Diseases of the Visceral Circulation

Edited by

GEORGE GEROULAKOS, MD, PhD, FRCS, DIC
Consultant Vascular Surgeon, Charing Cross and Ealing Hospital, Senior Lecturer in Vascular Surgery, Imperial College School of Medicine, London, UK

and

KENNETH J. CHERRY Jr, MD
Consultant, Division of Vascular Surgery, Mayo Clinic and Mayo Foundation, Professor of Surgery, Mayo Medical School, Rochester, MN, USA

ARNOLD

A member of the Hodder Headline Group
LONDON

First published in Great Britain in 2002 by
Arnold, a member of the Hodder Headline Group,
338 Euston Road, London NW1 3BH

http://www.arnoldpublishers.com

Distributed in the United States of America by
Oxford University Press Inc.,
198 Madison Avenue, New York, NY 10016
Oxford is a registered trademark of Oxford University Press

British Library Cataloguing in Publication Data
A catalogue record for this book is available from the British Library

Library of Congress Cataloging-in-Publication Data
A catalog record for this book is available from the Library of Congress

ISBN 0 340 80722 9 (hb)

1 2 3 4 5 6 7 8 9 10

Publisher: Nick Dunton
Development Editor: Michael Lax
Production Editor: Rada Radojicic
Production Controller: Martin Kerans
Cover Design: Mouse Mat

Typeset in 10/12 Minion by Integra Software Services Pvt. Ltd, India
Printed and bound in Italy by Giunti Industrie Grafiche

This book is dedicated to the memory of my father Dr Constantine Geroulakos and his brother Dr Demetrius Geroulakos who established and ran, for more than 50 years, the Geroulakos Clinic of Sparta in Greece. The staff of this institution offered an enormous service to the population of Sparta at a time when there was no National Health Service in Greece.

Contents

Contributors

Gianni Belcaro, MD PhD
Professor of Angiology, Department of Biomedical Sciences, Chieti University, Italy

Anna-Maria Belli, FRCR
Consultant Radiologist and Senior Lecturer, Department of Radiology, St George's Hospital, London, UK

David Bergqvist, MD, PhD
Professor of Vascular Surgery, Department of Surgical Sciences, University Hospital, Uppsala, Sweden

Marc Borge, MD
Associate Professor of Radiology and Chief, Section of Interventional Radiology
Department of Radiology, Loyola University Medical Center, Maywood, IL, USA

Sandra C. Carr, MD
Assistant Professor of Surgery, University of Wisconsin Medical Center, Madison, WI, USA

Kenneth J. Cherry Jr, MD
Consultant, Division of Vascular Surgery, Mayo Clinic and Mayo Foundation, Professor of Surgery, Mayo Medical School, Rochester, MN, USA

Harold Ellis, CBE, MCh, FRCS
Professor of Applied Clinical Anatomy, Division of Anatomy, Guy's, Kings and St Thomas' School of Biomedical Sciences, London, UK

John R. Galloway, MD
Associate Professor of Surgery, Department of Surgery, Emory University School of Medicine, Emory University Hospital, Atlanta, GA, USA

George Geroulakos, MD, PhD, FRCS, DIC
Consultant Vascular Surgeon, Charing Cross and Ealing Hospital, Senior Lecturer in Vascular Surgery, Imperial College School of Medicine, London, UK

Peter Gloviczki, MD
Professor of Surgery, Mayo Medical School, Chair, Division of Vascular Surgery, Mayo Clinic and Mayo Foundation, Rochester, MI, USA

Robert D. Goldin, MD, FRCPath
Consultant Histopathologist, Department of Histopathology, Imperial College Faculty of Medicine at St Mary's, London, UK

Ulf Haglund, MD
Professor of Surgery, Department of Surgery, University Hospital, Uppsala, Sweden

Nadey S. Hakim, MD, PhD, FRCS, FACS, FICS
Consultant Surgeon and Surgical Director, The Transplant Unit, St Mary's Hospital, London, UK

Corey J. Jost, MD
Vascular Fellow, Division of Vascular Surgery, Mayo Clinic and Mayo Foundation, Rochester, MN, USA

Ajay K. Kakkar, BSc, MBBS, PhD, FRCS
Senior Lecturer and Consultant Upper Gastrointestinal Surgeon, Department of Surgery, Imperial College School of Medicine, Hammersmith Hospital, London, UK

Nicos Labropoulos, BSc, PhD, DIC, RVT
Assistant Professor of Surgery, Department of Surgery, Loyola University Medical Center, Maywood, IL, USA

Nick J.M. London, MB, ChB, MD, FRCS, FRCP (Ed)
Professor of Surgery, Department of Surgery, University of Leicester, UK

Robert Morgan, FRCR
Consultant Radiologist, Department of Radiology, St George's Hospital, London, UK

William H. Pearce, MD
Professor of Surgery and Chief, Division of Vascular Surgery, Northwestern University Medical School, Chicago, IL, USA

Mario J. Perko, MD
Department of Thoracic Surgery, Rigshospitalet, University of Copenhagen, Denmark

Peter Robless, FRCS
Department of Vascular Surgery, Ealing Hospital, London, UK

Atef Salam, MD
The Emory Clinic, Atlanta, GA, USA

Torben V. Schroeder, MD, DMSc
Professor of Surgery and Consultant Vascular Surgeon, Department of Vascular Surgery, Rigshospitalet, University of Copenhagen, Denmark

Chitranjan J. Shukla, BSc, MBBS, MRCS (Eng), MRCS (Ed)
Surgical Trainee, Norfolk and Norwich University Hospital, Norwich, UK

William L. Smead, MD
Associate Professor of Surgery and Chief, Division of Vascular Surgery, The Ohio State University Hospital, Columbus, OH, USA

Gerard Stansby, MA (Cantab), MB, MChir, FRCS
Professor of Vascular Surgery, Northern Vascular Unit, Freeman Hospital, Newcastle upon Tyne, UK

Paul A. Thodiyil, FRCS (Gen Surg)
Senior Registrar in Surgery, Department of Surgery, Hammersmith Hospital, London, UK

Huw J.W. Thomas, MA, PhD, FRCP
Consultant Gastroenterologist and Senior Lecturer, St Mark's & St Mary's Hospitals, Imperial College, London, UK

Foreword

The visceral circulation, which irrigates the liver, spleen, pancreas and the whole of the alimentary tract from stomach to rectum, receives a fifth of the cardiac output and is subject to all manner of fluctuations in pressure and flow. Its major component, the intestinal circulation, is unique in that much of the tissue that it supplies is in contact with micro-organisms, and that critical reduction in blood flow results in infection and bacterial invasion. This in part explains why occlusion of its main vessel, the Superior Mesenteric Artery, is so particularly lethal. Furthermore, the recognition that the gut is not simply a propulsive tube, but a crucial endocrine organ, has focussed attention on its oxygen supply.

However, in contrast to the volume of published research related to the cardiac, cerebral and renal circulations, the visceral component has received comparatively little attention. This book redresses the balance and presents a timely update on current information in an important area. The introductory chapters lay the anatomical and physiological basis for the clinical syndromes to be described. The management of acute and chronic visceral ischemia is laid out in detail. The most important advance in the treatment of vascular disease over the last decade has been the spectacular progress achieved in imaging techniques and the resulting capacity to correct arterial occlusions and stenoses by non-operative means. A careful comparison is made between, on the one hand, the various operative approaches to the visceral vessels, and on the other, to the emerging range of techniques available to the vascular radiologist.

The venous system is not ignored, and mesenteric vein thrombosis and portal hypertension are discussed in detail. Rarer syndromes such as angiodysplasia, ischemic colitis and visceral artery aneurysms are briefly dealt with. Whether or not the 'celiac axis compression syndrome' is a clinical entity remains debatable, and the evidence for and against is fairly presented.

All in all, this book represents a gathering together of information which is scattered throughout specialist journals and hence not readily available to the practicing surgeon or gastroenterologist. The editors are to be congratulated on a most useful addition to the medical literature.

Adrian Marston
June 2002

Preface

This book provides the data currently available on the diseases of the visceral circulation to help the reader decide how to best manage patients with these conditions.

Critical reviews contributed by a multidisciplinary international team of experts discuss the normal anatomy and pathophysiology of the visceral circulation and focus on new ideas on the non-invasive diagnosis of visceral ischemia. A variety of surgical approaches are discussed that are useful for the management of acute and chronic visceral ischemia. The advantages and disadvantages of angioplasty and stenting for the management of symptomatic stenoses of the major visceral arteries are contrasted with the outcome of surgery. New ideas for diagnosis and management of mesenteric venous thrombosis, the celiac compression syndrome and visceral aneurysms are discussed. The introduction of endovascular techniques for the management of portal hypertension and bleeding from angiodysplasias of the gastrointestinal tract are also covered.

The book will be of help to vascular surgeons, gastrointestinal surgeons, gastroenterologists and vascular radiologists.

George Geroulakos and
Kenneth J. Cherry
June 2002

1

Normal and variant anatomy and collateral mesenteric circulation

CHITRANJAN J. SHUKLA AND HAROLD ELLIS

INTRODUCTION

In this chapter, we discuss the normal and variant anatomy of the mesenteric circulation and the collateral circulation that occurs. We will also highlight basic clinical relevance of the anatomy to surgical practice.

The mesenteric circulation comprises the celiac, superior mesenteric, and inferior mesenteric vessels, and all the viscera supplied by them. These include the gastrointestinal tract from the region of the lower part of the esophagus to the anal canal and the related viscera – the spleen, liver, biliary tract, pancreas, and omentum. The description of its anatomy is taken from numerous studies involving prosections and radiographic appearances.[1–32] There are numerous variations, many of which are not clinically relevant and therefore will be omitted or mentioned only briefly. Others, of course, are of considerable importance to the surgeon. They encompass variations in origin and in spatial relationships.

THE CELIAC ARTERY

The celiac artery, or trunk (Figure 1.1), is the first of the mesenteric vessels to arise from the ventral surface of the aorta. This occurs just as the aorta emerges from the aortic hiatus of the

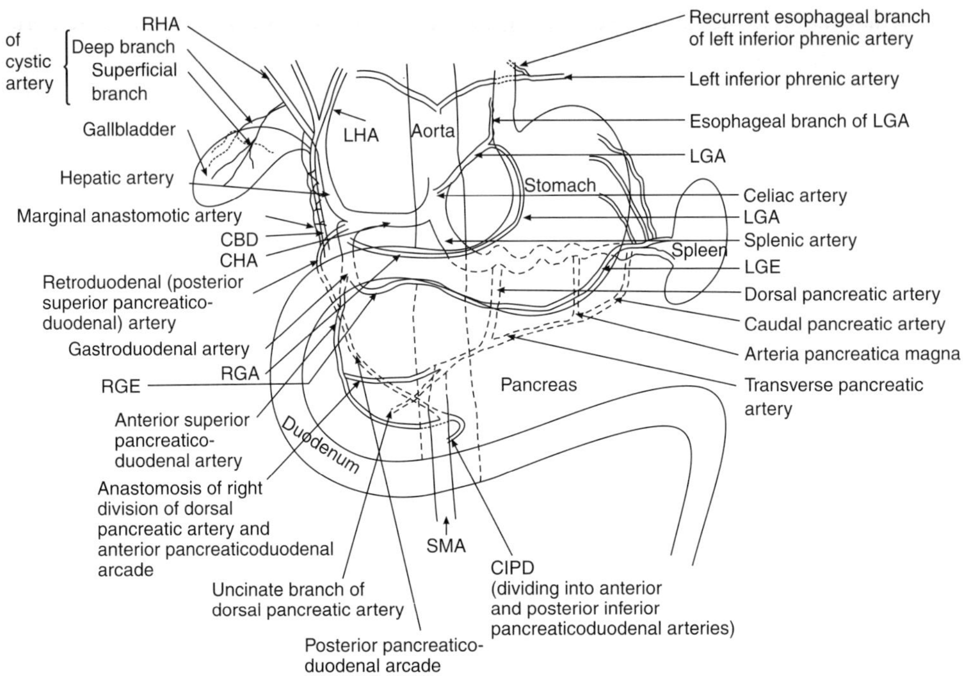

Figure 1.1 *The celiac trunk and its divisions. CBD, common bile duct; CHA, common hepatic artery; CIPD, common inferior pancreaticoduodenal artery; LGA, left gastric artery; LGE, left gastroepiploic artery; LHA, left hepatic artery; RGA, right gastric artery; RGE, right gastroepiploic artery; RHA, right hepatic artery; SMA, superior mesenteric artery.*

diaphragm at the level of the twelfth thoracic vertebra, behind the median arcuate ligament where the syndrome of celiac artery compression may occur.[11, 33, 34]

The artery then passes nearly horizontally forward and to the right, lying behind the lesser sac. As it passes to the right, it divides into its three main branches, the splenic artery, the left gastric artery (LGA), and the common hepatic artery (CHA).

Relations

The celiac plexus surrounds the axis at its origin, and produces branches that pass along the vessel and its subsequent branches. To the left of the artery lies the left crus of the diaphragm, the suspensory ligament of the duodenum and the gastric cardia. To the right lies the right crus of the diaphragm and the caudate process of the liver. Inferiorly lie the splenic vein and the upper border of the pancreas.[8, 35, 36]

Variations

The textbook description[30, 36] of the origin of the trunk is present in approximately 65–75% of dissections.[28] Variations include:

1 a true trifurcation of the celiac trunk into its three branches in 25% of cases;[28]
2 situations where the celiac and superior mesenteric vessels arise as a common celiaco-mesenteric trunk (1% of dissections). Here, this single artery is the sole source of

vascularization to the supramesocolic organs with collateral flow from inferior mesenteric, phrenic, esophageal, and retroperitoneal arteries;

3 situations where the branches vary in number or origin. There may be four or more branches of the celiac artery, such as the dorsal pancreatic, the inferior phrenic, the middle colic, or an esophageal branch;[21]
4 the celiac artery may give origin to the superior mesenteric artery (SMA), and the LGA may arise individually from the aorta. Such a vessel is called the lienohepatomesenteric trunk;
5 the lienomesenteric trunk arises as the splenic and SMA form a common trunk, leaving the CHA and LGA to arise as a separate trunk;
6 the hepatogastric trunk is one from which the LGA and CHA arise together, leaving the splenic artery to arise directly from the aorta.

THE SPLENIC ARTERY

This is the largest branch of the celiac artery in 82% of adults (in fetuses, it is the CHA).[36] It usually arises from the celiac axis, but rarely it may arise from the aorta, SMA, middle colic artery, LGA, left hepatic artery, or accessory right hepatic artery.[14] Initially, it takes a short route to the right and then courses to the left, lying retroperitoneally behind and superior to the pancreas, crossing the left crus of the diaphragm and laterally, the upper pole of the left kidney with the splenic vein lying below it. The splenic artery reaches the spleen, where it divides into two to four branches before plunging into the splenic hilum. Classically, its course is described as being tortuous, but this is not always so as fetuses and neonates have a straight splenic artery.[14, 30] Additionally, 15% of adults have a straight splenic artery, as discovered by Michels in his 100 dissections.[14] The degree of tortuosity is variable and has various possible etiologies and consequences.[9, 14] It can be harvested surgically after a splenectomy and used as an arterial bypass graft.[11] The splenic artery may also be used as a conduit for left renal artery revascularization in patients with atherosclerotic renal artery disease.[37]

Along its course to the spleen, the splenic artery gives off numerous branches to the stomach, pancreas, and greater omentum. The pancreatic branches include the dorsal pancreatic, the *arteria pancreatica magna*, and the caudal pancreatic vessels.

Branches to the stomach include the short gastric and the left gastroepiploic artery, as well as the cardio-esophageal artery and occasionally the LGA. Rarely mentioned is the posterior gastric artery, present in approximately 60% of people. This artery courses retroperitoneally behind the lesser sac and the bed of the stomach to pass to the posterior wall of the stomach via the gastrophrenic fold of peritoneum; it is a cause of significant hemorrhage during mobilization procedures of the stomach during partial gastrectomy procedures.[36]

Epiploic vessels to the greater omentum also arise from the splenic artery, as does the arch of Barkow indirectly (arising from the left and right gastroepiploic vessels).

The course of the splenic artery can be divided into four parts: suprapancreatic, pancreatic, prepancreatic, and prehilar. The artery divides into its terminal branches in the third or fourth segment. In 70% of cases, the vessel branches into its terminal branches in the second or, more commonly, the third segment, and is described as being distributive as more numerous branches are given off. In the cases where the artery branches in its fourth segment, it is described as being magistral, and the branches are less numerous.[14] These terminal arteries are end arteries.

The suprapancreatic portion

This is the first and shortest segment, consisting of one anticlockwise loop whose concavity points upwards and to the left. The clinical importance of this segment of the vessel is that occasionally it can be excessively long and may extend its loop into the lesser omentum. Here, it is susceptible to clamping or severance during partial gastrectomy.[31]

From this segment of the artery may arise other branches, which in order of frequency are:[14]

1 the left inferior phrenic artery;
2 the dorsal pancreatic artery, which arises from the splenic artery in 39% of individuals (in 69% it arises from the celiac, the SMA, or the right hepatic artery).[16] Located behind the confluence of the superior mesenteric and splenic veins, the artery descends to the dorsal surface of the pancreas, where the artery divides into two within the pancreas. A left branch, the transverse pancreatic artery, courses to the left of the pancreas to supply the body and tail of the pancreas. Two right branches travel to the right. One supplies the uncinate process of the pancreas, and the other anastomoses with the anterior superior pancreaticoduodenal, gastroduodenal, or right gastroepiploic artery. The middle colic artery may arise from an additional branch of the dorsal pancreatic artery, which passes behind the pancreas to anastomose with the SMA. This artery has been named the artery of Riolan;[8]
3 the superior polar artery to the spleen;
4 the cardio-esophageal artery, a long and slender vessel supplying the gastric cardia that may pass through the esophageal opening of the diaphragm to supply the lower esophagus;
5 an accessory gastric or hepatic artery;
6 the inferior mesenteric artery (IMA).

The pancreatic segment

This presents the highest degree of tortuosity of the splenic artery. It is usually suprapancreatic, although it may be prepancreatic, retropancreatic, or pancreatic relative to the pancreas. The close relation of this vessel to the pancreas explains the abundance of pancreatic vessels arising from this segment of the splenic artery, of which the most notable is the *arteria pancreatica magna*. This large vessel plunges into the posterior surface of the pancreas and splits into a left branch that courses within the gland to the left to anastomose with the caudal pancreatic artery, and a right branch that anastomoses with the transverse pancreatic artery.

The prepancreatic segment

As the pancreatic portion of the vessel leaves the upper border of the pancreas, it comes to lie on the anterior surface of the pancreas and branches into its terminal branches in 70% of individuals. These branches consist of an upper and lower (and occasionally a middle) terminal splenic artery. This segment also commonly gives origin to the short gastric and left gastroepiploic arteries that course to the right and traverse the gastrosplenic ligament to supply the fundus and cardia of the stomach (short gastric), and as it lies in the folds of greater omentum its greater curvature (left gastroepiploic). If one of the vessels accompanying the short gastric arteries is larger than usual and is distributed to the fundus of the stomach, it is called an accessory left gastric artery.

Epiploic vessels that supply blood to the greater omentum arise from an arch made of two constituent vessels arising from the left and right gastroepiploic arteries. Traversing in the posterior layers of the greater omentum below the transverse colon, these two vessels anastomose to form the arch of Barkow, which may afford additional blood supply to the transverse colon if the vessels anastomose with the middle colic artery.[8, 36]

The caudal pancreatic arteries to the tail of the pancreas also arise from this segment of the artery, or from its superior or inferior terminal branches, or indeed from the left gastroepiploic artery. The pancreas has prolific anastomoses within its structure between the dorsal pancreatic, arteria pancreatica magna, and the caudal pancreatic vessels, facilitated by the transverse pancreatic artery.

The prehilar segment

This usually consists of the terminal branches of the artery to the spleen traversing the space between the pancreatic tail and the splenic hilum, through the lienorenal ligament. In 30% of cases, the magistral pattern of vessel distribution occurs as the splenic artery terminates into its terminal branches in this region of the artery.

The splenic artery gives off two (superior and inferior) or occasionally three (middle) terminal end arteries to the spleen. In addition to the terminal arteries, the splenic artery gives off polar arteries to the spleen proximal to its terminal division. Surgically important are the superior and inferior polar arteries, which are end arteries and are occasionally given off as far proximally as the celiac trunk, a situation simulating a double splenic artery.

THE LEFT GASTRIC ARTERY

The LGA is often the smallest branch of the celiac artery, although it may occasionally arise from the aorta (with the rest of the divisions of the celiac trunk forming the lienohepato-mesenteric trunk with the SMA). It passes upwards and to the left, lying retroperitoneally behind the lesser sac to reach the bare area of the gastric cardia (so called because it is devoid of peritoneum as it lies apposed to the undersurface of the diaphragm). In reaching the bare area of the stomach, the artery raises a fold of peritoneum from the superior recess of the omental bursa. This fold of peritoneum is known as the gastropancreatic fold as it extends from the superior surface of the pancreas to the stomach. Having reached the cardia, the artery provides a few vessels to the gastric cardia and fundus before splitting into anterior and posterior divisions that turn sharply downwards, enveloped in the free fold of lesser omentum, following the lesser curvature of the stomach, and terminating by an often tenuous anastomosis with the anterior and posterior division of the right gastric artery (RGA).

In the region of the cardiac end of the stomach before its division, the LGA gives off an esophageal branch, which passes upwards (lying to the right of the abdominal esophagus) and through the esophageal hiatus of the diaphragm to supply the lower esophagus. The left inferior phrenic may also be a branch of the LGA before its division.

The blood supply of the lower esophagus is augmented by esophageal branches of the short gastric arteries near the fundus of the stomach and posteriorly by a recurrent esophageal branch of the left inferior phrenic artery, as the artery passes behind the esophagus on its way from the aorta to the undersurface of the left hemidiaphragm. Other sources of blood supply to the esophagus may be from:[22]

1 an aberrant (replaced or accessory) left hepatic from the left gastric (as occurs in 25% of cases)[8] or an accessory left gastric from the left hepatic;
2 cardio-esophageal arteries from the splenic trunk or its superior polar, terminal divisions, the short gastrics, or the previously mentioned posterior gastric artery;
3 a lone, tenuous cardio-esophageal artery of aortic, celiac, or splenic origin.

The clinical significance of the blood supply to the lower esophagus is in mobilization procedures of the upper end of the stomach when these esophageal branches are sacrificed by clamping of the left gastric, posterior gastric, or inferior phrenic vessels. This is not usually a problem due to a copious anastomosis providing for the abdominal esophagus, but it may be a problem if this anastomosis is meager. In this situation, an esophagogastric anastomosis may break down.

THE COMMON HEPATIC ARTERY

Typically, the CHA originates from the celiac artery, but it may be partially or wholly replaced from the SMA in 18.5% of cases.[8] It is directed anteriorly and to the right, passing retroperitoneally behind the lesser sac along the upper border of the head of the pancreas. It reaches the upper surface of the first part of the duodenum, passing below the epiploic foramen within the right hepatopancreatic fold. Here, it gives off the hepatic artery, which runs superiorly in the anterior fold of peritoneum lining the epiploic foramen of Winslow.

Its branches include the hepatic arteries to the liver and gallbladder, the gastroduodenal artery with its branches to the stomach, pancreas, and duodenum, and the RGA to the stomach.

THE HEPATIC ARTERY

As the hepatic artery (Figure 1.2) courses to the porta hepatis in this fold of peritoneum, it usually lies anterior to the portal vein and to the left of the common bile duct (CBD). At the porta hepatis, it divides into a left, a right, and a middle branch to supply the parenchyma of the liver.

The left branch of the hepatic artery supplies the left lobe of the liver. It is usually (75% of cases) a single artery from the hepatic artery, passing caudal to the left branch of the portal vein, along the fossa for the ligamentum venosum en route to supply the left lobe of the liver.[15] The middle hepatic artery usually arises from the left hepatic artery, although it may arise from the common hepatic or right hepatic, and in some instances it is the only vessel from the celiac trunk (the left and right hepatic arteries being replaced by branches from the LGA and SMA). The middle hepatic artery characteristically supplies the quadrate lobe of the liver, often augmented by branches from the left hepatic artery. The right hepatic artery supplies the right lobe of the liver. Before doing so, it often makes a caterpillar-like loop within the cystohepatic angle of Calot. Here, it may be mistaken for the cystic artery (usually a branch of the right hepatic artery from a lower level) if the loop is too low and is thus susceptible to damage

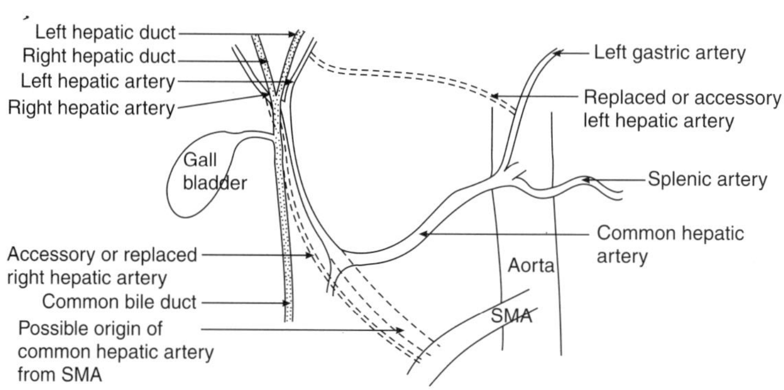

Figure 1.2 *Clinically important variations in the origin of the hepatic arteries. SMA, superior mesenteric artery.*

in laparoscopic cholecystectomies and other procedures in this region, resulting in infarction of the lobe it supplies.[8, 15, 21] This highlights an important point, namely that the hepatic arteries are end arteries.[8, 21] It is possible for a lobe of the liver to survive if the hepatic artery to it is damaged due to the presence of intrahepatic translobar and extrahepatic collaterals,[21] such as those from the inferior phrenic, right gastric, and gastroduodenal arteries.

This pattern of blood supply to the liver by the CHA is present in approximately 75% of dissections.[6] Variations are common and diverse, and are of clinical significance, particularly in biliary tract surgery, liver transplantation,[21, 29] and embolization or intra-arterial infusion of cytotoxic agents.[21] Anomalies of the hepatic arterial anatomy occur in up to one-third of individuals and compromise outcome if multiple anastomoses or direct anastomosis to the recipient aorta are required.[29]

The variations may result from replaced hepatic arteries (where one or both hepatic arteries arises from an abnormal source) or an accessory hepatic artery (where the aberrant hepatic artery is an additional blood supply to the normal pattern). The incidence of these variations depends on the series being studied.[6, 8, 15, 28]

In fact, the CHA may itself originate directly from the aorta (either alone or in common with the left gastric trunk in <1% of cases) or the SMA in 1.5–2.5% of cases, depending on the series of studies.[6] The SMA may give origin to a replaced right hepatic artery (11–16%) or an accessory right hepatic artery (6%), or a combination of both. Thus, it is important to preserve the SMA in operations about the duodenum and the pancreas, for the ligation of a common or replaced hepatic artery will result in complete necrosis of the entire liver or one of its lobes.

The left hepatic artery may take origin from the LGA (10%) or have an accessory left hepatic artery from the LGA (8%). It is also possible to get a combination of a replaced or accessory artery to any lobe in any permutation with a replaced or accessory artery to the other lobe of the liver. The surgical significance of a replaced or accessory left hepatic artery of left gastric origin is in procedures involving division of the hepatogastric ligament, in which the artery runs and is susceptible to division. Such procedures include partial hepatectomy or gastrectomy procedures.

An important branch of the CHA is the RGA, which arises in equal proportions from the common hepatic or left hepatic arteries (40% each),[8] although it may arise from the gastroduodenal artery (8%), or the middle or left hepatic artery (5%). This is a small vessel that descends to the gastric pylorus, before turning left into the lesser omentum on its way to anastomose (often tenuously) with the LGA. In the region of the pylorus, it gives origin to the pyloric artery, not to be confused with the supraduodenal artery (of Wilkie),[32] which usually arises from the gastroduodenal artery.

BLOOD SUPPLY OF THE GALLBLADDER AND THE COMMON BILE DUCT

Having passed from behind the common hepatic duct on its way to the liver, the right hepatic artery gives off a single (75% of cases) cystic artery in Calot's triangle,[15] where it bifurcates into a superficial and a deep cystic branch. The superficial cystic branch of the cystic artery is distributed anteriorly onto the free peritoneal surface of the gallbladder, whereas the deep cystic branch courses behind the gallbladder onto its bed on the liver. Here, the deep branch gives off variably sized arteries to the liver. These are a cause of hemorrhage during cholecystectomy, when attempts are made to free the gallbladder from its attachments to the liver.

Variations are common in the origin and course of the cystic artery, which, if not truly appreciated, may make a laparoscopic cholecystectomy difficult and cause significant morbidity to a patient.

Dual cystic arteries are common and occur in 25% of cases, the most frequent instances being those in which the superficial and deep branch originate from outside the cystohepatic angle from the same right hepatic artery or from two right hepatic arteries within the triangle. The two arteries may have a variable source, arising from a combination from the right, middle, or left hepatic artery, or from some other artery in the vicinity, such as the gastroduodenal artery. In such instances, the artery courses below the cystic duct instead of within Calot's triangle, important during cholecystectomy. A common pattern is one in which the deep cystic artery arises within the triangle from the right hepatic artery, leaving the superficial cystic to arise from the middle or left hepatic or gastroduodenal artery from outside the triangle.[15] It is important to clamp and divide both cystic arteries of different origins during cholecystectomy to prevent hemorrhage by avulsion of an unclamped vessel.

The location of the cystic artery is very important during laparoscopic cholecystectomies. The artery is usually found in Calot's triangle having originated from the right hepatic artery as it crosses behind the common hepatic duct. In 25% of situations, the cystic artery will cross anteriorly to the hepatic duct, as may the right hepatic artery in 12–26% of cases.[21]

In 20% of Michels' 200 dissections,[15] the cystic artery originated from outside the cystic triangle, stemming from the right hepatic artery to the left of the hepatic duct (13%), or from the middle, left, or common hepatic trunk (5%), or from the gastroduodenal artery and its retroduodenal branch (4%). A common occurrence (11%) was the situation in which the cystic artery arose from a replaced hepatic artery of superior mesenteric origin.

The CBD (Figure 1.3) consists of two sections, supraduodenal and infraduodenal. The upper part of the supraduodenal portion lies adjacent to the right hepatic artery and as a consequence receives blood from it. The lower part of the supraduodenal portion and the infraduodenal part of the CBD receive important vessels in varying patterns from the retroduodenal artery with minor contributions from the gastroduodenal and the SMA. What is evident is that there are extensive anastomoses ensheathing the CBD, with two to five main vessels arising from the retroduodenal artery ascending up the supraduodenal portion of the duct to anastomose with branches from the right hepatic or cystic arteries. The anastomosis of these vessels forms a plexus, excessive stripping of which may cause a stricture or necrosis

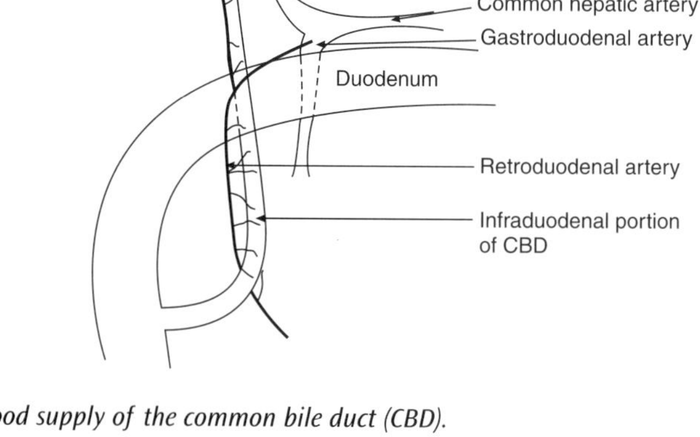

Figure 1.3 *The blood supply of the common bile duct (CBD).*

of the CBD and possible perforation. One particular vessel, the marginal anastomotic artery, is prominent in this epidochal plexus. It arises from the retroduodenal artery as it courses anterior to the CBD to lie on its right anterolateral aspect.[16, 26] It is from this marginal artery that the important vessels to the CBD arise as the vessel ascends to anastomose with branches of the right hepatic artery.

As the retroduodenal artery crosses the bile duct anteriorly first and then posteriorly (see below), it gives forth more vessels to the extra- and intrapancreatic portions (respectively) of the CBD. At the bile duct termination near the ampulla of Vater, the anterior superior pancreatic artery provides it with a few vessels. It is thus seen that the infraduodenal portion of the CBD has a more extensive blood supply than the supraduodenal portion.

The common variations to the blood supply of the CBD relate to the sparsity or lack of the marginal artery, or a significant anastomosis along the CBD. In these situations, the right hepatic artery or a cystic artery augments the blood supply and may arise from a lower level to do so. Thus, a clamp placed mistakenly on such a low-lying cystic artery may cause necrosis of the duct wall and result in postoperative biliary leak.

Another vascular pattern that may cause such a postoperative biliary leak is one in which neither marginal artery nor an epidochal plexus exist. Instead, two or three large vessels connect the right hepatic, cystic, and retroduodenal arteries, and course tortuously at a distance supplying the CBD with small vessels. Clamping of these vessels may result in necrosis of a segment of the CBD. Bile duct necrosis is not as common as would be expected due to the presence of an intramural plexus formed within the wall of the CBD by the vessels derived from the epidochal plexus. These vessels run along the wall of the CBD and at regular intervals give small branches that pierce the fibrous connective sheath to anastomose with other similar branches, and then remerge to anastomose with other main vessels on the surface of the duct. From this first sheath anastomosis, more branches percolate through to the lamina propria to form a second subepithelial plexus.[26] Thus, the plexus so formed provides a collateral in case of inadvertent ligation of a main vessel. It has been proposed that the complications of necrosis, failure of anastomosis, fistula formation, and stricture occur as a result of a rise in intramural pressure secondary to mucosal inflammation, compressing the intramural plexus against its envelope of a non-expansile fibrous sheath.[26] This would lead to ischemic necrosis of this segment of the CBD.

THE GASTRODUODENAL ARTERY

This is the second terminal branch of the CHA. It may arise less commonly from the left or right hepatic arteries, or from the SMA from a replaced hepatic trunk. It may be absent, in which case its branches arise from the right hepatic artery of superior mesenteric origin.

The gastroduodenal artery passes behind the first part of the duodenum, lying anterior to the portal vein and to the left of the CBD (although it may lie anterior to it). It descends to the inferior border of the first part of the duodenum, where it bifurcates into its terminal branches, namely the right gastroepiploic artery and the anterior superior pancreaticoduodenal arteries. Behind the first part of the duodenum, it may be penetrated by a posterior duodenal ulcer, with resultant severe hemorrhage.

The gastroduodenal artery may be used effectively as a source of arterial inflow to the right renal in approximately 50% of cases in patients with atherosclerotic renal artery disease. Before using the gastroduodenal artery for right renal artery revascularization, it is essential to rule out with an imaging technique the presence of celiac axis stenosis. Additionally, in the presence of SMA stenosis, the gastroduodenal artery has the capacity to dilate and is a major collateral pathway between the celiac axis and the SMA. In these cases, the gastroduodenal artery cannot be used as a conduit for right renal artery revascularization.[37]

One of the branches of the gastroduodenal artery is the retroduodenal artery, also known as the posterior superior pancreaticoduodenal artery. In approximately 90% of cases,[16] it is the first branch of the gastroduodenal artery, arising from its left. In remaining cases, it arises from the hepatic, right hepatic, or dorsal pancreatic artery. Its importance is in that it courses retroperitoneally behind the duodenum and the pancreas to form the posterior pancreaticoduodenal arcade of vessels that supplies the posterior aspect of the duodenum and the head of the pancreas.

From its 'cryptic origin' (due to dense fibrous and pancreatic tissue) above the duodenum and the head of the pancreas, the retroduodenal artery assumes a tortuous course, descending to the left of the CBD and crossing it anteriorly above the first part of the duodenum. It then lies to its right before spiraling down posterior to the intrapancreatic portion of the CBD to participate in the posterior pancreaticoduodenal arcade, a U- or V-shaped arcade lying behind the head of the pancreas.[16]

The retroduodenal artery terminates within the arcade by anastomosing with the posterior inferior pancreaticoduodenal artery originating from the SMA. The artery comes into full view during Kocherization procedures of the duodenum, where it is covered by Toldt's fascia, a condensation of primitively fused mesoduodenum.

The posterior pancreaticoduodenal arcade lies behind the pancreas and is more superior in position to the anterior arcade formed similarly by the anterior superior and inferior pancreaticoduodenal arteries from the gastroduodenal and superior mesenteric arteries, respectively.

Branches of the retroduodenal artery or the posterior pancreaticoduodenal arcade include:

1. duodenal branches, one of which may be the supraduodenal artery of Wilkie.[32] Duodenal branches are usually vasa recta to the posterior surface of all segments of the duodenum;
2. pancreatic branches to the head of the pancreas;
3. branches to the extrahepatic biliary tract, in particular branches to the CBD, and occasionally cystic arteries to the cystic duct.

A similar anterior pancreaticoduodenal arcade is formed by the anterior superior pancreaticoduodenal artery, one of the terminal branches of the gastroduodenal arteries, emerging from behind the first part of the duodenum and coursing inferiorly and anterior to the pancreas in the duodenopancreatic groove in the form of an arc on the anterior surface of the pancreas before plunging into the substance of the head of the pancreas to anastomose with the anterior inferior pancreaticoduodenal artery or the common inferior pancreaticoduodenal artery (CIPD) (whose posterior branch receives the posterior arcade).

Both anterior and posterior pancreaticoduodenal arcades give off vasa recta to the anterior and posterior surfaces of the duodenum, respectively. These are end arteries, interruption of which can lead to perforation of the duodenum. Also given off are pancreatic branches to the head of the pancreas, which freely anastomose with intrapancreatic branches derived from the dorsal pancreatic artery. Within the pancreas, the arteries anastomose to give the gland a rich vascular supply.

THE SUPERIOR MESENTERIC ARTERY

This vessel (Figure 1.4) originates from the ventral aspect of the aorta at the level of the first lumbar vertebra, with the splenic vein and the body of the pancreas lying anteriorly to it. It in turn lies anterior to the left renal vein, the uncinate process of the pancreas, and the third part of the duodenum, as the artery passes over this. This is the site where the SMA may rarely compress the third part of the duodenum and give rise to superior mesenteric artery

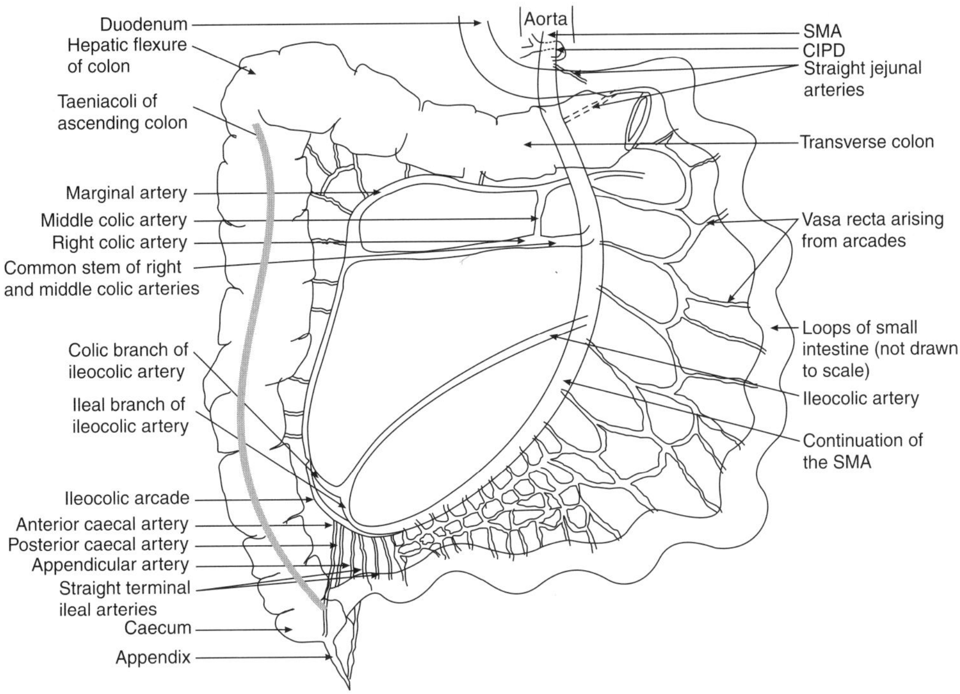

Figure 1.4 *The superior mesenteric artery (SMA) and the distribution of its branches. CIPD, common inferior pancreaticoduodenal artery.*

compression syndrome (of Wilkie).[38] Rather than arising in isolation, the SMA may originate in whole or partial combination with the celiac artery or its branches, the most common and certainly of considerable clinical importance is the hepatomesenteric trunk (in 16%), where the common hepatic or an aberrant right hepatic (replaced or accessory) arises with the SMA.

From the anterior aspect of the duodenum, the SMA courses in the small bowel mesentery with the superior mesenteric vein (SMV) lying to its right. Within the mesentery, the artery continues in an arc with its convexity forwards and inferiorly to the left, following the ileum to terminate at the ileocolic junction by anastomosing with the ileocolic artery, a branch of the parent artery. In its course, the SMA crosses the inferior vena cava, the ureter, and the right psoas major muscle.[11, 12, 17, 35, 36]

Along its course, the SMA gives off numerous vessels that supply the mid-gut (the small intestine from the second part of the duodenum distal to the ampulla of Vater to the junction of the proximal two-thirds and distal one-third of the transverse colon), sections of the pancreas, and occasionally parts of the liver by accessory or replaced hepatic arteries.

BRANCHES

The anterior and posterior inferior pancreaticoduodenal arteries arise individually (in 60%) or as a CIPD artery (in 40%) from the right side of the SMA (in 50%), from its first jejunal branch (in 48%), or less commonly from its second jejunal branch (2%).[16, 17] From here, they ascend on the posterior aspect of the uncinate process of the pancreas, from where the anterior division passes anteriorly to anastomose with the anterior superior pancreaticoduodenal vessel, leaving the posterior division to continue retropancreatically to anastomose with the retroduodenal artery.[16, 17]

Jejunal and ileal branches run parallel to each other and form increasing numbers of arcades progressively along the small intestine. From these arcades arise straight vessels, the vasa recta, to the jejunum and ileum. These vessels are distributed alternately to the anterior and posterior aspects of the small intestine, or may bifurcate to supply both sides of the bowel in a complicated manner.[17] Occasionally, two arcades – one anterior and one posterior – are present and give forth anterior and posterior vasa recta, respectively. As a result of the progressively increasing number of arcades along the small intestine, the vasa recta to the jejunum are longer and less numerous than along the ileum. One clinical feature of this arrangement of vessels is that an embolus that lodges more proximal in the jejunal vasculature is more likely to infarct a larger area of bowel than one that lodges at the same level in the ileal arcade. The ileal tissue is protected by a collateral circulation that will supply the area of bowel that would otherwise be devoid of blood. In practical terms, the implication is that for a small, isolated section of the ileum to be infarcted, the embolus would have to be small enough to lodge distally. In contrast, infarction of the jejunum can be achieved by a fairly large embolus placed proximally (and, as such, is accessible for embolectomy). The terminal arcades within the small intestine form a continuous link from the region of the first jejunal branch to the terminal ileal branch to form a vessel that runs parallel to the mesenteric border of the small bowel. It is from this vessel that the vasa recta to the small intestine arise. This vessel, described by Dwight (1898) and known as the vessel of Dwight,[17] is an important link and source of the collateral circulation from the pancreatic (via the inferior pancreaticoduodenal vessels) to the colonic circulation (via the ileocecal arcade)(see below).

The middle colic artery arises from the SMA just below the level of the pancreas. Its mode of origin varies. It may arise as a common trunk with the right colic artery (52%), as an isolated vessel from the SMA (44%), from the celiac artery (0.5%), or as a common trunk with the right and ileocolic arteries (0.5%). Nelson *et al.* reported an example of the middle colic artery arising from the left colic and cases of right and middle colic arteries arising from the ileocolic artery, in which the ileocolic supplied blood to the colon down to the splenic flexure.[21] The middle colic artery is absent in 3% of cases, and an accessory middle colic may be present in 8% of all cases.[17] From its origin, the middle colic artery enters the transverse mesocolon and divides into left and right branches, which anastomose with the left colic and right colic arteries, respectively to aid in the formation of the marginal artery (of Drummond).[1]

The right colic artery arises from the middle of the concavity of the SMA. It commonly originates as a common trunk with the middle colic (52%), as a separate vessel (38%), or as a common trunk with the ileocolic artery (8%).[17,21] It is absent in 8% of cases. It passes retroperitoneally towards the ileocecal junction, crossing the right gonadal vessels, the right ureter, and the right psoas major. It occasionally comes off more proximally from the middle colic above the duodenum, in which case it crosses the third part of the duodenum and the lower part of the right kidney. Upon reaching the right side of the colon, it divides into an ascending branch that courses superiorly to anastomose with the middle colic artery, and a descending branch that anastomoses with a branch of the ileocolic artery. The right colic artery supplies the upper two-thirds of the ascending colon.

The ileocolic artery is the only constant and the last major branch of the SMA.[17] The ileocolic artery is retroperitoneal as it courses to the right iliac fossa and divides into superior and inferior divisions that anastomose with the descending branch of the right colic and the terminal part of the SMA, respectively. It gives off the anterior and posterior cecal arteries to the cecum and the appendicular artery to the appendix. It is therefore the blood supply to the lower part of the ascending colon, the cecum, the appendix, and the terminal portion of the ileum.

The ileocolic arcade is an arcade of arteries that links the ileal and colic branches of the ileocolic artery and is present in 76% of individuals. It is an important source of collateral blood flow between the ascending colon and the small bowel.

The appendicular artery is very variable in site, size, and origin, mainly because of the variation in length and position of the appendix. It originates most commonly from an ileal branch within the ileocolic arcade (38%), but it may arise directly from the ileocolic artery (21%), from the cecal arteries (11%), or from the ileocolic trunk (11%).[17] The artery descends behind the terminal ileum and gives off a recurrent branch that anastomoses with the posterior cecal artery and then continues within the meso-appendix, in the free margin initially, and then in the terminal part lies in contact with the appendicular wall. It supplies the free margin of the appendix, the root usually being supplied by a small arterial twig from the posterior cecal artery. In 8% of cases, a double appendicular artery is present.

The anterior and posterior cecal arteries arise as separate branches (64%) or as a common trunk (36%) from the ileocecal arcade, colic, or ileal branches, or from the main ileocolic trunk (in decreasing order of frequency).[17, 19] The two arteries pass to the anterior or posterior surface of the cecum and frequently form arcades between themselves, which in turn give off terminal cecal vessels. The posterior cecal artery additionally gives off a small twig to the root of the appendix and may be a possible bleeder during appendectomies. The anterior cecal artery also gives off one or more terminal ileal arteries that pass horizontally to the left to supply the terminal portion of the ileum.

The blood supply to the terminal portion of the ileum is from the ileal branches of the ileocolic artery, the ileocolic arcade (formed by an ileal and colic vessel anastomosing), and a recurrent ileal artery. Present in 39% of the population, this artery arises from the cecal branch or the ileocolic arcade and courses horizontally along the terminal ileum to anastomose with the ileal branch of the ileocolic artery. It forms a pathway for extrinsic blood supply to the terminal ileum, with vasa recta arising from this artery. When absent, in most cases neighboring vessels overlap to furnish the terminal ileum. In the 16% of cases in which no overlap occurs, a section of 3–5 cm of terminal ileum is devoid of vasa recta. This section of the terminal ileum then has a precarious blood supply, which depends on straight arteries from the cecal arteries.

The recurrent ileal artery and the ileocolic arcade are imperative to complete the collateral link between the parallel vessel of Dwight, which supplies the small intestines, and the marginal artery of Drummond, which supplies the large bowel.[17, 19]

THE INFERIOR MESENTERIC ARTERY

The IMA (Figure 1.5) supplies the large intestine from the left half to one-third of the transverse colon to the anal canal. It arises from the ventral aspect of the aorta at the level of the third lumbar vertebra, just proximal to the lower border of the third part of the duodenum. It is also retroperitoneal in route, and courses to the left, crossing the left common iliac vessels and lying medial to the left ureter. It continues over the pelvic brim as the superior rectal artery within the sigmoid mesocolon, and passes into the lesser pelvis. Along its course it has the inferior mesenteric vein on its left. Its main branches are the left colic artery, the sigmoid arteries, and superior rectal artery.[11, 12, 17, 35, 36] The branches may be described as being ascending (left colic artery), middle, descending, and the superior rectal artery, with contributory branches from these vessels constituting the sigmoid artery.[17]

The *left colic artery*, also known as the ascending artery by previous authors,[5, 18] arises from the IMA and runs superiorly and retroperitoneally over the psoas major muscle to divide into two branches. The main trunk or its branches cross the gonadal vessels and the left ureter and, in 86% of people, reach the splenic flexure (or the middle of the descending colon in 14%, according to standard textbook descriptions[35, 36]). At the splenic flexure, the artery bifurcates into a right branch, which anastomoses with the left branch of the middle colic artery, and a

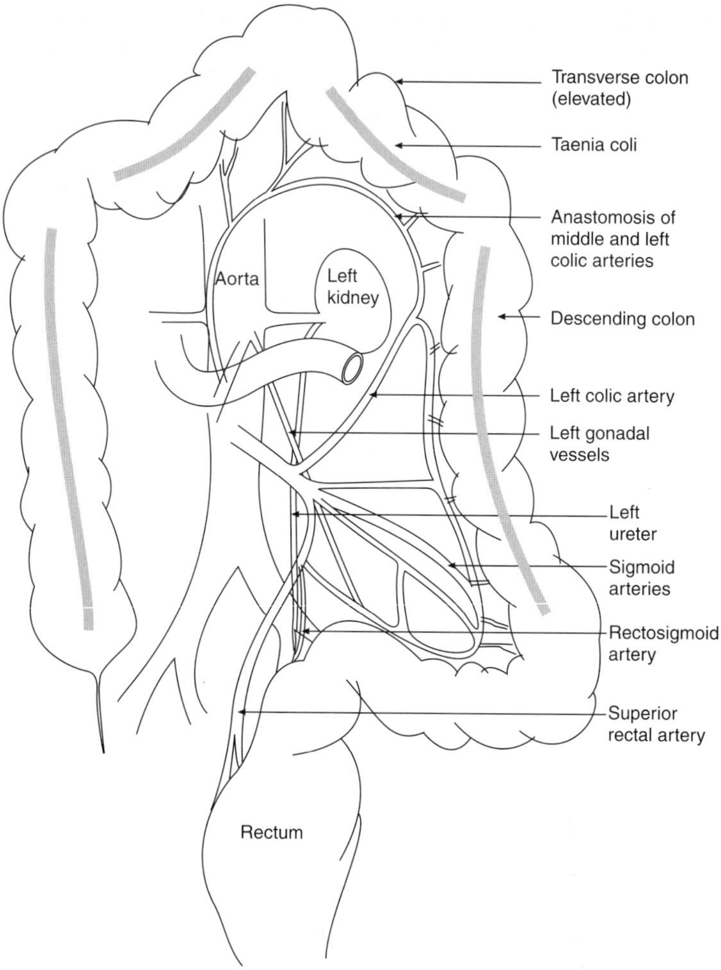

Figure 1.5 *The inferior mesenteric artery (IMA) and its branches.*

left branch, which descends along the descending colon to constitute and complete the marginal artery. Usually, the bifurcation occurs close to the splenic flexure, thus preserving the normal pattern of the marginal artery, with vessels to the colon and splenic flexure arising from this bifurcation. However, in a significant number of people, the bifurcation of the left colic artery may occur more proximally, thus placing the blood supply of the splenic flexure and the neighboring large bowel in a precarious position as the two branches act as the anastomotic link along the marginal artery. If they were to be ligated distal to the bifurcation, the marginal artery would be effectively interrupted, causing necrosis of the region of bowel supplied. This point in the vasculature is named Griffiths' point.[5, 13, 17–19] Its clamping has accounted for anastomotic leaks following left hemicolectomy, and necrosis of the descending colon during other procedures in which the artery had been tied distal to bifurcation. The anastomosis between the left and right branches of the left colic artery is usually substantial or adequate in approximately 61%, tenuous in 32%, and absent in 7% of cases.[17] Frequently, the blood supply to the splenic flexure is augmented by the accessory middle colic artery arising from the celiac trunk (celiocolic trunk), splenic artery (splenocolic trunk), an aberrant right hepatic from the SMA (hepatocolic trunk), dorsal or transverse pancreatic artery (pancreaticocolic trunk), gastroduodenal artery (gastrocolic trunk), or an omentocolic artery, where the blood supply to the splenic flexure is augmented by posterior epiploic arteries from the arc of Barkow. These vessels assist in the completion of the marginal artery and, if not preserved, may interfere with the blood supply of this region.

Sigmoid arteries vary in number from one to five, although usually two or three are present. The number of sigmoid arteries and their lengths depend on the length of the sigmoid mesocolon. If the mesocolon is long and wide, the arteries are derived from the ascending and descending branches of the IMA. If it is short and narrow, they are usually derived from the middle and descending branches of the IMA. The usual pattern is to have one artery derived from the ascending middle branches and two from the descending branch.[17]

The *rectosigmoid arteries* are those that are derived from the superior rectal artery with or after the last sigmoid artery but before its terminal bifurcation at the rectum. Present in 64% of individuals, and one to three in number, they course parallel to the rectum and supply the rectosigmoid junction. When absent (in 36%), the rectosigmoid junction is supplied by paired superior rectal arteries.[17]

The *superior rectal artery* is a continuation of the IMA. It acquires its name as it crosses the pelvic brim across the common or external iliac vessels. It courses into the pelvic mesocolon hidden by dense fat, and divides into two usually unequal-sized vessels at the level of the third sacral vertebra, below and behind the pouch of Douglas. This bifurcation is marked by numerous important lymph nodes, which are often involved in malignant disease and which require clearance during operative procedures in rectal carcinomas.

The branching of the superior rectal artery into its terminal branches can take the form of numerous patterns. In most cases (81%), the artery bifurcates into a larger right and smaller left. The larger branch is considered to be a continuation of the IMA and is distributed to the posterolateral surface of the rectum by its branches. The smaller left branch, considered to be a collateral of the right branch, angles to the left to supply the anterior and lateral surface of the rectal ampulla.

Instead of bifurcating, the IMA may trifurcate into three branches of equal size (13%), divide into more than three branches (4%), or course in the midline on the posterior aspect of the rectum to form anastomotic loops.[17]

The *intramural plexuses* consist of subserous, muscular, submucosal, and mucosal plexuses within the colon, rectosigmoid junction, and upper rectum. They are formed by the vasa recta, which first pierce the serosa to form a subserous plexus, and continue to the submucosa to form the submucous plexus. Two terminal sets of arteries leave the submucous plexus. One passes outwards (or centrifugally) to form the muscular (or myenteric) plexus, which consists of interwoven arterioles laminated between longitudinal and circular layers of the muscles. The other set courses inwards (centripetally) subepithelially to form the mucosal plexus.[17–19] The plexus provides for a collateral route for blood flow should one of the main vessels contributing to the blood supply of the colon and rectum be compromised, for example during surgical procedures.

In 52% of cases, a grossly visible anastomosis is present between the rectosigmoid or last sigmoid and the superior rectal arteries. It was originally suggested that a failure of gross anastomosis might lead to a compromise of blood supply in this region should one of the main contributing arteries be sacrificed. This region was called Sudeck's critical point, which has now been proven not to be of clinical significance due to the presence of the aforementioned plexus.[11, 12, 17, 18, 36]

The *middle rectal arteries* are not branches of the mesenteric arteries but are mentioned for the sake of completion. They arise from the internal iliac arteries or their branches. Consisting of one to three branches (usually one) on each side, they divide into three or four and are distributed asymmetrically onto the anterolateral and posterolateral surfaces of the rectal ampulla. A gross anastomosis with the superior rectal artery was observed in only 58% of cases,[18] but nonetheless contributed to the intra- and extramural plexus.

The *inferior rectal artery* arises from the internal pudendal artery below the level of the levator ani and crosses the ischiorectal fossa to divide into three branches, which anastomose with the superior and middle rectal arteries to contribute to the intramural plexus. This anastomosis is grossly visible in 14% of cases.[18] The artery supplies the canal, internal and external anal sphincters, and the rectum as high up as the peritoneal reflection.

The *median sacral artery* arises as a terminal division of the aorta, and contributes minimally to the plexus, but it is important in that it tends to be a source of hemorrhage during resections of the rectum.

Clinical significance of the mesenteric circulation of the colon

1. Variations in colonic and marginal arteries need to be borne in mind during replacement of esophagus or bladder segments using a section of the colon on a vascular pedicle.[21]
2. Resections of the colon or rectum during treatment of carcinoma of the colon or rectum must ensure adequate clearance of tumor, preserve the blood supply to the anastomotic margins, and maintain the marginal anastomosis. An example is too distal a ligation of the left colic artery in instances where no communication exists between its left and right branches. This compromising Griffiths' critical point during left hemicolectomies causes postoperative anastomotic leakage.
3. Operations involving the supposed avascular area of the transverse mesocolon can compromise an anastomosis between the left colic and middle colic arteries (replaced or accessory), and can compromise the blood supply to the splenic flexure.

THE COLLATERAL ARTERIAL CIRCULATION

Numerous channels exist as potentials for collateral circulation within and between the three main mesenteric vessels.[8, 17, 19, 28] They are summarized here briefly. Their importance lies in the ability to dilate and accommodate a gradually occluding vessel and allow for viability in the event of chronic ischemic occlusion. Thus, acute occlusion may not be tolerated, as a collateral channel may not have had a chance to dilate.

It is important to note that anatomic collaterals do exist as seen in the dissection room, but they may not be indicative of occlusive disease, whereas radiological signs are more evident of occlusive disease.[8]

The routes of collateral circulation are numerous, and may be classified as follows:

1. routes of anastomoses within each of the three main mesenteric vessels;
2. routes of collateral flow between the three main mesenteric vessels;
3. routes of collateral flow between the mesenteric circulation and the parietal or body wall vessels.

Collaterals within the mesenteric vessels

Within the *celiac axis* are numerous channels that act as an anastomotic link between the branches of its three main branches:

- Around the stomach are the anastomoses between the left and right gastric and gastroepiploic vessels. Also present at the region of the fundus is an anastomosis between the left gastric and the short gastric arteries of splenic origin.

- Within the pancreas are the anastomotic links between the pancreatic branches of gastroduodenal and splenic origin (aided by the SMA branches).
- The LGA may give origin to the accessory left hepatic artery just as the left hepatic artery may give origin to an accessory left gastric artery, which form a collateral flow to and from the liver. At the region of the porta hepatis, the left and right hepatic arteries may form an anastomosis before supplying the liver.

Within the *superior mesenteric circulation*, the main channels of collateral flow are:

- links between the inferior pancreaticoduodenal and jejunal vessels, the absence of which is the first weak point in the collateral circulation;
- links between the first and second jejunal vessels, the absence of which is the second weak point in the collateral circulation;
- anastomoses between the right, middle, and ileocolic arteries, which aid in the formation of the central anastomotic artery and the marginal artery (see below);
- the anastomosis between the ileocolic artery and the SMA;
- the ileocolic arcade and the recurrent ileal artery, which link the parallel vessel of Dwight (another important source of collateral blood flow) to the colic vasculature. An interruption of the ileal arcade or the recurrent artery will lead to interruption of the link between the small and large intestines.[17, 19]

Within the *inferior mesenteric circulation* are:

- the central anastomotic artery and the marginal artery linking the three main branches;
- links between the sigmoid, rectosigmoid, and superior rectal arteries;
- the intramural plexus (see above).

Collaterals between the mesenteric arteries

Linking the celiac and SMA vessels are:

- the arch of Bühler, an embryological remnant of a vertical vessel that linked the proximal parts of the celiac artery and the SMA. This artery is present in 2% of angiograms,[8] and is important when an obstruction exists in either main artery proximal to the insertion of the artery of Bühler;
- the pancreatic arcade, which facilitates an anastomosis between the superior and inferior pancreaticoduodenal arteries of celiac and superior mesenteric arteries, respectively;
- the transverse pancreatic artery, which affords an anastomosis between the various pancreatic arteries. Thus, a dorsal pancreatic artery of superior mesenteric origin establishes an anastomosis with the transverse pancreatic artery and hence with the *arteria pancreatic magna* and the caudal pancreatic artery of celiac origin;
- the middle colic artery, which may originate from the celiac artery or its branches and thus links the celiac artery and the SMA via its anastomosis with the right colic artery (and the IMA via the left colic artery).

The link between the SMA and the IMA is fraught with confusion in the terminology of the vessels.[1, 3, 5, 20, 28] This is due to the fact that there may be up to three sets of anastomotic vessels linking the blood supply of the large intestines, which have been given different sets of names by different researchers. The classification produced by Kornblith *et al.* aids as a simplification of the complexity of the terminology.[8] Three main sets of vessels exist:

1 A small, calibered artery placed peripherally near the mesenteric border of the large bowel from which vasa recta arise is known as the marginal artery. These vasa recta may be distributed to the colon in a complicated permutation.[18]
2 A larger, more centrally placed artery, which links the middle and left colic arteries, is known as the central anastomotic artery, the loop of Riolan, or the meandering mesenteric artery (Figure 1.6) of the colon. This gives vessels that anastomose to form the marginal artery. The presence of this vessel in radiological studies is indicative of vascular occlusion; the occluded vessel (SMA or IMA) can be deduced from the pattern of filling of contrast within this vessel.[20]
3 The arc or artery of Riolan is a vessel that links the proximal part of the middle colic artery to the point where the middle colic and the ascending branch of the left colic artery join.

The first two arteries have often been referred to as the arterial complex of Drummond.

Collaterals between the mesenteric and parietal arterial circulation

This is a link between the mesenteric circulation and that of the body wall. The main anastomoses are:

1 between the branches of the celiac artery and the parietal circulation, such as:

- between the esophageal branches of the inferior phrenic artery and the esophageal branches of the left gastric and short gastric arteries. The inferior phrenic artery may actually originate from the celiac axis and may anastomose with the superior phrenic arteries;

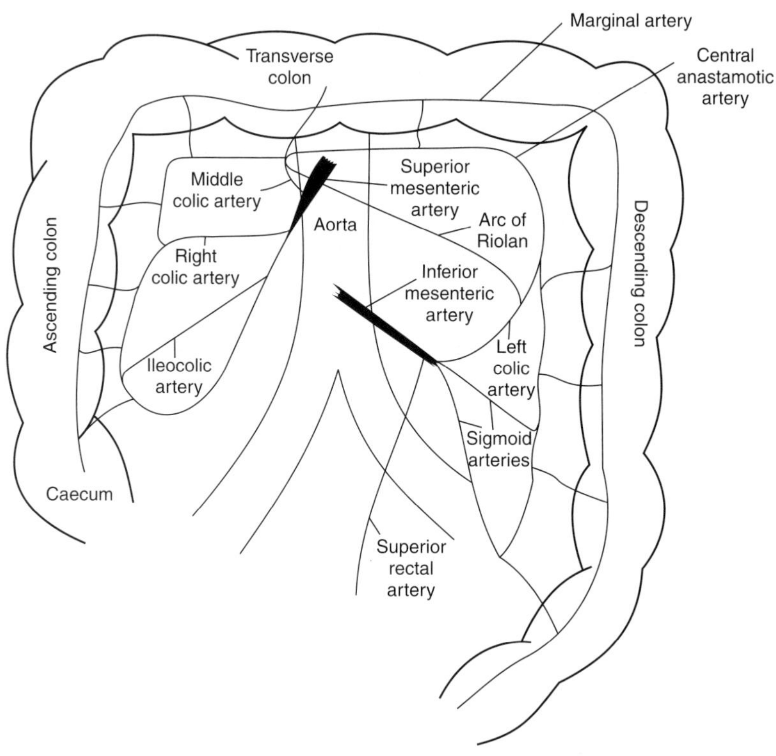

Figure 1.6 *The course of the meandering mesenteric artery.*

- between the hepatic arteries and the internal thoracic and superior epigastric arteries via the falciform ligament.[8, 12] Other sources of collateral blood flow to the liver are via the intercostal arteries and the posterior abdominal wall arteries. These collaterals have been shown to be an important source of collateral blood flow in patients with coarctation of the aorta;[7, 10]

2 between the IMA and the parietal circulation, which would occur where:

- the superior rectal artery anastomoses with the middle sacral, middle, and inferior rectal arteries to form a perirectal plexus, which may be linked, for instance, to the inferior vesical plexus. The anastomoses between the rectal arteries allow the IMA to supply the lower limbs via the iliac arteries in patients with obstruction in the lower abdominal aorta;
- the retroperitoneal plexus of Turner, which consists of vessels such as those in the renal capsule and communicates with twigs from the marginal artery;[17, 18]
- there is communication between the lumbar arteries on the posterior abdominal wall and the marginal arteries or its branches.

This brief account illustrates the extensive routes of collateral circulation available should one or more of the main mesenteric arteries or the lower abdominal aorta become occluded.[7, 10, 27]

THE MESENTERIC VEINS

The venous return from the mesenteric circulation consists of veins that usually follow their respective arteries to terminate in the hepatic portal vein, which then forms a second set of capillaries within the liver before draining into the inferior vena cava via the hepatic veins. As a result, two sets of exchange vessels exist in the mesenteric circulation.[35, 36]

The portal vein is formed behind the neck of the pancreas by the confluence of the superior mesenteric and splenic veins, which join roughly at a right angle. From here, it ascends behind the pancreas and the first part of the duodenum to enter the anterior free margin of the lesser omentum at the epiploic foramen (foramen of Winslow). It accompanies the CBD and the hepatic artery, which lie anterior to it, to reach the porta hepatis. Here, the vein divides into its terminal left and right branches, which enter the liver. The right division often receives the cystic veins before supplying the right lobe of the liver, whereas the left division supplies the left, caudate, and quadrate lobes. The left division is often joined to the para-umbilical veins via veins within the ligamentum teres (which are remnants of the left umbilical vein), and is itself linked to the inferior vena cava by the ligamentum venosum (a remnant of the fetal ductus venosus). The portal vein drains the gastrointestinal tract from the region of the lower esophagus to the lower part of the anal canal, together with the spleen, pancreas, and gallbladder.

Tributaries

The splenic vein arises from the splenic hilum, courses through the lienorenal ligament, and runs behind the pancreas and inferior to the splenic artery. It crosses the left renal hilum and the lower part of the suprarenal gland and the left renal vessels (which separates it from the left crus of the diaphragm and the left sympathetic trunk), and crosses the aorta separated from it by the SMA and left renal vein. It terminates behind the neck of the pancreas by joining the SMV at a more or less right angle to form the portal vein. Draining into the splenic vein are:

- the short gastric veins, which drain the fundus and the upper part of the greater curvature of the stomach, and traverse the gastrosplenic ligament to drain into the splenic vein;

- the left gastroepiploic vein, which drains both surfaces of the greater curvature and, like the short gastric veins, courses in the gastrosplenic ligament to join the proximal part of the splenic vein;
- pancreatic veins, which may be of small caliber and numerous, in which case they usually drain directly into the splenic vein from the posterior aspect of the body and tail of the pancreas, or of large caliber and few in number, in which case they drain into the anterior and posterior pancreaticoduodenal arcades. These arcades are venous arcades, very much like the arterial arcades. The upper end of the anterior aspect of the head is drained by the anterior superior pancreatic vein, which is formed in the anterior pancreaticoduodenal sulcus. It passes upwards, leaving the sulcus to drain into the gastrocolic trunk (see below), which ultimately drains into the SMV. The lower part of the anterior aspect of the pancreatic head drains into the anterior inferior pancreaticoduodenal vein, which is formed at the duodenal sulcus near the origin of the anterior superior pancreaticoduodenal artery. It then courses medially behind the uncinate process of the pancreas and the SMV, and into the first jejunal tributary of the SMV or the vein itself. The posterior aspect of the head of the pancreas is drained by the posterior arcade of veins formed by the posterior superior pancreaticoduodenal vein (which drains the upper part of the posterior aspect of the head of the pancreas and courses superiorly with the artery to pass behind the CBD and drain into the posterolateral aspect of the portal vein) and the posterior inferior pancreaticoduodenal vein (which drains into the SMV via either its first jejunal tributary or a common inferior pancreaticoduodenal vein);
- the left gastric vein, which drains the upper part of the lesser curvature of the stomach and courses superiorly to reach the esophageal hiatus of the diaphragm, where it receives the esophageal veins. It then passes inferiorly behind the lesser sac, accompanying its named artery, and terminates into the portal vein directly, into the terminal part of the splenic vein, or into its junction with the portal vein;
- the inferior mesenteric vein, which is formed in the region of the anal canal by the superior rectal vein, which arises from the rectal venous plexus, where it communicates with the middle and inferior rectal veins. It accompanies the named artery and passes upwards across the pelvic brim, crossing the left common iliac vessels medial to the ureter. It lies to the left of the IMA and receives the sigmoid and left colic veins as it ascends across the psoas muscle and gonadal vessels to the region of the duodenojejunal junction. Here, it lies above and behind the junction, often in the anterior wall of the paraduodenal fossa (when present). It terminates by draining into the splenic vein close to its termination into the portal vein or directly into the portal vein.

The SMV begins in the ileocolic region and follows the SMA, lying to its right. It receives the venous tributaries of the named branches of the SMA (the middle colic, right colic, and ileocolic). The main vein is formed by the confluence of this right trunk, which lies to the right of the artery, and another left trunk formed by tributaries of the upper jejunum and the inferior pancreaticoduodenal veins. The left trunk lies to the left of the artery and passes behind or occasionally anterior to the artery to drain into the right trunk.[2] Having formed the main vein, the SMV ascends to pass over the third part of the duodenum, passing anterior to the uncinate process of the pancreas and behind the neck of the pancreas to form the portal vein. Just before it drains with the splenic vein to form the portal vein, it receives the gastrocolic trunk. This vessel is formed as the right gastroepiploic vein (which drains the right part of the greater curvature of the stomach) passes into the right extremity of the transverse mesocolon and arches to the left to receive the anterior superior pancreaticoduodenal vein. It then continues into the anterolateral aspect of the SMV as this vein disappears behind the pancreas.[2]

The right gastric vein drains the lower part of the lesser curvature of the stomach, receiving the prepyloric vein (which drains the anterior aspect of the pylorus) as it passes into the free margin of the lesser omentum to drain into the portal vein or the posterior superior pancreaticoduodenal vein.

The cystic veins drain the gallbladder. Those from the upper surface of the gallbladder enter directly into the substance of the liver to drain into the hepatic veins; the others form one or two main cystic veins at the region of the neck of the gallbladder to enter the substance of the liver or join the veins draining the upper part of the extrahepatic biliary tract. It is rare to find a single cystic vein that drains into the portal vein.

The para-umbilical veins drain into the left branch of the portal vein and usually have an obliterated lumen.

PORTOSYSTEMIC ANASTOMOSIS

Dilation of the portosystemic anastomoses occurs during portal hypertensive states. The sites of these anastomoses are:[35, 36]

- at the region of the lower esophagus: the left gastric vein of the portal system and the azygous veins are confluent and result in a mixing of portal and systemic blood. This anastomosis may dilate, causing esophageal varices, and may be a cause of hematemesis in cirrhotics;
- at the region of the lower anal canal: the superior rectal vein of the portal system and the middle and inferior rectal veins, tributaries of the iliac systemic circulation, are another point of portosystemic anastomosis;
- the lumbar veins and the colic veins may form an anastomotic channel on the posterior abdominal wall in the region of the Turner's avascular area on the left and a corresponding area on the right. The posterior abdominal veins may also anastomose with the portal circulation at the region of the bare area of the liver;
- para-umbilical veins link the epigastric veins draining the abdominal wall and the left branch of the portal vein and form an important portosystemic anastomosis, as they dilate in portal hypertensive states to give the clinical sign of caput medusae;
- very rarely, the ductus venosus remains patent and links the portal vein to the inferior vena cava.[36]

REFERENCES

1 Drummond H. Some points relating to the surgical anatomy of arterial supply of large intestine. Proc Roy Soc Med 1913; **7**: 185–93.

2 Falconer CWA, Griffiths E. The anatomy of blood vessels in the region of the pancreas. *Br J Surg* 1950; **37**: 334–44.

3 Fisher DF Jr, Fry WJ. Collateral mesenteric circulation. *Surg Gynecol Obstet* 1987; **164**: 487–92.

4 Flint ER. Abnormalities of the right hepatic, cystic and gastroduodenal arteries, and of the bile ducts. *Br J Surg* 1920; **10**: 509–17.

5 Griffiths JD. Surgical anatomy of the blood supply of the distal colon. *Ann R Coll Surg* 1956; **19**: 241–56.

6 Hiatt JR, Gabbay J, Busuttil RW. Surgical anatomy of the hepatic arteries in 1000 cases. *Ann Surg* 1994; **220**: 50–2.

7 Kahn P, Abrams HL. Inferior mesenteric arterial patterns: an angiographic study. *Radiology* 1964; **82**: 429–42.

8 Kornblith PL, Scott J, Boley SJ, Whitehouse BS. Anatomy of the splanchnic circulation. *Surg Clin North Am* 1992; **72**: 1–30.

9 Kupic EA, Marshall WD, Abrams HL. Splenic arterial patterns: angiographic analysis and review. *Invest Radiol* 1969; **2**: 70–98.

10 Lipchik EO, Rob CG, Schwartzberg S. Obstruction of the abdominal aorta above the level of the renal arteries. *Radiology* 1964; **82**: 443–6.

11 Marston A. Applied anatomy of the mesenteric circulation. In: Marston A. *Intestinal ischaemia*. London: Edward Arnold, 1977: 5–22.

12 Marston A, Pegington J. Macroscopic anatomy of the splanchnic anatomy. In: Marston A, Bulkley GB, Fiddian-Green RG (eds). *Splanchnic ischaemia and multiple organ failure*. London: Edward Arnold, 1989; 3–17.

13 Meyers MA. Griffiths' point: critical anastomosis at the splenic flexure. *Am J Radiol* 1976; **126**: 77–94.

14 Michels N. The variational anatomy of the spleen and the splenic artery. *Am J Anat* 1942; **70**: 21–72.

15 Michels N. Newer anatomy of liver-variant blood supply and collateral circulation. *JAMA* 1960; **172**: 125–32.

16 Michels N. The anatomic variations of the arterial pancreaticoduodenal arcades: their import in regional resections involving the gallbladder, bile ducts, liver, pancreas and part of the small and large intestines. *J Int Coll Surg* 1962; **37**: 13–40.

17 Michels N, Siddharth P, Kornblith P, Parke WW. The variant blood supply to the small and large intestines: its import in regional resections. *J Int Coll Surg* 1963; **39**: 127–70.

18 Michels N, Siddharth P, Kornblith P, Parke WW. The variant blood supply to the descending colon, rectosigmoid and rectum based on 400 dissections. Its importance in regional resections: a review of medical literature. *Dis Colon Rectum* 1965; **8**: 251–78.

19 Michels N, Siddharth P, Kornblith P, Parke WW. Routes of collateral circulation of the gastrointestinal tract as ascertained in a dissection of 500 bodies. *Int Surg* 1968; **49**: 8–28.

20 Moskowitz M, Zimmerman H, Felson B. The meandering mesenteric artery of the colon. *Am J Roentgenol* 1964; **92**: 1088–99.

21 Nelson TM, Pollak R, Jonasson O, Abcarian H. Anatomic variants of the celiac, superior mesenteric and inferior mesenteric arteries and their clinical relevance. *Clin Anat* 1988; **1**: 75–91.

22 Netter F. *The CIBA collection of medical illustrations: the upper digestive tract*. Vol. 3, part I. Summit, NJ: Ciba Pharmaceutical Products Inc., 1959.

23 Netter F. *The CIBA collection of medical illustrations: the lower digestive tract*. Vol. 3, part II. Summit, NJ: Ciba Pharmaceutical Company, 1962.

24 Netter F. *The CIBA collection of medical illustrations: the liver, biliary tract and pancreas*. Vol. 3, part III. USA: Colorpress, 1967.

25 Netter F. The abdomen. In: Netter F. *Atlas of human anatomy*. Summit, NJ: Ciba-Geigy USA, 1989; 282–303.

26 Parke WW, Michels N, Gosh GM. Blood supply of the common bile duct. *Surg Gynecol Obstet* 1963; **117**: 47–55.

27 Rob CG, Owen K. Ligation of both celiac axis and superior mesenteric artery with survival of the patient. *Br J Surg* 1956; **44**: 247–9.

28 Rosenblum JD, Boyle CM, Schwartz LB. The mesenteric circulation, anatomy and physiology. *Surg Clin North Am* 1997; **77**: 289–306.

29 Soin AS, Friend PJ, Rasmussen A, Saxena R, Tokat Y, Alexander GJM, *et al.* Donor arterial variations in liver transplantation: management and outcome of 527 consecutive grafts. *Br J Surg* 1996; **83**: 637–41.

30 Shukla CJ, Ellis H. Tortuosity of the splenic artery, human and animal studies. *Clin Anat* 1997; **10**: 214.

31 Waizer A, Baniel J, Dintsman M. Clinical implications of anatomic variations of the splenic artery. *Surg Gynecol Obstet* 1989; **168**: 57–8.

32 Wilkie DPD. The blood supply of the duodenum with special reference to the supraduodenal artery. *Surg Gynecol Obstet* 1911; **13**: 399–405.

33 Lord RSA, Stony RJ, Wyke EJ. Coeliac axis compression. *Lancet* 1968; **2**: 795–8.

34 BMJ. Compression of the coeliac axis [leading article]. *BMJ* 1971; **4**: 378–9.

35 Sinnatamby CS. The abdomen. In: Sinnatamby CS (ed.). *Lasts anatomy regional and applied.* Edinburgh: Churchill Livingstone, 1999; 215–81.

36 Gabella G. (ed.) The cardiovascular system. In: Williams PL, Bannister LH, Berry M, Dyson M, Dussek JE, Ferguson MWJ (eds). *Gray's anatomy.* New York: Churchill Livingstone, 1995; 1451–1626.

37 Geroulakos G, Wright JG, Anderson L, Tober CT, Smead W. Use of the splenic and hepatic artery for renal revascularisation in patients with atherosclerotic renal artery disease. *Ann Vasc Surg* 1997; 85–9.

38 Cheshire NJ, Glazer G. Stomach and duodenum. In: Zinner M, Schwarts S, Ellis H (eds). *Maingot's abdominal operations*, Vol. 1. Stanford: Appleton and Lange, 1997; 913–40.

2

Pathophysiology of the visceral circulation

ULF H. HAGLUND

INTRODUCTION

The splanchnic area receives approximately 25% of the cardiac output in resting man. The pathophysiological events caused in different disease conditions may therefore be of interest for this fact alone. Since there is growing evidence that the pathophysiological changes in the splanchnic area in different disease conditions may by themselves contribute significantly to the development of critical illness, there is good reason to pay interest to, and try to understand the underlying processes in, the visceral reactions to hypotension, hemorrhage, and sepsis.

Functional anatomy of the mesenteric circulation

The gastrointestinal vascular system is built up of several circuits coupled in parallel. These circuits are directed to the different layers of the gastrointestinal wall, such as the mucosa and the muscular layer. Each circuit is made up of segments coupled in series:[1] the resistance vessels, the precapillary sphincters, the exchange vessels, the post-capillary sphincters, and the capacitance vessels (Figure 2.1). The resistance vessels correspond anatomically mainly with the arterioles. The major part of the resistance to blood flow and consequently the pressure drop over the circuit takes place in this segment. The precapillary resistance segment or the precapillary sphincters do not contribute very much to the total vascular resistance of the segment, but the balance between the pre- and post-capillary sphincters determines the mean hydrostatic capillary pressure and hence the direction and the magnitude of any net fluid movement across the capillary wall. There are no anatomical sphincters identified pre- or post-capillary, and the word 'sphincter' in this context is used as a functional entity. The capillary or the exchange segment is the one in which the important exchange of fluid and metabolites take place. Finally, the capacitance segment is the one that contains the major part

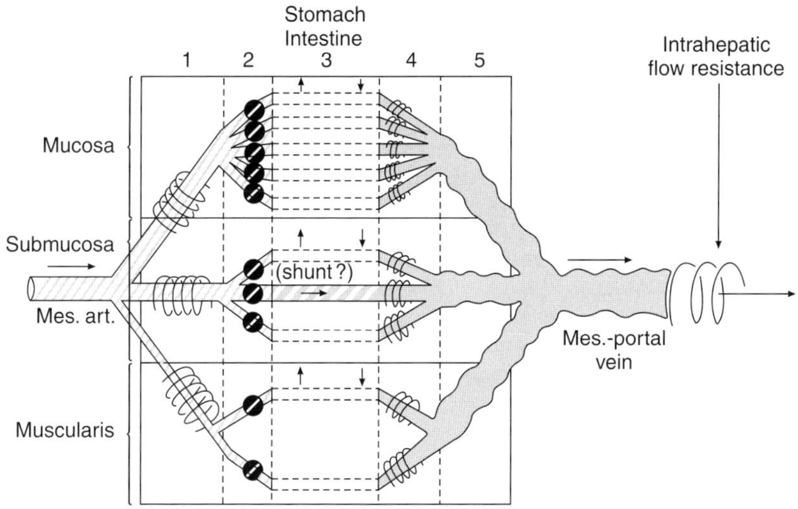

Figure 2.1 *The gastrointestinal vascular system is built up from circuits coupled in parallel, each of which consists of sections coupled in series. Reproduced from Folkow B. Regional adjustments of intestinal blood flow.* Gastroenterology *1967;* ***52****: 423–32.*

of the regional blood volume. This segment corresponds mainly to the veins and the larger venulae.

Capillary pressure and net fluid movements across the capillary walls

All blood directed to the splanchnic organs has to pass two capillary beds – that of the primary organ as well as the hepatic capillary bed – before returning to the heart. To maintain a driving force for blood in the secondary portal circulation, the venous pressure of the first capillary bed (e.g. the small intestine) has to be unusually high. However, the balance between the pre- and post-capillary resistance ('sphincters') is such that the mean hydrostatic capillary pressure prevents net movements of fluid over the capillary walls.[2] This is important since the huge capillary bed of the viscera and the much higher capillary permeability in these capillaries compared with, for example, skeletal muscle capillaries could create a potential for very significant fluid losses.

Visceral vascular response to sympathetic nerve stimulation

Neither stimulation of the sympathetic vasoconstrictor nerves nor intra-arterial infusion of noradrenaline and other vasoconstriction substances change the Starling equilibrium across the capillary wall.[2,3] Not even at very reduced blood pressure levels,[4] or in shock states induced by, for example, hemorrhage or sepsis,[5,6] is there any net movement of fluid across the gastrointestinal vascular bed, unlike that demonstrated in, for example, skeletal muscle (Figure 2.2).[7]

The gastrointestinal reaction upon sympathetic vasoconstrictor nerve stimulation (or intra-arterial infusion of noradrenaline) also differs in other respects from that of skeletal muscle. In the gut, sympathetic nerve activation induces a prompt vasoconstriction, which subsides after about a minute despite ongoing nerve activation.[3] A couple of minutes after

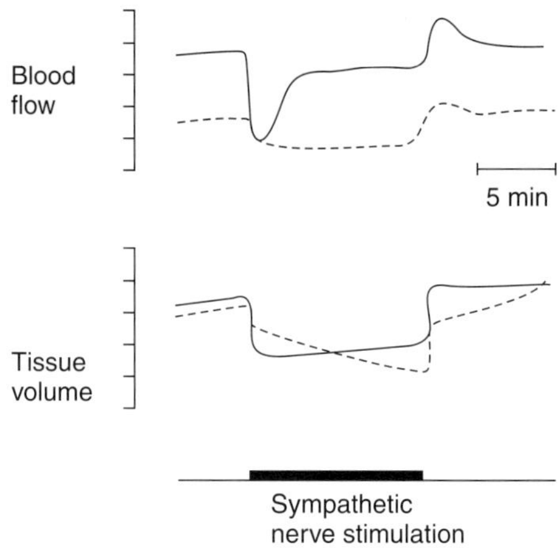

Figure 2.2 *The effects of sympathetic nerve stimulation on blood flow (resistance vessels) and tissue volume (capacitance vessels and changes in the pre- to post-capillary resistance ratio) in the intestinal vascular bed (solid line) compared with the reactions in skeletal muscle (interrupted line).*

initiating the vasoconstrictor nerve activation, a steady-state situation is reached for the remainder of the nerve activation period. During this period, blood flow is reduced only slightly, and resistance to blood flow is increased only slightly (5–10%), compared with the prestimulatory period. This effect is referred to as the gastrointestinal autoregulatory escape from the response to sympathetic nerve stimulation. Despite this autoregulatory escape, there is a paradoxical hyperemic response upon cessation of the nerve stimulation (or infusion of catecholamines): reactive hyperemia. The autoregulatory escape of the blood flow response following infusion of vasoconstrictors is sometimes also referred to as tachyphylaxis. It is of interest to note that tachyphylaxis has not been demonstrated for the vasoconstrictor response following infusion of angiotensin or vasopressin.[8]

The blood volume contained in the capacitance vessels (mainly venous vessels) is reduced by about 40% in response to maximal gastrointestinal vasoconstrictor nerve stimulation. This mobilization of blood participates in the autotransfusion by which man may mobilize about half a liter of blood as a first compensatory reaction upon hypovolemia.

Visceral vasoconstriction at hypovolemia

The natural response of smooth muscle is to relax when tension is reduced and to contract upon increased tension.[9] The smooth muscles of the visceral blood vessel walls (mainly the resistance vessels) are no exception. Consequently, the isolated visceral vascular bed would autoregulate its blood flow over a fairly wide blood pressure range.[10] This means that changes in the vasculature tend to guarantee a more or less constant blood flow, even if blood pressure varies.

Pathophysiology of the visceral circulation during hypovolemia and other conditions of critical illness, however, is dependent not only on the natural reactions of the vascular smooth muscle cells of the visceral vessels but even more on the influences the systemic (i.e. non-visceral) parts of the body exert on the visceral vasculature and its smooth muscle cells.

During hemorrhage and in septic states, gastrointestinal blood flow is reduced in proportion to the reduction in cardiac output, or sometimes more than what could be expected from the change in cardiac output. Especially following high-volume losses (bleeding of more than 35% of the blood volume), there is a very significant and disproportionate visceral vasoconstriction.[11] It follows from the discussion above that the mediators of this visceral vasoconstrictor response are not the effects of the sympathetic nervous influence; neither is it due to catecholamine-dependent mechanisms, since it is not influenced to any significant extent by alpha-adrenergic receptor blockade. Instead, data indicate the renin–angiotensin axis to be of significant importance for the intense visceral vasoconstriction seen in shock. This response is blocked effectively by, for example, the renin–angiotensin axis blocker enalapril (or prior nephrectomy). Vasopressin has also been discussed as an important factor contributing to visceral vasoconstriction in hypovolemia, but recent data indicate that confirmed blockade to vasopressin did not influence this response significantly.[11]

In this chapter, we discuss some particularly important aspects of gut pathophysiology in response to reduced blood pressure, hemorrhage or sepsis. Particular emphasis will be put on the changes taking place in the gut and how these may have systemic influences, i.e. on the non-mesenteric part of the body.

REACTIONS WITHIN THE GUT

The main event in the gut as a consequence of hypotension, hypovolemia or sepsis is tissue ischemia. The part of the intestinal wall most sensitive to ischemia, and the part first suffering from ischemia, is the mucosa, particularly the superficial part of the mucosa. When blood flow to the intestine is reduced by partial obstruction of the superior mesenteric artery blood flow or by inducing sepsis, blood flow is reduced to all layers of the intestine. However, blood flow to the mucosa is preferred, and a larger fraction of the total intestinal blood flow is directed to this layer.

Visceral tissue oxygenation

In most conditions, gastrointestinal tissue ischemia is caused by a reduced delivery of oxygen, which in turn is usually caused by inadequate blood flow. Following an embolus in the main trunk of the superior mesenteric artery or hypovolemia due to heavy bleeding, the cause of the reduced oxygen delivery is obvious. In addition, the vascular anatomy of the superficial part of the small-intestinal mucosa, the villi, is such that it creates a countercurrent exchange, exaggerating the tissue hypoxia.[12] The villi are supplied by one arterial vessel running in the middle. This vessel does not branch until it reaches the tip of the villus. The blood is transported back from the tip in a network of capillaries and venulae running in the periphery of the villus. The distance between the two sets of vessels running in parallel with mainly opposite directions of flow is less than 20 μg, which allows diffusion equilibrium of easily diffusible substances such as oxygen (Figure 2.3).[9] The countercurrent mechanism provides the prerequisites for a short-circuiting of oxygen at the base of the villi, causing hypoxia at the tips despite more than adequate blood flow. In support of this concept, experimental data have demonstrated a shorter distance to travel from the arterial to the venous side of the small intestine for oxygen than for hemoglobin.[13] The latter cannot short-circuit the vascular loop of the villus. Furthermore, measurements of tissue pO_2 have revealed very low values at the tip of the villus in normal conditions.[14] In low flow conditions, the number of perfused villi is increased and the blood flow velocity becomes markedly reduced.[15] Consequently, there will

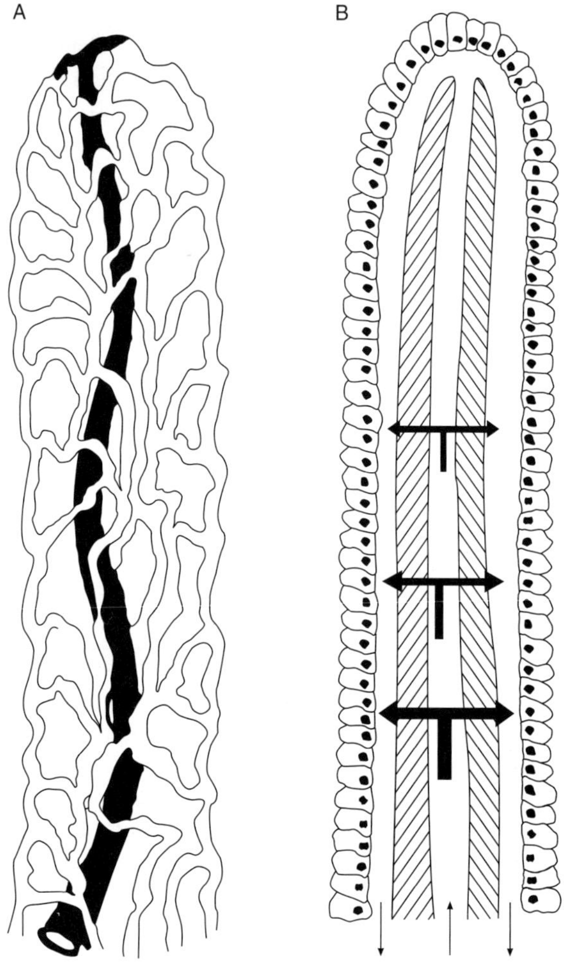

Figure 2.3 *Schematic illustration of the vascular arrangement of the small intestinal villus. Note that the short distance between the centrally located supplying, arterial vessel and the peripheral draining vessels constitutes the prerequisite for a counter-current exchange of easily diffusible substances. Adapted from Lundgren O. Studies on blood flow distribution and countercurrent exchange in the small intestine.* Acta Physiol. Scand. Suppl. *1967;* ***303****: 1–42.*

be more time for diffusion, hence the possibility of the countercurrent system to create a more effective short-circuiting of oxygen from the tips of the villi increases.[16] Therefore, in hypotensive states the tips of the villi may become almost anoxic despite a fairly normal volume of blood flowing through the intestines.

In sepsis, the situation regarding tissue oxygenation in the viscera is complicated by an increased tissue demand for oxygen,[17] and at the same time an impaired ability to extract and utilize the available oxygen.[18] The main bulk of information obtained in this field of research is from experiments in which the oxygenation of the superficial part of the mucosa has been followed using a tonometric technique. Fiddian-Green and co-workers originally described the use of this method to monitor tissue oxygenation in the gut of critically ill patients.[19, 20] The tonometer allows measurements of luminal pCO_2. This has been demonstrated experimentally to be in equilibrium with the pCO_2 of the superficial part of the mucosa.[21] Provided enough time has been given to allow for equilibrium, luminal pCO_2 can be used to measure tissue pCO_2.

This value can then be used to calculate the tissue pH using the Henderson–Hasselbalch equation. If so, the bicarbonate concentration of the mucosal tissue is approximated as equal to arterial bicarbonate. This technique has been validated for normal conditions, for endotoxemia, and for reduced blood flow caused by partial occlusion of the superior mesenteric artery.[22] However, the tissue pH obtained by this technique following total occlusion of the superior mesenteric artery will be an underestimation of the metabolic consequences.[22] Because of this and other factors, many would argue that measuring the difference between tissue pCO_2 and end-tidal pCO_2 (pCO_2 gap) or arterial pH and tissue pH (pH gap) would reflect more accurately the metabolic situation and tissue oxygenation in some circumstances.[23] However, most of our clinical knowledge regarding correlation between gut mucosal oxygenation and outcome of crtical illness has been based on calculations of tissue pH (see below).

The blood flow of the gut is designed to guarantee not only adequate supply of oxygen to the gastrointestinal tract and tissue survival but the absorptive function of the gastrointestinal tract.[10] Consequently, visceral blood flow can be reduced to about half of its normal level without any change of tissue pO_2 or any detectable biochemical or morphological evidence of tissue injury.[24] The consequences of further reduction of the gastrointestinal blood flow are dependent on the extent of ischemia induced and its duration.[25] If blood flow to the gut is reduced to less than 50% but more than 30%, there will be detectable changes in capillary and mucosal permeability apparent within an hour. If blood flow becomes reduced to about 20–25% of control, morphological injury at the tip of the villi will be detectable after 1–2 h. Injury caused by partial ischemia does not induce injury to layers of the intestinal wall deeper than the superficial part of the mucosal layer. If ischemia is total or near total, tissue injury may be detectable sooner (after 30–40 min), and the extent of injury may be deeper. After 8–24 h, total ischemia will cause transmural gangrene. If venous congestion (strangulation) is added to or causing the reduced inflow of oxygenated blood, tissue injury is evident sooner and the degree of injury becomes more profound than that caused by a similar period of reduced inflow of blood alone.

Reperfusion injury

In addition to injury caused during ischemia, there may also be a reperfusion component of tissue injury apparent *after* ischemia at reperfusion (Table 2.1). Granger and co-workers, in experiments in which the mucosal permeability of the small intestine was measured following partial ischemia and reperfusion, first described the reperfusion injury.[26] Several mechanisms have been proposed as the pathophysiology of this reperfusion component of tissue injury.[27] During ischemia, energy-containing nucleotides are metabolized. The further metabolism to

Table 2.1 *Causes of ischemic gastrointestinal injury*

A. *Ischemic part*
Reduced oxygen delivery (>50% reduction)
Short-circuiting of oxygen in the villi
In sepsis:
Increased demand of oxygen
Impaired extraction of oxygen
Impaired utilization of oxygen
B. *Reperfusion part*
Increased generation of oxygen free radicals at reperfusion

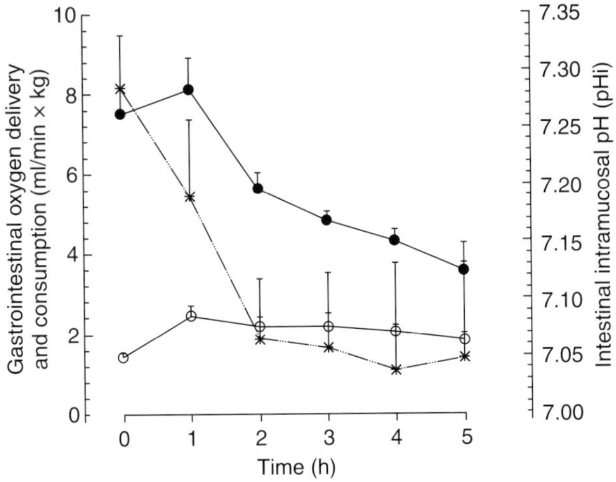

Figure 2.4 *Gastrointestinal oxygen delivery (solid circles), pH of the superficial mucosa of the ileum (asterisks) and gastrointestinal oxygen consumption (open circles) during experimental peritonitis induced by fecal instillation. Note that, despite increased oxygen consumption, the mucosa rapidly becomes acidotic indicating an increased and unmet demand of oxygen during sepsis. Reproduced from Rasmussen I, Haglund U. Early gut ischemia in experimental fecal peritonitis.* Circ Shock *1992;* ***38****: 22–8.*

uric acid is halted at hypoxanthine, since the hypoxanthine-to-xanthine step requires the presence of oxygen, which is lacking by definition.[28] This step is catalyzed mainly by the enzyme xanthine dehydrogenase (D). During ischemia, D is converted rapidly to xanthine oxidase (O) in a proteolytic process (D to O conversion).[29] Xanthine oxidase catalyzes the conversion of hypoxanthine to xanthine in the presence of oxygen by adding an electron to oxygen, creating a superoxide anion (superoxide radical).[26] At reperfusion, when oxygen is available again, the substrate is accumulated and xanthine oxidase will create more superoxide anions than could be handled by the natural defense mechanisms.[28] With the creation of superoxide anions, a chain reaction is started, since superoxide radicals react rapidly with molecules in the vicinity, creating new radicals. Radicals may cause tissue injury by themselves, but it is considered more likely that the tissue injury seen in the reperfusion component is secondary to endothelial cell injury and activation of polymorphonuclear white blood cells by superoxide radicals.[27]

The reperfusion component of ischemic tissue injury is detectable only if the ischemia component leaves a space for further injury.[25, 30] Reperfusion injury in the gastrointestinal tract is detectable following partial ischemia with blood flow reduction of more than 80%. This holds true for permeability changes as well as morphologic injury of the villous layer. In fact, for this type of injury 2 h of partial ischemia followed by 1 h of reperfusion has been demonstrated to cause more severe injury than 3 h of partial ischemia. It should be noted that following periods of total or near-total ischemia, it is difficult to demonstrate any significant reperfusion component of tissue injury.

Special effects caused by sepsis

In sepsis, blood flow to the splanchnic area is unlikely to be reduced by more than 50%. Characteristic morphologic injury indicating tissue hypoxia was evident in animal experiments (anesthetized pigs) within 3–5 h following induction of peritonitis in more than 50% of the experiments.[31] In such experiments, it has been demonstrated that the oxygen consumption of the gut of the septic animals is higher than that found in controls (Figure 2.4).[32, 33] Similar findings have been reported from septic patients who were found to have a lower oxygen content in the hepatic venous blood compared with age-matched controls suffering from significant trauma but not sepsis.[17] In addition, it was found in a series of porcine experiments that if intestinal blood flow was reduced to 50% of control in one series of animals by infusion of endotoxin and in another series by mechanical partial occlusion of the superior mesenteric artery, decreased tissue pH and increased tissue permeability could be demonstrated only in the endotoxic animals.[34] This strongly indicates that sepsis causes more damage to the tissues than could be explained by reduced blood supply alone.[35] In other experimental series, it has been demonstrated that the characteristic ischemic intestinal tissue injury develops despite a high tissue oxygen partial pressure level,[36] indicating a pathologically impaired utilization of oxygen in sepsis. Other experiments have demonstrated a reduced capacity for the visceral tissue to extract oxygen from arterial blood.[18]

A special aspect of hepatic circulation

The liver obtains arterial blood by the hepatic artery and venous blood, normally with high oxygen content, by the portal vein. The oxygen delivered to the liver comes in approximately equal parts from these two sources. However, in conditions characterized by impaired gastro-intestinal blood flow, and in which the oxygen extraction by the gastrointestinal tract is increased significantly, such as in hypotension, the oxygen content of portal blood becomes very much reduced.[3] The portal circulation then helps only very little to supply the liver according to its oxygen demand. The liver then becomes more or less dependent on the arterial supply for tissue oxygenation.

SYSTEMIC EFFECTS OF VISCERAL ISCHEMIA

It was demonstrated some 50 years ago that if the intestines were perfused with oxygenated blood at normal blood pressure during a period in which the remainder of the body suffered from hemorrhagic hypotension, mortality was prevented.[37] A similar cross-perfusion of other parts of the body during hemorrhagic shock did not exert a similar preventive effect. Over the years, several mechanisms for the systemic effects of visceral ischemia have been proposed. Some of these are summarized in Table 2.2 and are discussed below.

Using the tonometric technique described above, it was demonstrated by several groups that a low intramucosal pH (pHi) at admission to an intensive care unit predicts a bad outcome.[38, 39] Patients that had low pHi at admission but turned normal after 12 or 24 h of intensive care treatment had significantly lower mortality than those that arrived with a low pHi and stayed low despite treatment. Along the same lines, those that had normal pHi and stayed normal had significantly lower mortality compared with those that developed a low

Table 2.2 *Proposed mechanisms for systemic effects of intestinal ischemia*

Release of cardiotoxic substances to the blood
Release of other toxic substances, including mediators
Increased gut mucosal permeability
Translocation of bacteria and endotoxin
Effects on the gut as an immune organ
Hepatic consequences of gut ischemia

pHi despite treatment. In another study, a mixed group of medical and surgical intensive care unit patients were randomized to either best conventional intensive care treatment or, in addition, monitoring of gastric pHi and normalization of pHi as a goal of the treatment.[40] For those patients that had a normal pHi at admission, the treatment goal maintaining pHi normal was associated with increased survival. However, there was no difference in outcome between groups entering the study with abnormally low pHi.

Release of cardiotoxic substances from the viscera into the systemic circulation

Cardiac dysfunction is evident in many shock conditions, and it has been proposed that release of substances with a negative influence on the contractility of the heart from the ischemic viscera may be responsible. Several authors have demonstrated that plasma from patients in shock contains material that depresses the contractility of the heart in situ, ex vivo, or on cultured myocardial cells.[41, 42] The latter technique of assessing possible cardiotoxic effects has been most commonly employed. Furthermore, experimental data demonstrate similar effects in plasma from the visceral region following shock and hypotensive states. The small intestine has been proposed as the site of origin of cardiotoxic factors[43–45] as well as circulating factors causing lung injury.[46]

We were able to demonstrate a cardiotoxic effect in intestinal venous blood obtained during the first minutes of reperfusion following 2 h of regional intestinal hypotension simulating severe hemorrhagic shock (partial ischemia).[44] The effect on the heart was demonstrated in situ as well as in vitro preparations.[47] Filtering processes demonstrated that the activity was contained in a molecular mass of 500–1000 daltons,[48] i.e. similar to that described for myocardial depressant factor (MDF). It has been demonstrated recently that the effect on the heart of myocardial depressant substances could be blockade of the calcium channels of the myocardial cells.[49]

Release of toxic mediators

A number of plasma factors have been described from several laboratories as being released from the gastrointestinal tract following hypotension and shock causing pulmonary injury.[50, 51] Release of proteases from the splanchnic viscera has also been demonstrated.[52] Such release could exert harmful effects by itself as well as initiate the development of organ injury at distant sites. Other factors that have been shown to be released from the gastrointestinal tract following hypotensive states, and that have the potential to induce distant organ injury, include mediators such as tumor necrosis factor alpha (TNFα) and different interleukins.[53] Many of these mediators are produced by macrophages, and the gastrointestinal tract is the organ with the highest number of such cells. Many investigators in the field believe that multiple organ dysfunction develops because the defense mechanisms of the body cannot prevent a local release of such mediators from being converted to a systemic overflow.[54] If this hypothesis is true, the gastrointestinal tract represents a tissue with a very strong potential to induce such a systemic overflow.

Increased gut mucosal permeability

The intestinal mucosal injury characteristic of hypoxia is confined to the superficial part of the mucosa.[55–57] In its early stages, it is characterized as tissue edema, and the epithelial cells at the tips of the villi are lifted from the core. Longer duration and/or more extensive hypoxia creates a more advanced injury, with loss of the epithelial covering of the villi.[25] In a situation

Table 2.3 *Possible functional implications of superficial injury of the mucosa of the gastrointestinal tract*

Increased translocation of bacteria and endotoxins
Impaired absorption of fluids and food
Fluid losses into the gut lumen
Hemorrhage

characterized by extensive mucosal tissue injury, it is easy to assume that the permeability over the mucosal membrane is quite high and that the normal barrier between the exterior environment and the sterile interior is broken.

In the stomach, the superficial mucosal injury is known as stress ulceration.[58] Impaired mucosal oxygenation (reflected as low tissue pH) is one important risk factor for the development of bleeding stress ulceration in critical illness.[59] Other important risk factors include high acid secretion, as reflected in a low luminal pH, and reflux of duodenal contents into the gastric lumen.[56, 60] Superficial mucosal injury is demonstrated experimentally in the colon. However, it seems as if the colon is more resistant to ischemia than the small intestine and the stomach.

The development of superficial mucosal injury in the gastrointestinal tract has several possible implications (Table 2.3).

ABSORPTION OF FLUID AND NUTRIENTS

This has not been well studied during acute critical illness. However, from several investigations it seems that early enteral nutrition has a beneficial influence on the disease process, reduces the incidence of infectious complications, and shortens recovery in many critical surgical conditions. Parenteral nutrition itself may cause a mucosal injury very similar to that caused by ischemia.[61] Moreover, enteral nutrition, particularly if the nutrients are glutamine enriched, is proposed to prevent or ameliorate superficial mucosal injury caused by shock.

FLUID LOSSES TO THE ALIMENTARY TRACT

Such losses are not commonly regarded as a clinical problem following hemorrhage or sepsis, but this phenomenon could add to the clinical shock that develops following mesenteric vascular occlusion.

BLEEDING STRESS ULCERATIONS

Blood loss into the gastric lumen is the characteristic clinical symptom of stress ulcerations. Such bleeding can be profuse, and the development of stress ulcerations during critical illness increases the risks and mortality significantly.[58] The incidence of bleeding stress ulcerations has diminished over the last few decades. It is unlikely that this is due only to the fact that more effective acid-secretion blockers are available today. It is proposed that better general management of critically ill patients preventing tissue hypoxia is probably at least as important. Keeping luminal pH above 4 has been shown to be an effective prophylactic measure against bleeding stress ulcerations. It is likely that enteral feeding provides a very effective prophylactic measure.

Translocation of bacteria and endotoxin

This term originally referred to the passage of bacteria or bacterial products through the normal intact gastrointestinal mucosa. With time, the term has come to include passage through an injured mucosa. Nevertheless, it is still an open question whether translocation is related qualitatively and quantitatively to changes in the mucosal permeability or whether primarily it reflects something else. The concept does not represent a new idea, and the hypothesis that passage of bacteria and endotoxin may occur during shock states and be of significant importance for the outcome has been suggested by several authors. The main proponents were Jacob Fine and co-workers.[62] In the 1950s, this group of investigators suggested that absorption of bacteria and endotoxin from the gut lumen was the common complication in shock, thus converting all types of shock regardless of initial etiology into septic shock.[63] This was regarded by many as an oversimplification, and among other things it was demonstrated that germ-free animals also died following severe hemorrhage.[64] The concept was revitalized some decades later,[65] and it is now discussed as bacterial and endotoxin translocation. The concept is still controversial regarding the incidence of translocation and its relevance in clinical medicine. Furthermore, the discussion is not very different to that in the 1950s.

There is much evidence supporting the concept that translocation is common in small experimental animals, such as mice and rats.[65] Translocation has also been demonstrated in larger experimental animals, such as pigs and monkeys, but the phenomenon does not occur in such animals as frequently as it does in rodents. Experiments in pigs subjected to peritonitis have revealed that bacterial flora and their virulence are considerably more important for translocation to occur than the grade of physiological challenge.[66] The route for translocation could be direct invasion of the portal venous blood, transportation through the lymphatic route, or direct passage through the intestinal wall to the abdominal cavity. Experimental evidence suggests the lymphatic route to be the quantitatively most important one.[67] Therefore, culturing mesenteric lymph glands may be a better method to detect bacterial translocation than culturing portal venous blood.

In man, translocation to mesenteric lymph nodes has been reported in about 5–10% of all general surgical operations.[68, 69] Translocation was associated with no higher mortality but with higher incidence of infection complications.[69] Furthermore, translocation has been reported much more frequently among patients operated upon because of ileus or inflammatory bowel disease. Furthermore, early enteral nutrition was associated with fewer infectious complications than parenteral nutrition in trauma patients.[70] Moreover, selective gut decontamination using non-absorbable antibiotics reduced the complication rate and the mortality in patients with severe pancreatitis.[71] On the other hand, translocation was not commonly found in the portal venous blood of trauma patients.[72]

It seems logical to conclude that bacterial translocation may occur in man, and that this phenomenon can be of potential harm to some surgical patients. However, in other patients, bacterial translocation does not seem to be an important pathophysiological part of the disease process. We do not yet understand the differences between these two groups of patients; indeed, the difference may be related more to the bacteriology than the patients.

Effects on the gut as an immune organ

Gut ischemia may induce worse functional changes than the obvious permeability changes and the loss of the intestinal mucosa as the barrier. The gut contains the largest accumulation of tissue-bound macrophages and lymphocytes in the body, and it is likely that these have a role to play there. Recent data indicate that the cellular components of the immune system of the gut change significantly during development of shock.[73] In this study, a reduced number of lymphocytes and an increased number of macrophages were demonstrated in the small-intestinal mucosa of rats 24 h after induction of peritonitis with the cecal ligation and puncture technique. It is too early to comment on any functional implications of this cellular shift, but it could be speculated that bacterial translocation and septic complications could be explained more by changes in the resistance to bacteria than to increased permeability of the mucosa.

Hepatic consequences of gut ischemia

As discussed above, the liver is supplied with oxygen from two roughly equal sources: the hepatic artery and the portal vein. In situations characterized by inadequate supply of oxygen to the gut, this organ increases the extraction of oxygen from the perfusing blood. In hemorrhagic shock, this extraction can be up to 75% of the supplied oxygen content. However, in sepsis, the ability to extract oxygen is impaired and the intestines cannot compensate for reduced delivery.[37] Still, the increased extraction of oxygen by the gut means less oxygen is available downstream for the liver. Following hemorrhage, therefore, the hepatic oxygen supply (DO_2) becomes more impaired than that of the gut, and flow-dependent ischemia is likely to develop early during the course of shock.[32] This may have profound effects for the further development of the shock state, since a maintained functional capacity of the liver is important for defense reactions and recovery. It is therefore possible that, to a large extent, the pathophysiological explanation for an important role of the gut in shock may be that the compensatory defense reactions to impaired blood supply as a side effect cause hepatic ischemia.

REFERENCES

1 Folkow B. Regional adjustments of intestinal blood flow. *Gastroenterology* 1967; **52**: 423–32.

2 Folkow B, Lewis DH, Lundgren O, Mellander S, Wallentin I. The effect of the sympathetic vasoconstrictor fibres on the distribution of capillary blood flow in the intestine. *Acta Physiol Scand* 1964; **61**: 458–66.

3 Folkow B, Lewis DH, Lundgren O, Mellander S, Wallentin I. The effect of graded vasoconstrictor fibre stimulation on the intestinal resistance and capacitance vessels. *Acta Physiol Scand* 1964; **61**: 445–57.

4 Haglund U, Lundgren O. Reactions within the consecutive vascular sections of the small intestine of the cat during and after experimentally induced local hypotension. A preliminary report. *Rev Surg* 1972; **29**: 59–62.

5 Haglund U. Vascular reactions in the small intestine of the cat during hemorrhage. *Acta Physiol Scand* 1973; **89**: 129–41.

6 Falk A, Kaijser B, Myrvold H, Haglund U. Intestinal vascular and central hemodynamic responses following i.v. infusion of live *E. coli* bacteria. *Circ Shock* 1980; **7**: 239–50.

7 Lewis DH, Mellander S. Competitive effects of sympathetic control and tissue metabolites on resistance and capacitance vessels and capillary filtration in skeletal muscle. *Acta Physiol Scand* 1962; **56**: 162–88.

8 McNeill JR, Stark RD, Greenway CV. Intestinal vasoconstriction after hemorrhage: roles of vasopressin and angiotensin. *Am J Physiol* 1970; **219**: 1342–7.

9 Johnson PC, Hanson KM. Effect of arterial pressure on arterial and venous resistance of intestine. *J Appl Physiol* 1962; **17**: 503–8.

10 Folkow B, Neil E (eds). *Circulation*. London: Oxford University Press, 1971.

11 Toung T, Reilly PM, Fuh KC, Ferris R, Bulkley GB. Mesenteric vasoconstriction in response to hemorrhagic shock. *Shock* 2000; **13**: 267–73.

12 Wallentin I. Intestinal circulation. *Acta Physiol Scand* 1966; **69** (suppl 279): 3–38.

13 Kampp M, Lundgren O, Nilsson NJ. Extravascular shunting of oxygen in the small intestine of the cat. *Acta Physiol Scand* 1968; **72**: 396–403.

14 Bohlen HG. Intestinal tissue PO_2 and microvascular responses during glucose exposure. *Am J Physiol* 1980; **238**: H164–71.
15 Lundgren O, Svanvik J. Mucosal hemodynamics in the small intestine of the cat during reduced perfusion pressure. *Acta Physiol Scand* 1973; **88**: 551–63.
16 Haglund U, Jodal M, Lundgren O. The small bowel in arterial hypotension and shock. In: Shepherd AP, Granger DN (eds). *Physiology of the intestinal circulation*. New York: Raven Press, 1984: 305–19.
17 Dahn M, Lange P, Lobdell K, Bijaya H, Jacobs L, Mitchell R. Splanchnic and total body oxygen consumption differences in septic and injured patients. *Surgery* 1987; **101**: 69–80.
18 Antonsson JB, Haglund UH. Gut intramucosal pH and intraluminal pO_2 in a porcine model of peritonitis of hemorrhage. *Gut* 1995; **37**: 791–7.
19 Grum CM, Fiddian-Green RG, Pittenger GL, Grant BJB, Rothman ED, Dantzker DR. Adequacy of tissue oxygenation in intact dog intestine. *J Appl Physiol* 1984; **56**: 1065–9.
20 Fiddian-Green R. Should routine measurements of tissue pH and PO_2 be included in the routine monitoring of intensive care unit patients? *Crit Care Med* 1991; **19**: 141–3.
21 Bergofsky EH. Determination of tissue O_2 tensions by hollow visceral tonometers: effect of breathing enriched O_2 mixtures. *J Clin Invest* 1964; **43**: 193–200.
22 Antonsson J, Boyle CC III, Kruithoff KL, Wang H, Sacristan E, Rothschild HR, Fink MP. Validation of tonometric measurement of gut intramural pH during endotoxemia and mesenteric occlusion in pigs. *Am J Physiol* 1990; **259**: G519–23.
23 Van der Meer TJ, Wang H, Fink M. Endotoxemia causes ileal mucosal acidosis in the absence of mucosal hypoxia in a normodynamic porcine model of septic shock. *Crit Care Med* 1995; **23**: 1217–26.
24 Bulkley GB, Kvietys PR, Parks DA, Perry MA, Granger DN. Relationship of blood flow and oxygen consumption to ischemic injury in the canine small intestine. *Gastroenterology* 1985; **89**: 852–7.
25 Haglund U, Bulkley GB, Granger DN. On the pathophysiology of intestinal ischemic injury. *Acta Chir Scand* 1987; **153**: 321–4.
26 Granger DN, Rutili G, McCord JM. Superoxide radicals in feline intestinal ischemia. *Gastroenterology* 1981; **81**: 22–9.
27 Bulkley GB. Reactive oxygen metabolites and reperfusion injury: aberrant triggering of reticuloendothelial function. *Lancet* 1994; **344**: 934–6.
28 Schoenberg M, Muhl E, Sellin D, Younes M, Schildberg FW, Haglund U. Posthypotensive generation of superoxide free radicals – possible role in the pathogenesis of the intestinal mucosal damage. *Acta Chir Scand* 1984; **150**: 301–9.
29 Parks DA, Williams TK, Beckman JS. Conversion of xanthine dehydrogenase to oxidase in ischemic rat intestine: a reevaluation. *Am J Physiol* 1988; **254**: G768–74.
30 Bulkley GB. Free radical-mediated reperfusion injury: a selective review. *Br J Cancer* 1987; **55** (suppl VIII): 66–73.
31 Wollert S, Antonsson J, Gerdin B, Lundberg D, Rasmussen I, Haglund U. Intestinal mucosal injury during porcine faecal peritonitis. *Eur J Surg* 1995; **161**: 741–50.
32 Arvidsson D, Rasmussen I, Almqvist P, Niklasson F, Haglund U. Splanchnic oxygen consumption in septic and hemorrhagic shock. *Surgery* 1991; **109**: 190–7.
33 Rasmussen I, Haglund U. Early gut ischemia in experimental fecal peritonitis. *Circ Shock* 1992; **38**: 22–8.
34 Fink MP, Antonsson JB, Wang H, Rothschild HR. Increased intestinal permeability in endotoxic pigs. *Ann Surg* 1991; **126**: 211–18.
35 Fink M. Cytopathic hypoxia in sepsis. *Acta Anaesthesiol Scand* 1997; **110** (Suppl):87–95.
36 Antonsson JB, Engström L, Rasmussen I, Wollert S, Haglund UH. Changes in gut intramucosal pH and gut oxygen extraction ratio in a porcine model of peritonitis and hemorrhage. *Crit Care Med* 1995; **23**: 1872–81.

37 Lillehei RC. The intestinal factor in irreversible hemorrhagic shock. *Surgery* 1957; **42**: 1043–54.
38 Doglio GR, Pusajo JF, Egurrola MA, Bonfigli GC, Parra C, Vetere L, *et al.* Gastric mucosal pH as a prognostic index of mortality in critically ill patients. *Crit Care Med* 1991; **19**: 1037–40.
39 Maynard N, Bihari D, Beale R, Smithies M, Baldock G, Mason R, McColl I. Assessment of splanchnic oxygenation by gastric tonometry in patients with acute circulatory failure. *JAMA* 1993; **270**: 1203–10.
40 Gutierrez G, Palizas F, Doglio G, Wainsztein N, Gallesio A, Pacin J, *et al.* Gastric intramucosal pH as a therapeutic index of tissue oxygenation in critically ill patients. *Lancet* 1992; **339**: 195–9.
41 Carli A, Auclair M-C, Vernimmen C, Jourdoun P. Reversal by calcium of rat heart cell dysfunction induced by human sera in septic shock. *Circ Shock* 1979; **6**: 147–57.
42 Parillo JE, Burch C, Shelhamer JH, Parker MM, Natanson C, Schuette W. A circulating myocardial depressant substance in humans with septic shock. *J Clin Invest* 1985; **76**: 1539–53.
43 Williams LF, Jr, Goldberg AH, Polansky BJ, Byrne JJ. Myocardial effects of acute intestinal ischemia. *Surgery* 1969; **66**: 138–44.
44 Haglund U, Lundgren O. Cardiovascular effects of blood borne material released from the cat small intestine during simulated shock conditions. *Acta Physiol Scand* 1973; **89**: 558–70.
45 McConn R, Greineder JK, Wasserman F, Clowes GHA, Jr. Is there a humoral factor which depresses ventricular function in sepsis? *Circ Shock Suppl* 1979; **1**: 9–22.
46 Magnotti LJ, Upperman JS, Xu D-Z, Lu Q, Deitch EA. Gut-derived mesenteric lymph but not portal blood increases endothelial cell permeability and promotes lung injury after hemorrhagic shock. *Ann Surg* 1998; **228**: 518–27.
47 Lundgren O, Haglund U, Isaksson O, Abe T. Effects on myocardial contractility of blood borne material released from the feline small intestine in simulated shock. *Circ Res* 1976: **38**: 307–15.
48 Lundgren O, Haglund U. On the chemical nature of the blood borne cardiotoxic material released from the feline small bowel in regional shock. *Acta Physiol Scand* 1978; **103**: 59–70.
49 Hallström S, Koidl B, Müller U, Werdan K, Schlag G. A cardiodepressant factor isolated from blood blocks Ca^{2+} current in cardiomyocytes. *Am J Physiol* 1991; **260**: H869–76.
50 Farrington GH, Saravis CA, Cossette GR, Miller DA, Clowes GHA, Jr. Blood-borne factors in the pulmonary response to sepsis (acute experimental peritonitis). *Surgery* 1970; **68**: 136–44.
51 Hashimoto E, Thal AP. The lung lesion produced by materials released from the superior mesenteric vein after superior mesenteric arterial occlusion. *Jpn Circ J* 1971; **35**: 1071–80.
52 Haglund U, Lundholm K, Lundgren O, Scherstén T. The relationship between the intestinal release of lysosomal enzymes and cardiotoxic material during regional simulated shock. An experimental study in the cat. *Circ Shock* 1975; **2**: 49–56.
53 Deitch ED, Xu D, Franko L, Ayala A, Chaudry IH. Evidence favouring the role of the gut as a cytokine-generating organ in rats subjected to hemorrhagic shock. *Shock* 1994; **1**: 141–6.
54 Adrine C, Pinsky MR. The inflammatory balance in human sepsis. *Intensive Care Med* 2000; **26**: 364–75.
55 Chiu C-J, McArdle AH, Brown R, Scott HJ, Gurd FN. Intestinal mucosal lesion in low-flow states. *Arch Surg* 1970; **101**: 478–83.
56 Bounous G. Acute necrosis of the intestinal mucosa. *Gastroenterology* 1982; **82**: 1457–67.
57 Park PO, Haglund U, Bulkely GB, Fält K. The sequence of development of intestinal tissue injury following strangulation ischemia and reperfusion. *Surgery* 1990; **107**: 574–80.
58 Marrone GC, Silen W. Pathogenesis, diagnosis and treatment of acute gastric mucosal lesions. *Clin Gastroenterol* 1984; **13**: 635–50.
59 Fiddian-Green RG, McGough E, Pittenger G, Rothman E. Predictive value of intramural pH and other risk factors for massive bleeding from stress ulceration. *Gastroenterology* 1983; **85**: 613–20.
60 Arvidsson S, Haglund U. On the pathogenesis of acute gastric mucosal lesions in septic shock. *Scand J Gastroenterol* 1984; **19** (suppl 105): 67–70.
61 Haglund U. Gut ischaemia. *Gut* 1994; **1**: S73–6.

62 Fine J, Frank ED, Ravin HA, Rutenberg SH, Schweinburg FB. The bacterial factor in traumatic shock. *New Engl J Med* 1959; **260**: 214–20.

63 Fine J. The intestinal circulation in shock. *Gastroenterology* 1967; **52**: 454–8.

64 Carter D, Einheber A. Intestinal ischemic shock in germ-free animals. *Surgery* 1966; **122**: 66–76.

65 Deitch EA. Multiple organ failure. Pathophysiology and potential future therapy. *Ann Surg* 1992; **216**: 117–34.

66 Ljungdahl M, Lundholm M, Katouli M, Rasmussen I, Engstrand L, Haglund U. Bacterial translocation in experimental shock is dependent on the strains in the intestinal flora. *Scand J Gastroenterol* 2000; **35**: 389–97.

67 Olofsson P, Nylander G, Olsson P. Endotoxin routes of transport in experimental peritonitis. *Am J Surg* 1986; **151**: 443–6.

68 Brathwaite CEM, Ross SE, Nagele R, Mure AJ, O'Malley KF, Garcia-Perez FA. Bacterial translocation occurs in humans after traumatic injury: evidence using immunofluorescence. *J Trauma* 1993; **34**: 586–90.

69 Sedman PC, Macfie J, Sager P, Mitchell CJ, May J, Mancey-Jones B, Johnstone D. The prevalence of gut translocation in humans. *Gastroenterology* 1994; **107**: 643–9.

70 Kudsk KA, Croce MA, Fabian TC, Minard G, Tolley EA, Poret Ha, *et al.* Enteral versus parenteral feeding. Effects on septic morbidity after blunt and penetrating abdominal trauma. *Ann Surg* 1992; **215**: 503–11.

71 Luiten EJ, Hop WC, Lange JF, Bruining HA. Controlled clinical trial of selective decontamination for the treatment of severe acute pancreatitis. *Ann Surg* 1995: **222**: 57–65.

72 Moore FA, Moore EE, Pogetti R, McAnena OJ, Peterson VM, Abernathy CM, Parsons PE. Gut bacterial translocation via the portal vein: a clinical perspective with major torso trauma. *J Trauma* 1991; **31**: 629–36.

73 Österberg J, Johnsson C, Gannedahl G, Westlund A, Haglund U. Alterations in mucosal immune cell distribution in septic rats. *Shock* 1997; **7**: 182–5.

3

Imaging for visceral artery stenosis

NICOS LABROPOULOS AND MARC BORGE

INTRODUCTION

Stenosis in the visceral arteries can be diagnosed accurately with various invasive and non-invasive techniques. These methods include ultrasound, computerized tomography (CT) (conventional or spiral), magnetic resonance angiography (MRA), and contrast angiography. Pressure gradient across a stenosis is the true hemodynamic standard for evaluating a lesion. However, since pressures are not measured routinely in the visceral arteries, contrast angiography has been the accepted gold standard to which all the other methods are being validated. In this chapter, we will describe the precise methodology, diagnostic accuracy, advantages, and limitations of the above techniques.

COLOR FLOW DUPLEX SCANNING

Color flow duplex scanning (CFDS) is performed routinely for peripheral arterial and venous systems. Its use has replaced venography for the detection of thrombosis in the extremities, and it has reduced significantly the need for contrast angiography in the evaluation of the carotid arteries. CFDS has limited use in the imaging of the visceral arteries because its role in acute and chronic mesenteric ischemia has not been determined adequately. Furthermore, the prevalence of mesenteric ischemia is low, therefore due to the limited number of examinations, only a few places have the expertise to accurately detect disease in these vessels. In a recent editorial[1] and a review paper,[2] it was concluded that CFDS is an accurate method for screening patients with suspected visceral ischemia.

Method of examination

Patients are examined in the morning following overnight fasting to minimize the effect of intestinal gas during imaging. They are placed in the supine position with the head elevated by 20–30 degrees. Imaging is performed with a 2–3.5-MHz phased-array or curve–linear-array transducer. Initially, the transducer is placed in the sub-xyphoid area with its head facing the liver (Figure 3.1). Then the aorta is imaged in the long view, and a reference velocity is obtained at the level of the celiac artery (CA) (Plate 1). At this view, the long axis of the CA and superior mesenteric artery (SMA) are seen (Figure 3.2; Plates 2 and 3). In a cross-sectional view, the long axis of the CA and the common hepatic and splenic arteries can be seen (Plate 4). The inferior mesenteric artery (IMA) is found above the aortic bifurcation and is best seen in a transverse or oblique view (Plate 5). The angle of insonation is set at 60 degrees,[3] and a large

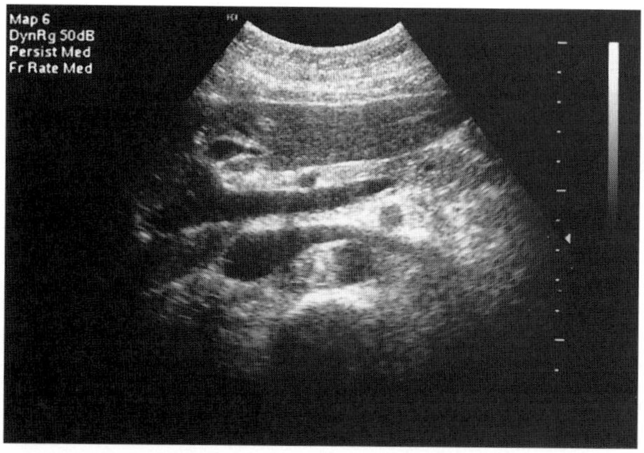

Figure 3.1 *Transverse view of the abdomen at the level of the left renal vein. From top to bottom are the left lobe of the liver, the extrahepatic segment of the portal vein, the superior mesenteric artery, the left renal vein uniting the inferior vena cava, the aorta, and the spine.*

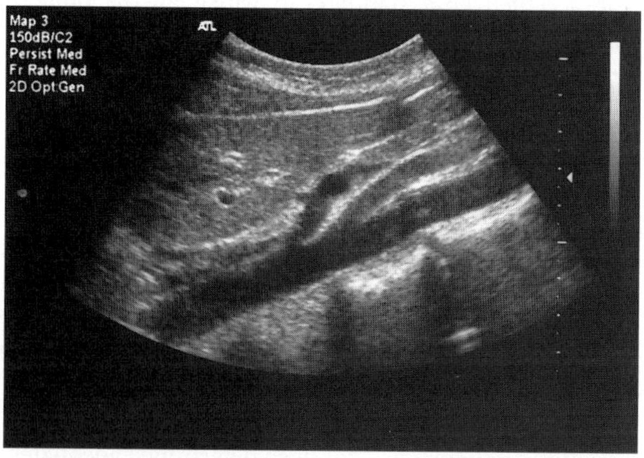

Figure 3.2 *Ultrasound image of normal visceral artery. B-mode imaging of the celiac artery (CA) and superior mesenteric artery in the long view. The long axis of the CA, common hepatic, splenic arteries, and inferior mesenteric artery are often best seen in the cross-sectional view.*

sample volume is used. Because steering is not possible with the low-frequency transducer, transverse, oblique, and longitudinal views in relation to the aorta are obtained to facilitate insonation of the visceral vessels at an angle of 60 degrees. A high-pass filter is used to reduce noise from the wall motion of the vessels. The peak systolic velocity (PSV) and end diastolic velocity (EDV) are taken at the origin of CA and up to the first 3 cm in the SMA and IMA.

Normal velocities and waveform patterns

The CA supplies the liver, the spleen, and part of the stomach. The majority of this blood volume goes to the first two organs, which are low-resistance vascular beds, therefore the EDV of the CA is high (Plate 6). In contrast, during fasting the SMA and IMA supply a high-resistance vascular bed because most of the mesenteric capillaries are shut down. During fasting, both arteries have a triphasic waveform, with low EDV and a reversed-flow component present (Plate 7). However, when the right hepatic artery originates from the SMA, the latter has a higher EDV.[2] Normal PSV in the CA and SMA is usually 200 cm/s. The mean values for PSV in healthy volunteers and in patients with atherosclerosis but without disease in the visceral arteries range from 120 to 160 cm/s. Normal EDVs in both vessels are 45 cm/s. The IMA has similar velocities to the SMA, but there are no studies comparing CFDS with contrast arteriography to determine cut-off values for significant stenosis.

After a meal, the PSV and EDV in the SMA increase significantly, reaching a peak 15–45 min postprandially depending on the food ingested.[4, 5] The velocities can be higher compared with baseline values for at least 90 min after ingesting the meal.[5] The velocities in the CA do not change postprandially. It has been shown that direct feeding from the duodenum causes dilation of the SMA but reduces the diameter of the CA.[6] This indicates that intestinal blood flow is dependent on the food stimulation site.

Detection of significant stenosis

Criteria for >50% diameter reduction in the CA and SMA include mainly increases in the PSV and EDV with post-stenotic turbulence being present (Plates 8 and 9). Reversed flow in the hepatic artery indicates significant CA stenosis or occlusion. Measurements during inspiration and expiration are diagnostic for the median arcuate ligament compression, known as the Dunbar syndrome (Plates 10 and 11). The above measurements have >80% diagnostic accuracy for detecting significant stenosis compared with contrast arteriography. Although the imaging of the visceral arteries may be limited by obesity and intestinal gas in the general population, this problem is not encountered in patients with chronic visceral ischemia. Occasionally, the presence of intestinal gas and the inability of a good angle of insonation may limit the diagnosis.

Significant stenosis in a visceral artery using duplex scanning was first reported by Jager *et al.*[7] They found increased velocities in the SMA in a female with postprandial epigastric pain, which was confirmed by arteriography. Qamar *et al.* determined the blood flow in the CA and SMA using ultrasound; they were the first to show increases in the SMA flow after glucose ingestion by this method.[8–10] Nicholls *et al.* described three patients with >90% stenosis in both the CA and the SMA, and a fourth patient with >90% CA and >65% SMA stenosis.[11] The coefficient of variation of PSV in the CA and the SMA in nine healthy volunteers was 19% and 12%, respectively.[12] Although these measurements are reliable in controls, such studies have not been performed in patients with stenosis. Larger retrospective and prospective series comparing CFDS with contrast arteriography appeared after 1990.[13–22] According to these studies, PSV > 200 cm/s in the CA and PSV > 275 cm/s in the SMA indicates significant stenosis. Using the EDV > 55 cm/s for the CA and EDV > 45 cm/s for the SMA, the diagnostic accuracy is at least as good as the PSV criteria.

The IMA (Plate 12) is the smallest of the three arteries and becomes important only when the CA and SMA have tight stenosis or occlusion. Currently, there are no published criteria for determining significant IMA stenosis. In our institution, we have had only seven patients with IMA stenosis >50% documented by both CFDS and arteriography. In all seven patients, the PSV was >290 cm/s and the EDV was >52 cm/s (Plate 13). In 10 normal volunteers, the PSV was 170 cm/s and the EDV was 25 cm/s. If we were to apply the SMA criteria to the IMA for the above patients, we would have obtained excellent results. However, more work is needed before we make such assumptions.

Investigators with adequate training in visceral imaging should be able to screen patients with suspected chronic visceral ischemia. The diagnostic accuracy is going to be around 90%. Although it may be feasible to initiate treatment in many of these patients based on the ultrasound findings alone, such an approach has not been performed prospectively. Therefore, all patients who are candidates for an intervention in the visceral arteries currently undergo arteriography.

VISCERAL ANGIOGRAPHY

Conventional catheter-based angiography is still considered by many to be the gold standard for visceral artery imaging. However, the technique is used less frequently due to the considerable advances in non-invasive methods of visceral arterial imaging (CT, magnetic resonance imaging (MRI), and duplex ultrasound). Nonetheless, each of these other imaging modalities has shortcomings, and visceral angiography still plays a role in evaluation of mesenteric vascular disease.[23] Most recent published reports continue to use conventional arteriography images or surgery as the gold standard for comparison of newer non-invasive imaging techniques.

In spite of the advances in non-invasive methods of visceral artery imaging, the techniques for catheter-based angiography have continued to evolve and improve.[24] The advent of computerized digital subtraction angiography (DSA) and improved contrast agents has improved the safety and efficacy of visceral angiography tremendously.

Technique

Historically, visceral artery angiography has been performed with film-screen or 'cut-film' imaging systems (Figure 3.3).[24, 25] This technique has been replaced almost completely by DSA (Figure 3.4). Film-screen systems provide high spatial resolution at approximately six line pairs per millimeter. Current high-resolution DSA systems with a 1024 × 1024 imaging matrix and 40-cm wide image intensifier have a spatial resolution of approximately three line pairs per millimeter, resulting in a definite loss of image resolution. However, DSA imaging provides considerable improvement in image contrast when subtracted images are utilized.

An additional advantage of DSA imaging is the rapidity of image acquisition and presentation. Computer-generated images are available for evaluation immediately without the need for dark-room development of film-screen cassettes. This can be particularly helpful in unstable patients or during prolonged arterial interventions.

The development of newer, low-osmolar contrast media (Ioxaglate, Iopamidol, Iohexol) have made visceral angiography considerably safer and better tolerated by patients. The older,

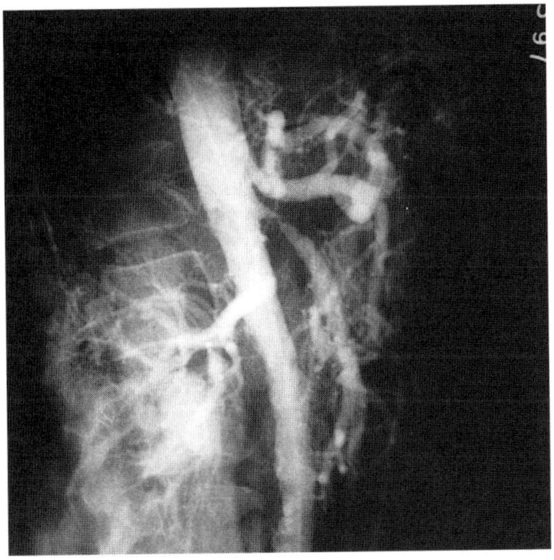

Figure 3.3 *Visceral artery stenosis. Cut-film lateral aortography demonstrates high-grade stenosis of the proximal superior mesenteric artery. The celiac axis is patent.*

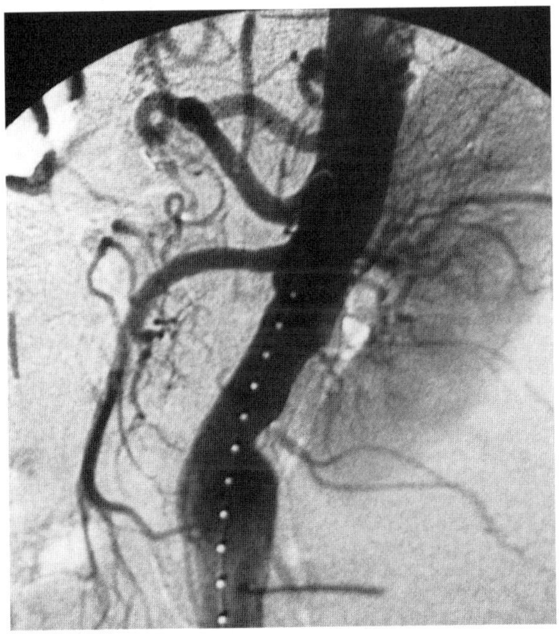

Figure 3.4 *Digital subtraction angiography lateral aortogram demonstrates profile of widely patent celiac axis and superior mesenteric artery.*

high-osmolar ionic contrast agents often caused considerable discomfort during injection, resulting in suboptimal images due to patient motion. With the newer contrast agents, prolonged arterial injections can be performed with minimal patient sensation. In most instances, a visceral artery angiogram for suspected visceral artery stenosis should begin with anterior-posterior (AP) and lateral aortography, with a flush catheter positioned in the region of the CA and the SMA. The AP view will delineate gross flow patterns as well as anomalous anatomy (Figure 3.5). The lateral view profiles the proximal CA and proximal SMA optimally.

For evaluation of the IMA, a flush catheter can be withdrawn to the L3 vertebral level and flush aortography performed in right anterior oblique (RAO) projection in order to profile the origin of the IMA arising anterolateral to the left. Flush aortography can often delineate proximal visceral artery stenosis. However, selective catheterization of visceral arteries is often required to further delineate proximal or distal lesions (Figure 3.6). Various catheter shapes can be used to cannulate the visceral arteries safely. Selective

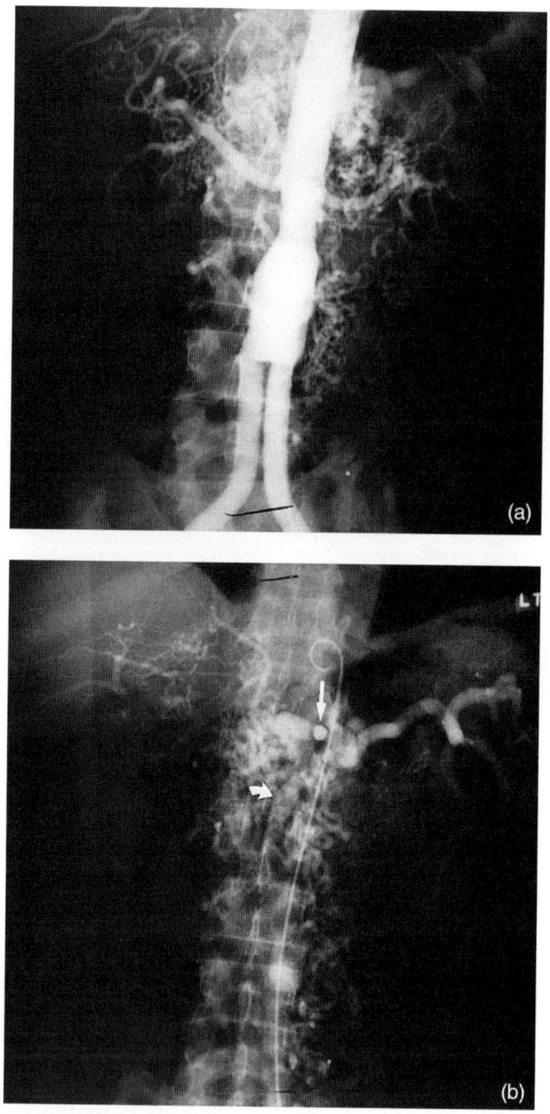

Figure 3.5 *Visceral artery collaterals. (a) Asymptomatic patient with complete occlusion of all three main visceral arteries. The patient has undergone aortofemoral bypass graft. (b) Delayed image demonstrates reconstitution of celiac axis (white arrow) and distal superior mesenteric artery (curved arrow) via collaterals.*

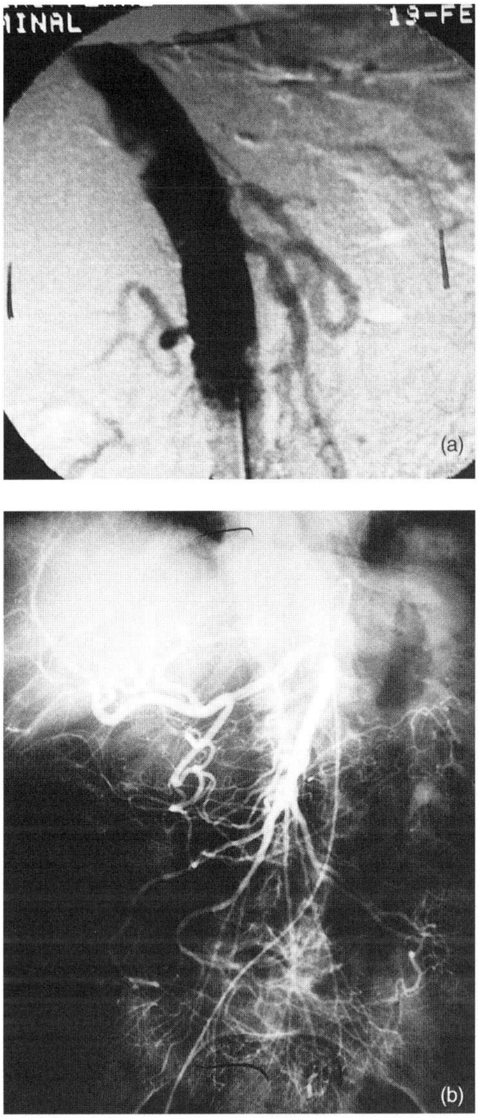

Figure 3.6 *Visceral artery collaterals. (a) Lateral aortogram demonstrates high-grade stenosis at origin of celiac axis. The superior mesenteric artery (SMA) is patent. (b) Selective SMA injection demonstrates collateral flow via gastroduodenal artery to the hepatic arteries.*

injection can further delineate the degree of stenosis, and can help demonstrate distal arterial disease and define collateral pathways.

Pharmacoangiography

Various medications can be used to optimize visceral arterial images. Glucagon 1 mg IV slow push given about 1 min before arterial imaging can help to decrease bowel peristalsis. This can be particularly useful in DSA, as the bowel peristalsis causes misregistration on the subtraction process.

Vasodilating drugs, particular papaverine and tolazoline, can be injected directly into visceral arteries before angiography in an effort to promote contrast transit through the visceral arterial system and optimize mesenteric venous enhancement.

VISCERAL ARTERY COMPUTED TOMOGRAPHIC ARTERIOGRAPHY

Visceral artery imaging has evolved considerably with the advent of CT and MRI imaging. Historically, catheter-based conventional angiography was required to evaluate visceral arterial disease. However, recent advances in CT have made it much more effective for visceral arterial imaging.[26] Earlier-generation 'single-slice' CT can provide considerable information regarding the status of visceral arteries. Arterial patency, as well as anatomy, calcification, and aneurysms can be demonstrated by conventional CT in a non-invasive and more cost-effective manner than conventional arteriography. However, the spatial resolution of conventional CT limits its utility in the morphologic evaluation of visceral arteries. The advent of helical CT and, more recently, multi-slice, multi-array helical CT has improved considerably the ability of CT to delineate visceral artery anatomy.

Helical CT can provide considerably better three-dimensional spatial resolution than single-slice CT, allowing visualization of visceral arteries in a more optimal fashion (Figure 3.7).[27] This can be accomplished because helical CT acquires image data as a volume as opposed to individual sections.[28] Conventional CT is accomplished with rotation of the X-ray tube through a single arc acquiring a single slice of information. With helical technology, continuous prolonged X-ray exposure is performed during continuous rota-

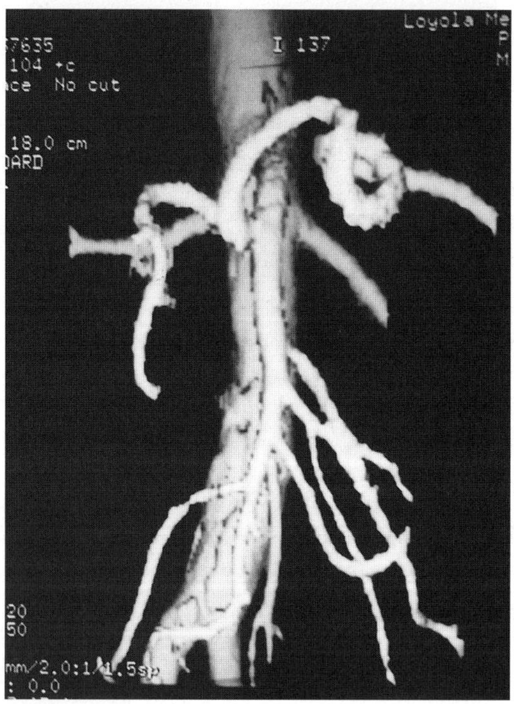

Figure 3.7 *Anomalous anatomy. Shaded-surface display rendering of three-dimensional computed tomographic angiography demonstrates celiac axis arising from the superior mesenteric artery.*

tion of the X-ray tube and detector system, with simultaneous movement of the patient through the slip-ring gantry. Helical CT therefore provides a continuous volume of data with spatial resolution in three planes dependent on the imaging parameters set. Image detail in the *x* and *y* planes is limited by the X-ray tube/detector scanning system. Image detail in the *z* axis (longitudinal) can be adjusted by changing the collimation or pitch ratio.

The helical scan technology also allows for more rapid scanning, which improves vascular imaging.[29] Conventional CT can take several minutes to scan the abdomen. This makes optimal arterial opacification with intravenous contrast difficult due to the prolonged timeframe. With helical scanning, images can be acquired during the time of peak arterial opacification after intravenous contrast administration.

Early helical CT scanners with a single X-ray tube and detector array were limited in their ability to provide broad scan coverage in the *z* axis, being unable, for example, to image the entire abdominal aorta and pelvic arterial system in one breath-hold. This can be performed with conventional helical scanners by adjusting the collimation and pitch to obtain relatively thick sections over the extended area of interest at the expense of image resolution.

The latest helical CT technology, including helical dual-slice, quad scanners, and multi-slice/multi-array scanners, allows imaging of a larger volume or longer area of interest, with considerably greater speed and detail (Figure 3.8).[30] These newer scanner technologies incorporate multiple X-ray tubes and detectors working in tandem, as though multiple helical CT scanners are combined in a focal area. This allows tighter collimation and considerably greater longitudinal imaging coverage in the *z* (longitudinal) axis during a single breath-hold or over a single contrast bolus. Images acquired with these types of scanners can then be reconstructed at very thin sections, allowing improved detail and spatial resolution.

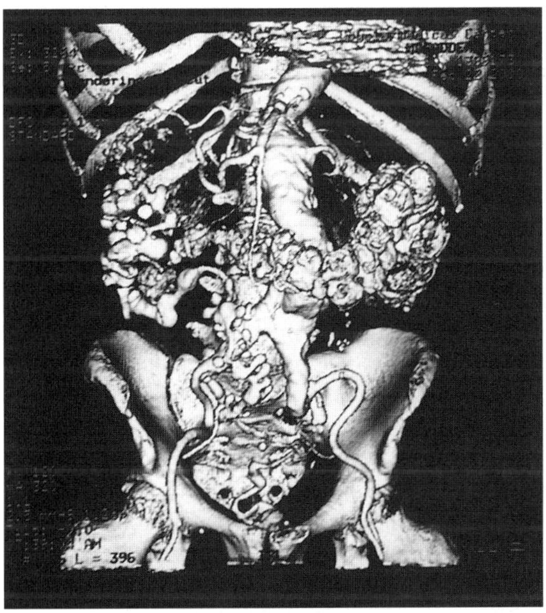

Figure 3.8 *Shaded-surface display image demonstrates the common hepatic artery arising from the superior mesenteric artery. Note the bones and high-density colonic contrast have been included in the three-dimensional reconstruction volume.*

Image analysis

The volumetric data set obtained can be displayed in axial sections or in three-dimensional models. The axial imaging sections are similar in appearance to images from conventional CT. Depending on the scanning collimation, spiral pitch, and image slice thickness, considerable detail can be achieved. These images demonstrate all of the data obtained in the volume. Compared with conventional CT there is a considerably higher level of intravascular contrast due to the speed of acquisition. A large number of scans can be obtained due to the thin slices obtained. These can be reviewed easily at a workstation (Figure 3.9). Frequently, a subset of axial images is printed on film.

The high level of intravascular contrast relative to background tissue facilitates computer-generated three-dimensional images. These three-dimensional models constructed from the volumetric data can be very helpful for evaluation of vascular structures. While it is important to review the axial images for specific details, the three-dimensional images allow considerably easier orientation in areas of complex anatomy (Figure 3.10). Three-dimensional reconstruction involves one of several techniques, including maximum-intensity projection (MIP), shaded-surface display (SSD), or volume rendering (Figure 3.11).[31, 32] All of these techniques take advantage of the fact that the vessels imaged after intravenous contrast administration demonstrate high attenuation relative to surrounding tissues.

The MIP image technique depicts the highest pixel value along the observer's imaginary line of sight through the volume set. Overlying dense structures, such as bone, must be removed electronically from the image set. MIP is the only three-dimensional rendering technique that allows for differentiation between calcification and contrast material.

The SSD technique depicts the outer contour of the column of contrast material as an opaque surface. Obscuring overlying structures are removed by setting a threshold range of attenuation below which structures will not be imaged. The final image demonstrates a three-dimensional appearance. Limitations of SSD images include loss of a large portion of the available data. Information about structures inside or behind the displayed vessel surface is lost. Post-processing changes in threshold may inadvertently lose a portion of vascular anatomy.

The third technique of three-dimensional imaging is volume rendering (Figure 3.12). This technique requires considerably more computer hardware and software power in order to incorporate the entire image data set into a three-dimensional volume image. This technique allows simultaneous evaluation of both the vasculature and adjacent organs. Theoretically, it is advantageous relative to MIP and SSD because the relative voxel attenuation is related to a gray scale in the final image, thereby yielding images that are more accurate than those rendered with SSD.[33] Interaction at a workstation yields optimal depiction of the area of interest. Modification of the reconstruction parameters changes conspicuity and resolution, and allows distinction between adjacent structures or between intravascular contrast and calcification.

At our institution visceral artery, computed tomographic angiography (CTA) is performed on a GE light-speed quad detector scanner. Oral contrast is not used routinely. Glucagon (1 mg) is given routinely before scanning. A large-caliber intravenous line is placed in a large vein, preferably the antecubital. The intravenous line is tested manually to ensure it will withstand a large rate/volume injection. Non-ionic contrast is injected at a rate of up to 5 cm^3 per second, and the injection is monitored by technical staff to prevent subcutaneous extravagation.

Images are obtained during breath-hold for up to 30–40 s. Coaching, hyperventilation, and supplemental oxygen are utilized when necessary.

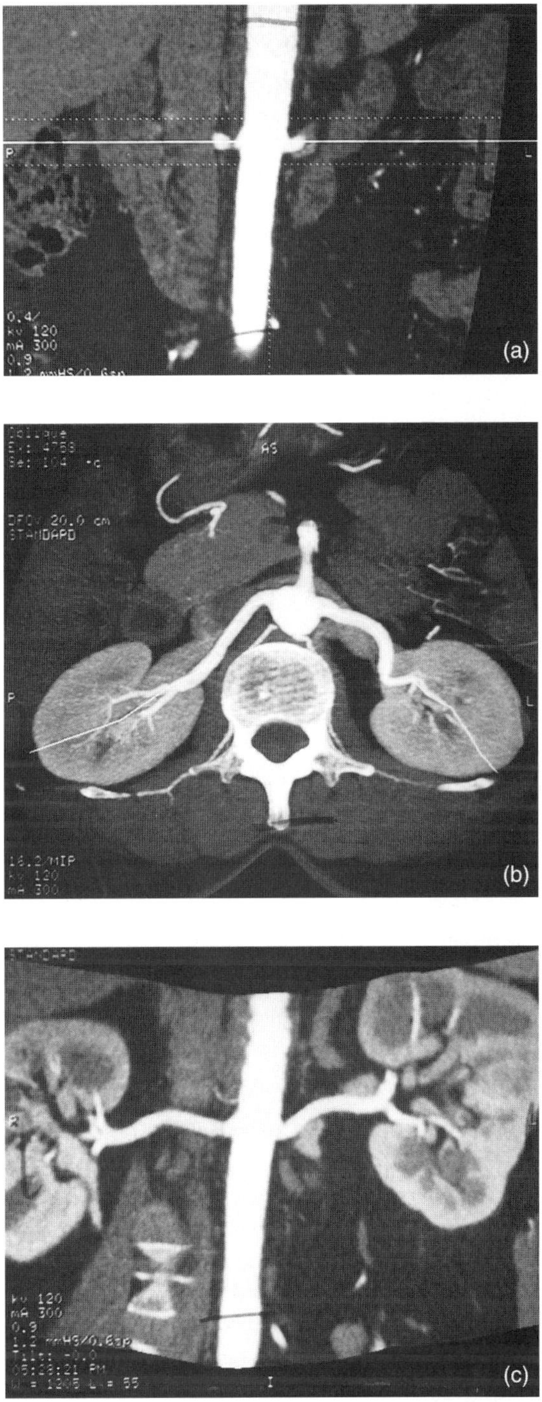

Figure 3.9 *Multiplanar reformatting. (a) Coronal localizer shows volume to be imaged for renal arteries. (b) Axial images through renal arteries with image reconstruction plane drawn through the renal arteries. (c) Multiplanar image of renal arteries constructed from the orientation in (b).*

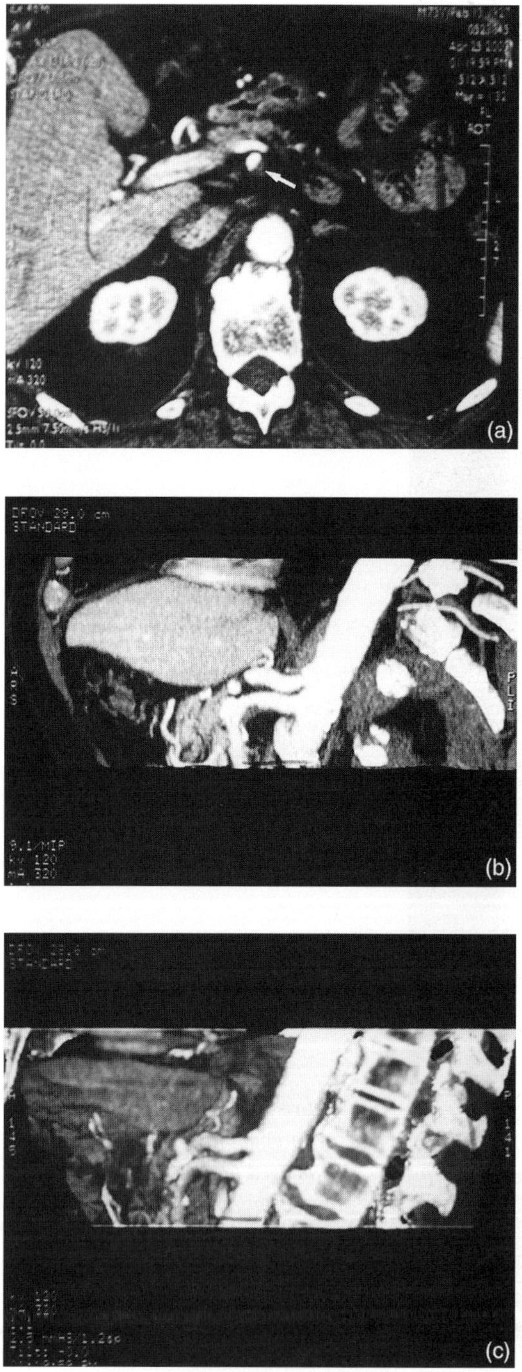

Figure 3.10 *Partially occlusive thrombus in superior mesenteric artery (SMA). (a) Thrombus is seen in the proximal SMA on contrast enhanced axial image (arrow). (b) Maximum-intensity projection. (c) Volume rendering images do not show the thrombus well. Review of axial source images is important to characterize the SMA narrowing on three-dimensional images as intraluminal clot.*

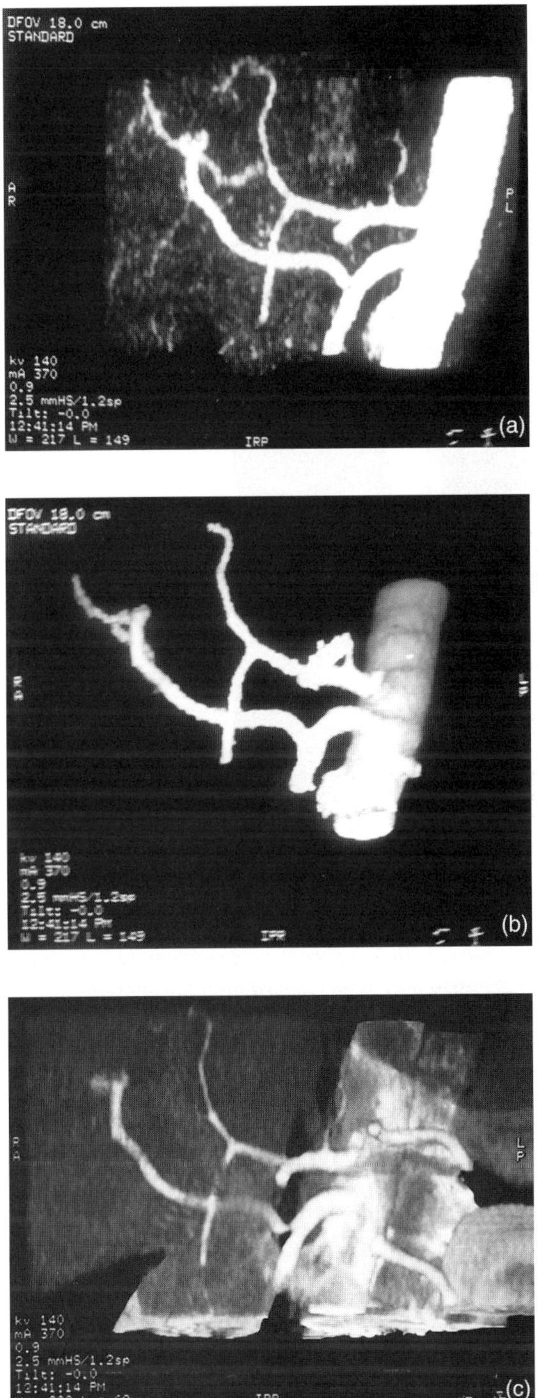

Figure 3.11 *Maximum-intensity projection (MIP), shaded-surface display (SSD), and volume rendering of superior mesenteric artery (SMA). (a) MIP image from lateral orientation demonstrates patent SMA with replaced right hepatic artery. (b) SSD image from left anterior oblique orientation. (c) Volume rendering image from left anterior oblique orientation. Note that stenosis at the origin of the right hepatic artery is best seen in (a) and (c).*

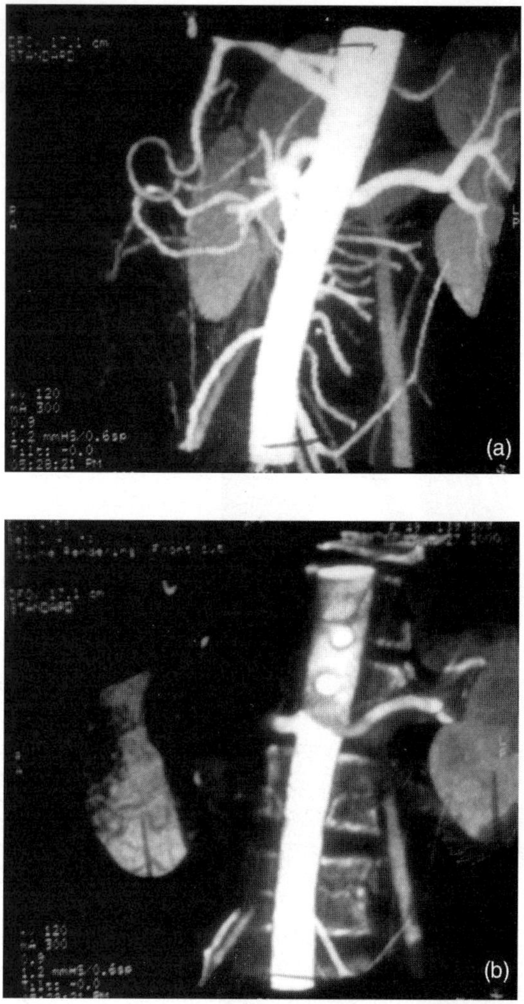

Figure 3.12 *Inferior mesenteric artery (IMA) and left colic artery. (a) Maximum-intensity projection image demonstrates good artery-to-background contrast. The left colic artery is seen coursing cephalic anterior to the gonadal vein. Note the subtle serrated contours due to a corrugation artifact. (b) Volume rendering image demonstrates the IMA and left colic artery with improved three-dimensional perspective. Note that the celiac axis and the superior mesenteric artery have been cropped from the image to improve visualization of the IMA.*

Ideally, a test bolus of contrast is administered and sequential images are obtained in the region of interest to evaluate time of peak enhancement (Figure 3.13). In patients in whom a test bolus cannot be administered, a slower injection rate is used and scan injection timing estimation is made.

Scan parameters are set to optimize *z*-axis coverage while staying within the limitations of tube heating. Such coverage depends on beam collimation, pitch, and scan duration. Axial images are obtained with a slice thickness of 2.5 mm. The reconstructed thin sections with overlap are ideal for evaluating small vessels. The doctors rather than the technical staff perform three-dimensional rendering because of the required knowledge of arterial anatomy and vascular variability. Three-dimensional reconstruction for a single patient routinely requires about 20 min of physician time.

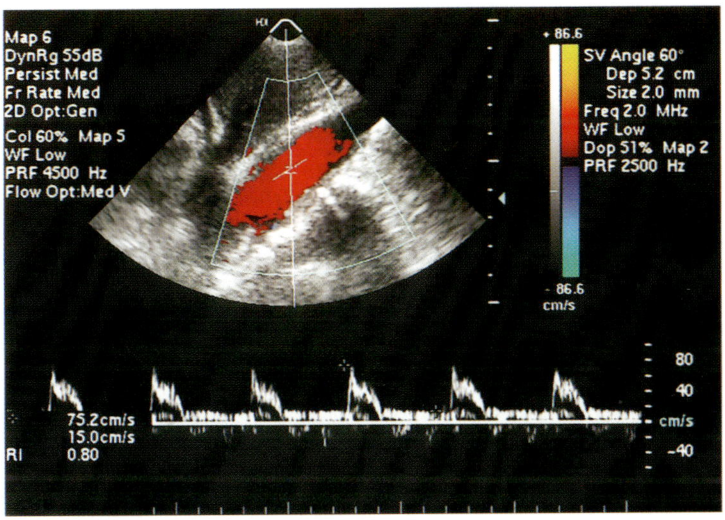

Plate 1 *Long view of the aorta by color flow imaging. A reference velocity is obtained in the vessel at the level of the visceral arteries.*

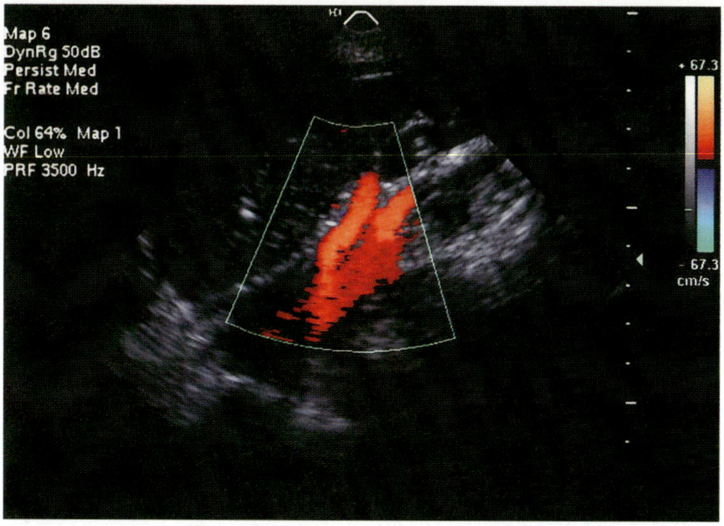

Plate 2 *Ultrasound image of normal visceral arteries. Color imaging of the celiac artery and superior mesenteric artery in the long view.*

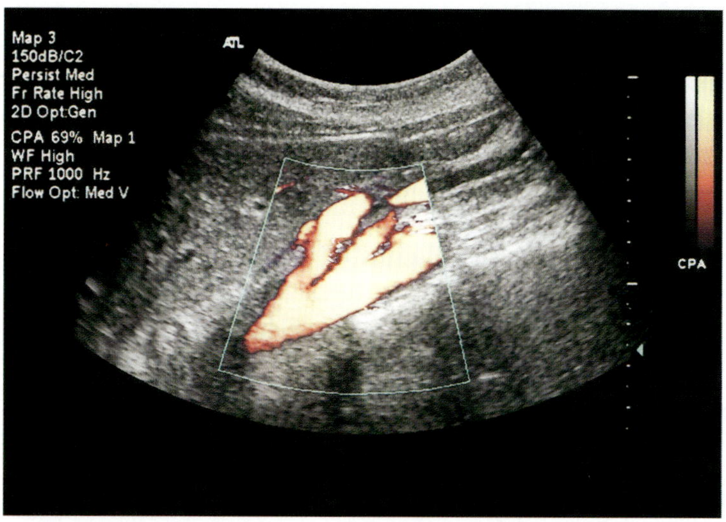

Plate 3 *Ultrasound image of normal visceral arteries. Doppler imaging of the celiac artery and superior mesenteric artery in the long view.*

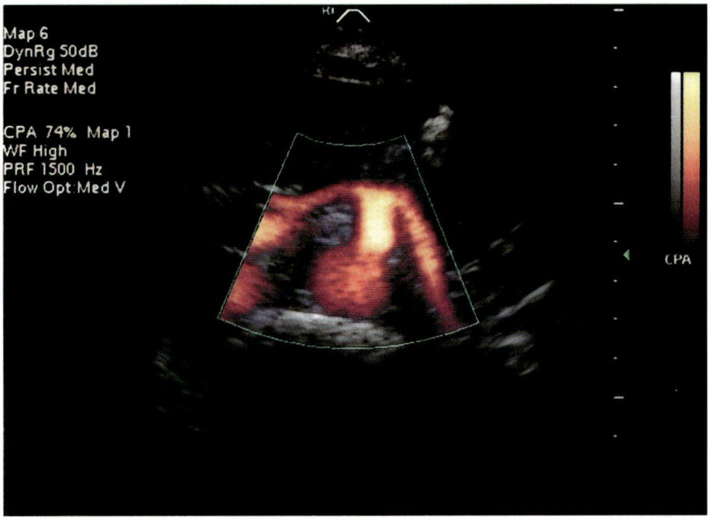

Plate 4 *Ultrasound image of normal visceral arteries. The long axis of the celiac artery, common hepatic artery, and splenic artery is often seen best in cross-sectional view.*

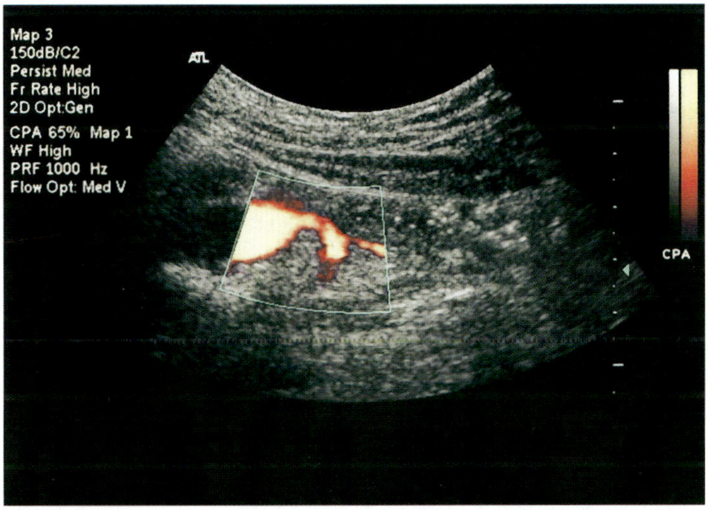

Plate 5 *Ultrasound image of normal visceral arteries. The long axis of the inferior mesenteric artery is often seen best in a transverse or oblique view.*

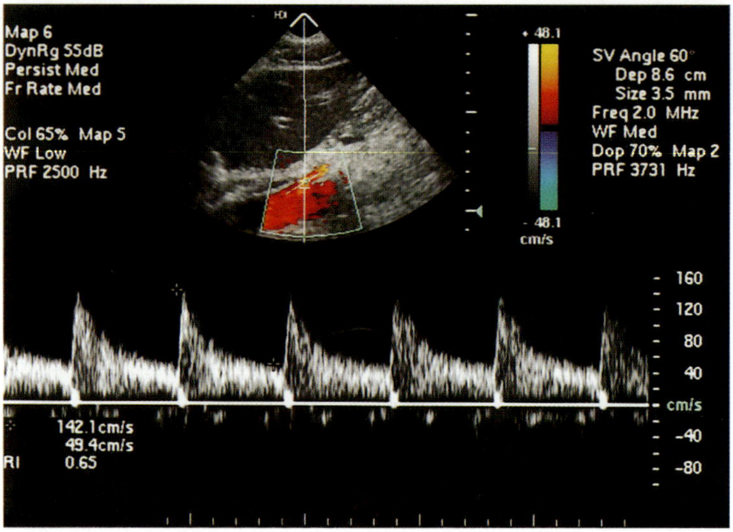

Plate 6 *Blood flow velocity waveforms from normal celiac artery. The celiac artery has a higher end diastolic velocity than the aorta, superior mesenteric artery (Plate 7), and inferior mesenteric artery (Plate 12) because it supplies low-resistance vascular beds with blood.*

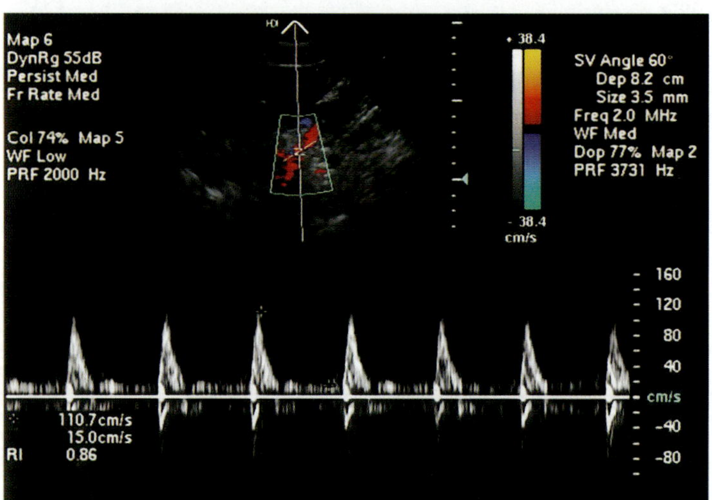

Plate 7 *Blood flow velocity waveforms from superior mesenteric artery.*

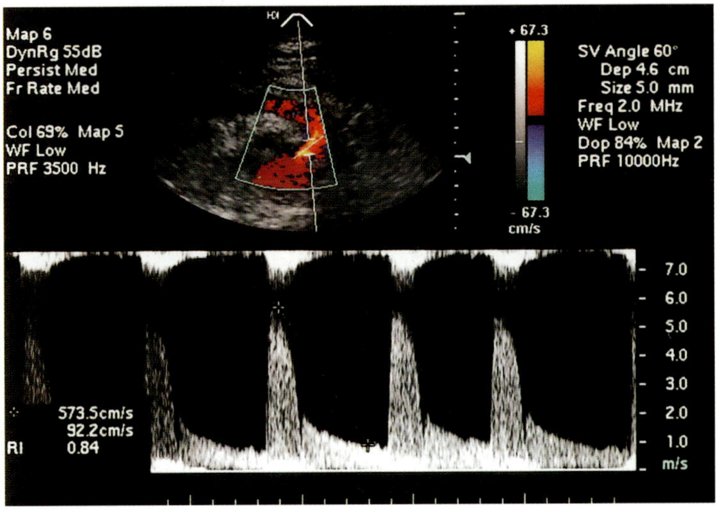

Plate 8 *Color flow image of the celiac artery with significant stenosis. The artery has post-stenotic turbulence, as seen by the mosaic color despite using a high pulse repetition frequency. Spectral waveforms indicate the presence of tight stenosis.*

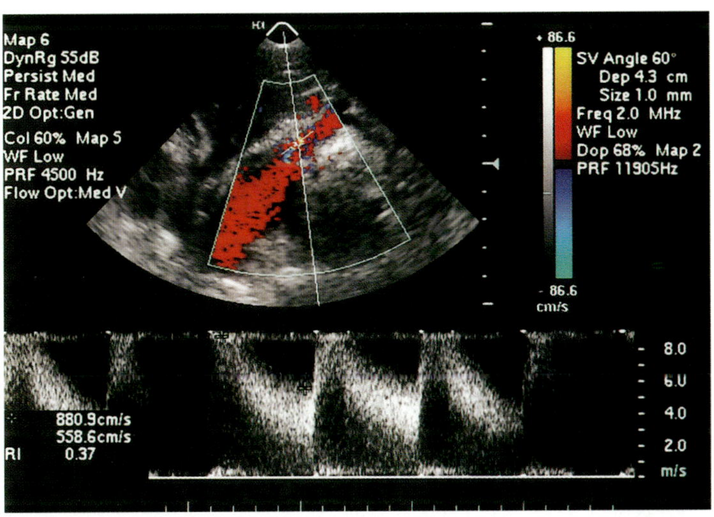

Plate 9 *Color flow image of the superior mesenteric artery with significant stenosis. The artery has post-stenotic turbulence, as seen by the mosaic color despite using a high pulse repetition frequency. Spectral waveforms indicate the presence of tight stenosis.*

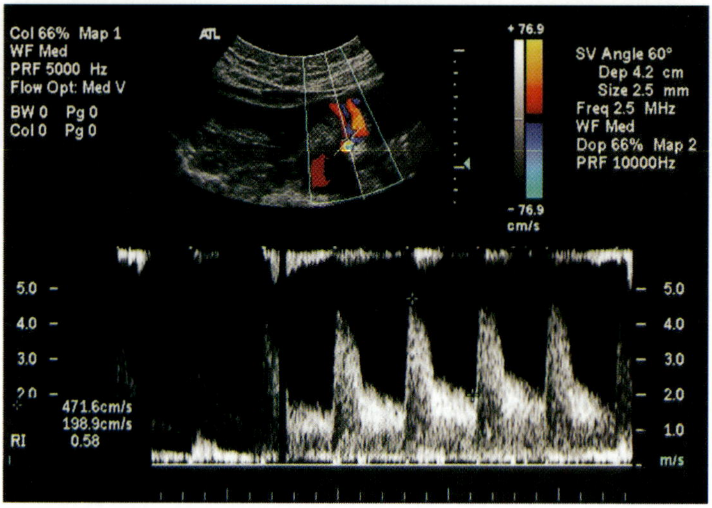

Plate 10 *Celiac axis stenosis from extrinsic compression by the median arcuate ligament. Peak systolic and end diastolic velocities are significantly increased.*

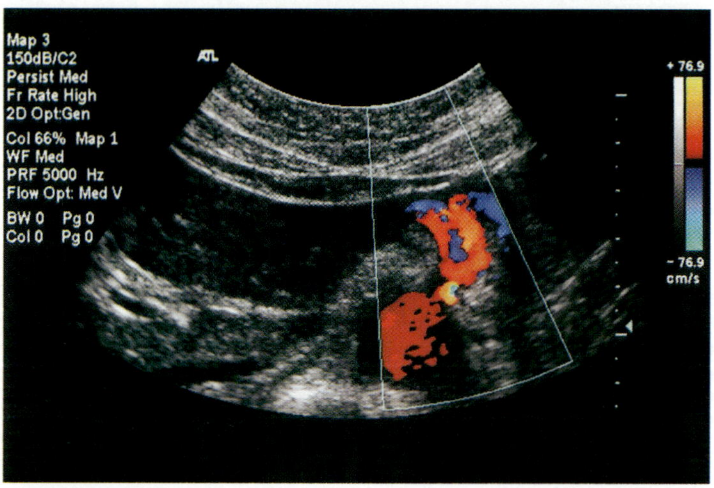

Plate 11 *There is a marked post-stenotic dilation in the celiac axis while the common hepatic and splenic artery are of normal size.*

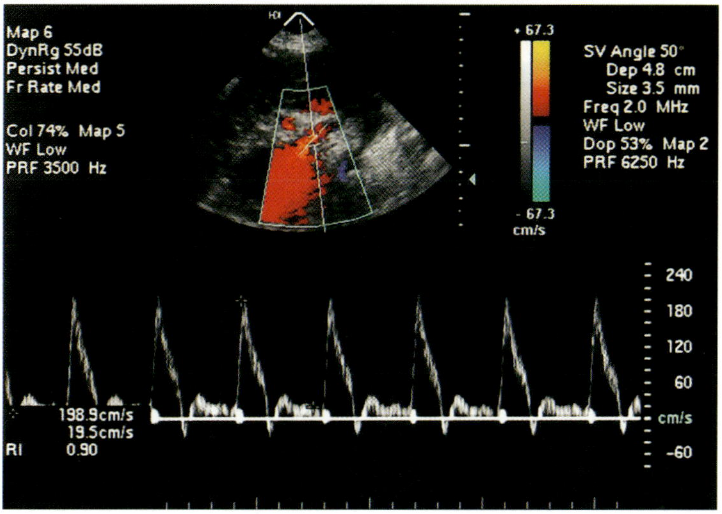

Plate 12 *Blood flow velocity waveforms from inferior mesenteric artery.*

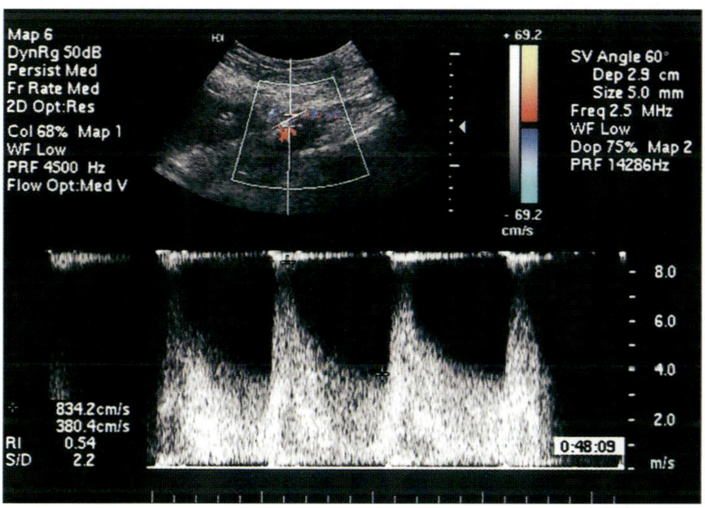

Plate 13 *Increased peak systolic and end diastolic velocities at the origin of the inferior mesenteric artery in the presence of a tight stenosis. Post-stenotic turbulence is seen (mosaic color) despite using a high pulse repetition frequency.*

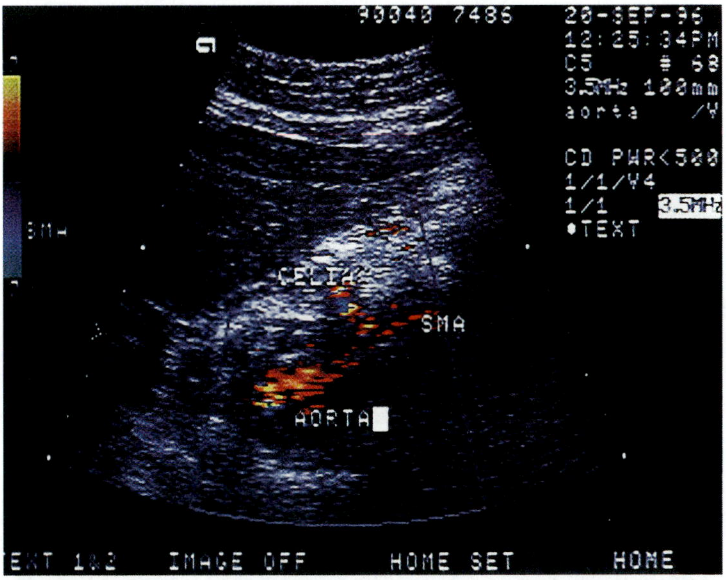

Plate 14 *Ultrasonic picture of an aortosuperior mesenteric bypass with implantation of the celiac artery on the side of the bypass.*

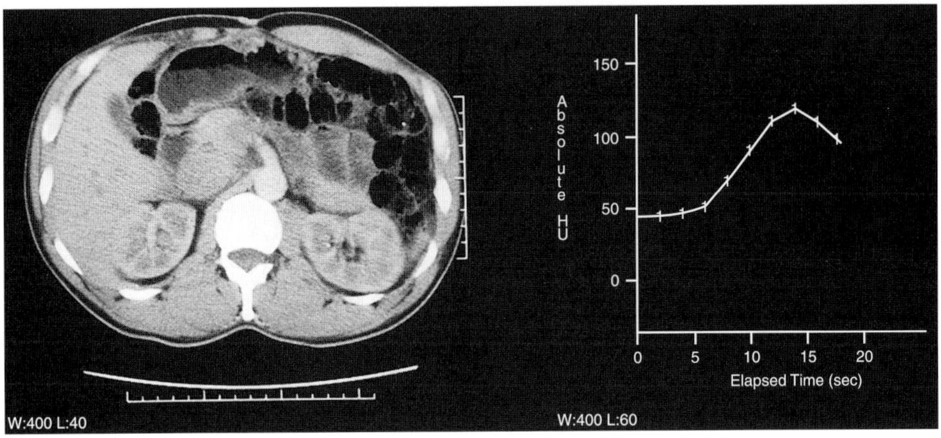

Figure 3.13 *Sequential image obtained at same level with ROI cursor over aorta after test bolus demonstrates peak arterial enhancement at approximately 14 s. Imaging for CTA would then be performed with a 14-s delay after contrast injection in this patient in order to optimize arterial opacification.*

Pitfalls

The potential for overestimation of arterial stenosis has been shown in clinical studies using CTA and MRA.[34, 35] In addition, critical stenosis can be misdiagnosed as occlusion. These studies pertain primarily to helical scanning. Assessment of stenosis with multi-detector or multi-array scanners may be more sensitive.

Artificial stenosis can be created due to improper thresholding or electronic disarticulation (Figure 3.14). Corrugation artifacts are often present (Figure 3.15).[36] This may make evaluation for assessment of stenosis difficult and assessment for fibromuscular dysplasia difficult or impossible.

CT angiography requires the use of intravenous contrast material. This can be a problem in patients with contrast allergy and/or renal insufficiency. CT image quality may be degraded in patients with extremely large body habitus or adjacent metallic foreign bodies.

Clinical applications

Winter *et al.* evaluated the ability of CT arteriography to delineate hepatic arterial anatomy in transplantation candidates.[37] They performed three-dimensional CT angiograms on 115 patients. Seventeen of the patients also underwent conventional angiography. Follow-up was also obtained in 16 patients who did not undergo angiography but underwent hepatic transplantation. Nine of the three-dimensional CTAs were considered technical failures. In the 17 patients with angiographic correlation, there was only one marked disagreement with three-dimensional CTAs. This involved demonstration of a replaced right hepatic artery at conventional angiography, which was not detected initially on the CTA. In retrospect, the artery was deemed to be visible. The authors emphasize that measurements of arterial diameter from major vessels should be made from axial images, as inaccuracies can be introduced in three-dimensional images due to threshold adjustment and corrugation artifacts. The authors conclude that three-dimensional CTA was as accurate as angiography in the assessment of hepatic arterial anatomy. Also, it was safer, more convenient, and more easily tolerated.

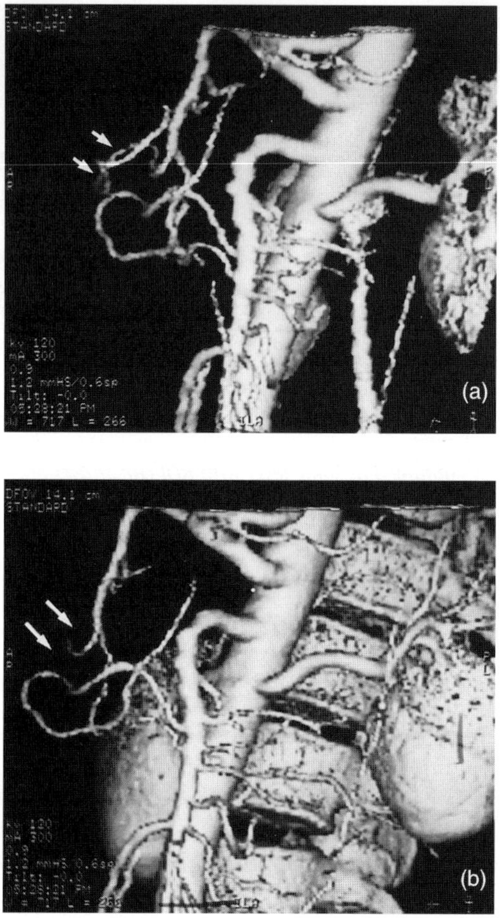

Figure 3.14 *Effect of changing the density threshold on arterial continuity. (a) Low-threshold image demonstrates intact and continuous pancreaticoduodenal arcade branches and mild superior mesenteric artery (SMA) stenosis (white arrows). (b) Higher-threshold image demonstrates artifactual discontinuity of pancreaticoduodenal branches and more prominent SMA stenosis (white arrows).*

Blomey *et al.* evaluated the utility of three-dimensional spiral CTA in pancreatic surgical planning using non-tailored protocols.[38] Specifically, they used protocols including oral contrast geared towards opacification of the pancreas and adjacent structures in addition to the visceral vasculature. They found that three-dimensional CTA was effective using narrow sections (collimation/table-feed protocol, 3 mm). Wider collimation/table-feed protocols failed to demonstrate one replaced right hepatic artery.

Johnson *et al.* evaluated patients using CTA with volume rendering.[33] They believe the volume rendering application is more versatile in that it allows visualization of vessels and adjacent organs or pathology. For example, volume rendering allows determination of relationships between hepatic tumors and liver vasculature for surgical planning. In addition, volume rendering conveys vessel depth and the interrelationships of vascular structures, which can be rotated at a workstation. The accuracy of the volume rendering technique for depiction of arterial anatomy and disease awaits a statistical comparison to convention angiography or other CTA techniques.

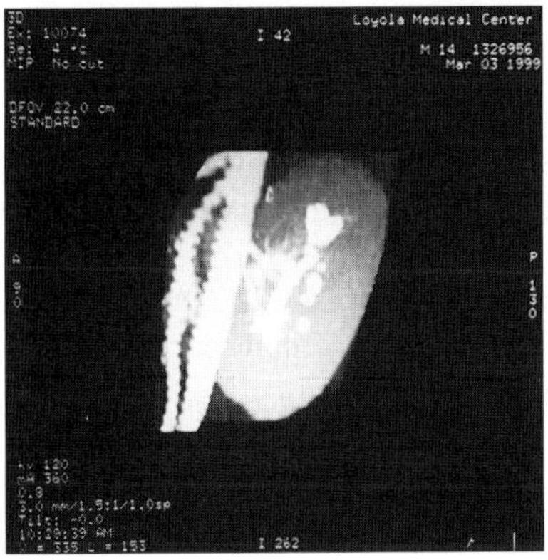

Figure 3.15 *Corrugation artifact. Maximum-intensity projection with lateral projection demonstrates regular narrowing and dilation of superior mesenteric artery and aorta due to corrugation artifact.*

In an early study of CTA, Ruben *et al.* evaluated CTA as an imaging modality for the abdomen aorta and its branches.[26] Patients were scanned during a 30-s breath-hold after intravenous contrast injection. Images were obtained to evaluate patients with aneurysms, aortic dissections, and mesenteric artery stenosis. The authors found good correlation with conventional arteriographic findings. In aneurysmal disease, the lumen, mural thrombus, and associated renal artery stenosis were visualized. True and false lumens in dissections and the anatomy of the visceral vessels were shown clearly in the latter two groups. The patency of visceral and renal reconstruction or stent placement was confirmed.

CTA using conventional helical imaging and dual-slice helical CT has been used to evaluate abdominal aortic aneurysms before repair. Gomes *et al.* evaluated abdominal aorta aneurysms with three-dimensional CTA.[39] Thirty-two patients were scanned using either 5-mm collimation or a combination of 3-mm collimation through the renal and mesenteric arteries and 7 mm collimation through the remainder of the abdominal aorta. The location, size, and extent of the aneurysm, as well as wall calcification and intra-aneurysm thrombus, were depicted well with the helical CT. Visceral aortic branches (including the detection of renal stenosis and accessory renal arteries) were seen consistently with the 3-mm/7-mm collimation protocol, which allowed detection of renal artery stenosis and accessory renal arteries.

Quanadli *et al.* evaluated abdominal aorta aneurysms preoperatively with dual-slice helical CTA.[30] In 47 consecutive patients, they found that dual-slice helical CT showed all main and accessory renal arteries with a sensitivity of 91% and a specificity of 100% for revealing renal arteries stenosis (>50%). Sensitivity and specificity of dual-slice helical CT for detecting stenosis greater than 75% in the CA and SMA were 100%. Notably, they found that dual-slice helical CT correlated better with operative findings than did DSA in determining IMA patency. They proposed that opacification of the IMA during angiography in supine patients with a large abdominal aneurysm may be difficult because the IMA originates from the anterior wall of the aneurysm.

Little is written about direct comparison of spiral CT versus conventional arteriography for evaluation of visceral arteries. Cikrit *et al.* evaluated 32 patients with CTA and conventional arteriography for visceral and renal artery disease.[40] Using conventional arteriography as the standard for comparison, they found that CT had a sensitivity of 67% and a specificity of 95% for depiction of ≥75% stenosis in the main renal artery. For detection of celiac stenosis of ≥75%, sensitivity was 75% and specificity was 100%. For detecting SMA stenosis, sensitivity was 100% and specificity was 91%. Significant difficulty was encountered due to calcification in the arterial wall. Cikrit *et al.* described at least two patients in their series in whom calcification in the aorta and the renal vessels was so dense that the degree of stenosis could not be determined.

Conclusion

In conclusion, CT angiography with three-dimensional rendering is a rapidly evolving means of imaging of visceral vessels. Current technology allows imaging of proximal visceral arteries. The newer CT technologies may allow more accurate visualization of more distal branches of the visceral circulation.

In comparison to conventional angiography, CTA is more rapid, non-invasive, and less expensive. At the present time, its use is limited in certain patients due to the requirements for intravascular contrast material.

VISCERAL ARTERY MAGNETIC RESONANCE ANGIOGRAPHY

Recent advances in magnetic resonance scanning technology have made MRI imaging of visceral arteries considerably more practical than in earlier studies. In early MRI imaging with standard T1 and T2 weighted spin echo (SE) or gradient echo (GRE) protocols, flowing blood in vessels was often considered a nuisance because of the artifacts it created. These early scan sequences allowed visualization of blood vessels and limited interpretation of anatomy.

More recent reports have described two-dimensional time-of-flight (TOF) MRA techniques and phase-contrast methods for assessment of visceral arteries.[41–43] The phase-contrast MRA acquired in either two-dimensional or three-dimensional approaches creates an image based on measured phase shifts within each imaging voxel, which are dependent on flow rates. These techniques have several shortcomings. Although two-dimensional TOF MRA does not require gadolinium, it is a time-consuming procedure that cannot be performed in a single breath-hold. The long-scan techniques are also susceptible to image degradation by respiratory and peristaltic motion. In addition, smaller vessels beyond the main trunks of the visceral vessel are poorly visualized.

More recent development of fast imaging techniques allows gadolinium-enhanced MRA during a single breath-hold (Figure 3.16). With these techniques, it is possible to obtain angiograms during separate arterial and venous phases. There are now multiple reports of these techniques being used to evaluate the mesenteric and renal vasculature.[44–54]

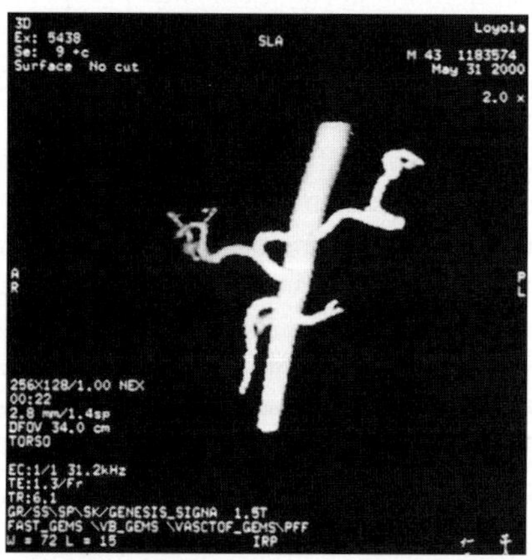

Figure 3.16 *Shaded-surface display rendering of three-dimensional magnetic resonance angiogram demonstrates patent celiac axis and superior mesenteric artery.*

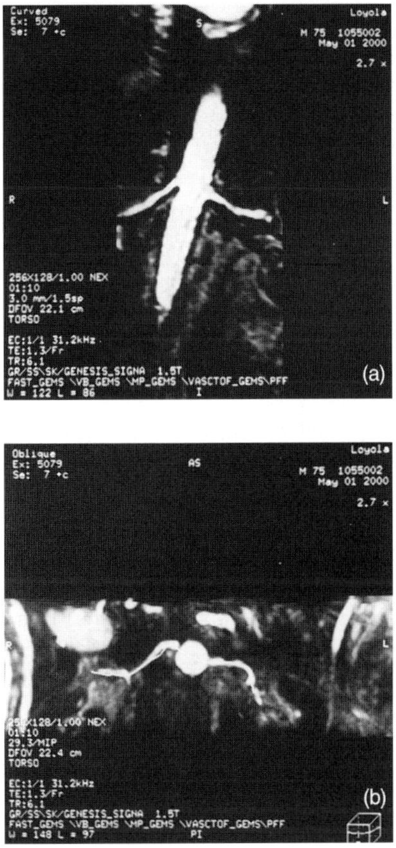

Figure 3.17 *(a) Frontal view of maximum-intensity projection reconstructed from subvolume of breath-hold ultrafast three-dimensional spoiled gradient echo magnetic resonance angiogram shows patent renal arteries. (b) Multiplanar image orientation used in (a).*

Image interpretation of MRA is made by viewing both the axial images and the three-dimensional reconstructions. Reconstructions can be performed at a workstation. Reformations can be obtained in flat two-dimensional planes and in curved planes (Figure 3.17). Images can be obtained in axial, sagittal and oblique planes in order to align the plane of reformation along the artery of interest. Slab thickness can be adjusted to encompass the entire artery targeted. Three-dimensional MIP and SSD images can be created with a similar appearance to three-dimensional CTA images (Figure 3.18).

Technique

Imaging is usually performed on a 1.5 T system with a high-performance gradient using a body or chest coil. A sagittal or axial localizer image is used to position the coronal three-dimensional gadolinium-enhanced sequence and to assess the patient's breath-holding capacity. The two-dimensional gradient echo imaging is performed with a flip angle of 30°, a time to echo (TE) of 13, and a time to repetition (TR) of 34.

Using this localizer sequence, the area of interest for the three-dimensional imaging can be targeted. The gadolinium-enhanced three-dimensional spoiled gradient echo sequence is performed using a vascular TOF sequence, such as three-dimensional fast spoiled gradient

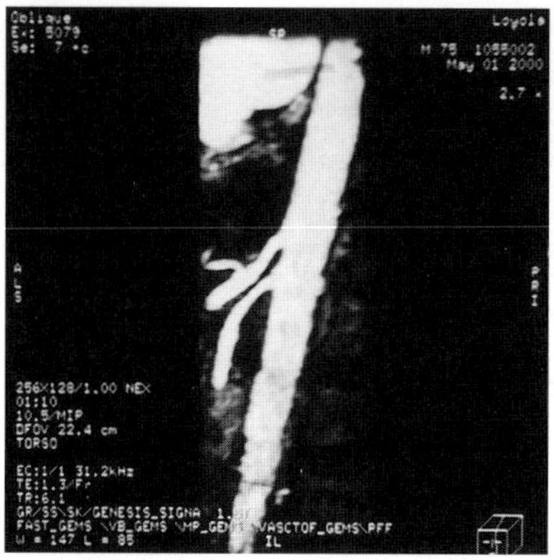

Figure 3.18 *Contrast-enhanced maximum-intensity projection magnetic resonance angiography image of patent celiac and superior mesenteric arteries.*

echo techniques (SPGR). A double dose of gadolinium is manually or power injected. Automated systems exist for optimizing peak contrast during image acquisition, such as GE's Smart PrepTM. This sequence has an image time of approximately 1 min using a flip angle of 45° and minimum TE.

Clinical use

While the effectiveness of visceral artery MRA is yet to be validated, multiple studies demonstrate the potential utility of three-dimensional gadolinium-enhanced MRA. Stafford-Johnson *et al.* evaluated the use of three-dimensional gadolinium-enhanced MRA in the detection of vascular complications of liver transplantation in 13 patients.[55] Ten vascular complications were identified in their series with conventional angiography or surgery. All 10 complications were diagnosed correctly with MRA. These included transplant hepatic artery thrombosis, stenosis, and portal vein stenosis or occlusions.

Prince *et al.* studied 63 patients with three-dimensional gadolinium-enhanced MRA during breath-hold.[47] Conventional angiography was performed in 19 patients. They found that renal artery, CA and SMA occlusive disease was appropriately graded in 15 of the 19 patients studied. Renal artery branches were visualized in 86 of 95 kidneys on breath-hold images compared with only 84 of 236 kidneys with free breathing. Overall, they found 25–50% greater signal-to-noise and 60–120% greater contrast-to-noise ratio with breath-holding compared with free breathing.

Ernst *et al.* prospectively performed MRA and digital subtraction angiography in 33 patients to evaluate hepatic, pancreatic, and mesenteric disease.[44] MRA was performed with three-dimensional acquisitions after administration of gadolinium. The authors described good or excellent agreement for visualization of the hepatic artery, SMA, and splenic artery. In addition, the portal vein, SMV, and splenic vein were well visualized. Agreement was poor for the intrahepatic arteries and branches of the SMA. In addition, the authors found that MRA depicted all arterial abnormalities and variations diagnosed with conventional angiography. These included replaced left and right hepatic arteries, an aneurysm of the hepatic artery, a gastroduodenal artery stenosis, a splenic artery stenosis, and an SMA thrombosis. The authors attribute the disagreement in visualization of small arteries to the differences in spatial resolution between the techniques.

Holland *et al.* prospectively studied the abdominal aorta and renal and visceral arteries in 63 consecutive patients using two-dimensional TOF and ultrafast three-dimensional SPGR

MRA.[46] Thirty-two of the patients had contrast angiography and/or surgery for comparison. Images were performed on a 1.5-T scanner. The breath-hold ultrafast three-dimensional SPGR correctly identified 31 of 31 stenoses of the renal arteries. Two-dimensional TOF detected 23 of the 31 renal artery stenoses. The ultrafast three-dimensional SPGR and two-dimensional TOF correctly identified 20 of 20 celiac, superior mesenteric and inferior mesenteric ostial stenoses or occlusions. Three Riolan's arcs were identified correctly by breath-hold three-dimensional SPGR7 but missed by two-dimensional TOF. Overall, the breath-hold ultrafast three-dimensional SPGR angiography correctly identified and graded 48 of 51 renal, celiac, superior mesenteric, and interior mesenteric artery stenoses or occlusions, while the two-dimensional TOF MRA correctly identified and graded 45 of 51 lesions. The authors emphasized the importance of breath-holding. Their technique was considerably better in breath-hold groups than in non-breath-hold groups.

Alternative magnetic resonance angiography techniques

Other investigators have attempted to assess the flow patterns present in the visceral vessels to evaluate disease.[56] Wasser *et al.* used systolic gating with phase-contrast MRA for imaging the mesenteric vessels.[41] The systolic gated three-dimensional phase-contrast angiography (PCA) studies demonstrated superiority over ungated studies in 10 patients with suspected mesenteric ischemia. However, their technique found a significant stenosis in only four of the six mesenteric vessels demonstrated to be stenotic by DSA. The authors found that by limiting data acquisition to the systolic phase of the cardiac cycle, they could decrease shift artifacts and maximize signal intensity in the vessels.

Li *et al.* have evaluated cine phase-contrast MRI to generate anatomic and physiologic information about visceral arteries.[57] They have used the data from PC MRI to evaluate flow velocity, volume flow rate, and displaced volume in dogs and humans. Their technique for non-invasive magnetic resonance oximetry relies on the fact that deoxyhemoglobin in red blood cells is paramagnetic, but oxyhemoglobin is not. Therefore, T2 relaxation time is correlated to oxygen saturation.

Hany *et al.* described the use of three-dimensional contrast-enhanced breath-hold MRA and segmented K-space two-dimensional contrast acquisitions before and after caloric stimulation in volunteers.[58] They report considerable increase in flow volume in the SMA after caloric stimulation. The ability to assess morphology and function may be helpful in the detection of chronic as well as acute mesenteric ischemia.

Kopka *et al.* evaluated dual-phase three-dimensional MRA in visceral arteries in 140 patients.[45] In 80 patients, the value of fat saturation, digital image subtraction, and anticholerigenic agents and high-choleric meals were evaluated. Their findings demonstrated that fat-saturated images were of significantly better quality than non-fat-saturated images. Digital imaging subtraction was useful in a subset of patients. The injection of an anticholerigenic agent was not helpful. High-calorific meals helped demonstrate the SMA and portal vein. MRA correctly identified the arterial blood supply in 57 of 60 patients. However, the authors noted that hemodynamic grading of CA stenosis could not be achieved with MRA because of the limited temporal resolution.

REFERENCES

1 Geelkerken RH, van Bockel JH. Duplex ultrasound examination of splanchnic vessels in the assessment of splanchnic ischaemic symptoms. *Eur J Vasc Endovasc Surg* 1999; **18**: 371–4.

2 Zwolak RM. Can duplex ultrasound replace arteriography in screening for mesenteric ischemia? *Semin Vasc Surg* 1999; **12**: 252–60.
3 Rizzo RJ, Sandager G, Astleford P, Payne K, Peterson-Kennedy L, Flinn WR, Yao JS. Mesenteric flow velocity variations as a function of angle of insonation. *J Vasc Surg* 1990; **11**: 688–94.
4 Qamar MI, Read AE. Effects of ingestion of carbohydrate, fat, protein, and water on the mesenteric blood flow in man. *Scand J Gastroenterol* 1988; **23**: 26–30.
5 Volteas N, Labropoulos N, Leon M, Kalodiki E, Chan P, Nicolaides AN. Detection of superior mesenteric and coeliac artery stenosis with colour flow Duplex imaging. *Eur J Vasc Surg* 1993; **7**: 616–20.
6 Geelkerken RH, Lamers CB, Delahunt TA, Hermans J, Zwijsen JH, van Bockel JH. Duodenal meal stimulation leads to coeliac artery vasoconstriction and superior mesenteric artery vasodilatation: an intra-abdominal ultrasound study. *Ultrasound Med Biol* 1998; **24**: 1351–6.
7 Jager KA, Fortner GS, Thiele BL, Strandness DE. Noninvasive diagnosis of intestinal angina. *J Clin Ultrasound* 1984; **12**: 588–91.
8 Qamar MI, Read AE, Skidmore R, Evans JM, Williamson RC. Transcutaneous Doppler ultrasound measurement of coeliac axis blood flow in man. *Br J Surg* 1985; **72**: 391–3.
9 Qamar MI, Read AE, Skidmore R, Evans JM, Wells PN. Transcutaneous Doppler ultrasound measurement of superior mesenteric artery blood flow in man. *Gut* 1986; **27**: 100–5.
10 Qamar MI, Read AE, Mountford R. Increased superior mesenteric artery blood flow after glucose but not lactulose ingestion. *QJM* 1986; **60**: 893–6.
11 Nicholls SC, Kohler TR, Martin RL, Strandness DE Jr. Use of hemodynamic parameters in the diagnosis of mesenteric insufficiency. *J Vasc Surg* 1986; **3**: 507–10.
12 Lilly MP, Harward TR, Flinn WR, Blackburn DR, Astleford PM, Yao JS. Duplex ultrasound measurement of changes in mesenteric flow velocity with pharmacologic and physiologic alteration of intestinal blood flow in man. *J Vasc Surg* 1989; **9**: 18–25.
13 Bowersox JC, Zwolak RM, Walsh DB, Schneider JR, Musson A, LaBombard FE, Cronenwett JL. Duplex ultrasonography in the diagnosis of celiac and mesenteric artery occlusive disease. *J Vasc Surg* 1991; **14**: 780–6.
14 Moneta GL, Yeager RA, Dalman R, Antonovic R, Hall LD, Porter JM. Duplex ultrasound criteria for diagnosis of splanchnic artery stenosis or occlusion. *J Vasc Surg* 1991; **14**: 511–18.
15 LaBombard FE, Musson A, Bowersox JC, Zwolak RM. Hepatic artery duplex as an adjunct in the evaluation of chronic mesenteric ischemia. *J Vasc Technol* 1992; **16**: 7–11.
16 Healy DA, Neumyer MM, Atnip RG, Thiele BL. Evaluation of celiac and mesenteric vascular disease with duplex ultrasonography. *J Ultrasound Med* 1992; **11**: 481–5.
17 Moneta GL, Lee RW, Yeager RA, Taylor LM, Jr, Porter JM. Mesenteric duplex scanning: a blinded prospective study. *J Vasc Surg* 1993; **17**: 79–84.
18 Harward TR, Smith S, Seeger JM. Detection of celiac axis and superior mesenteric artery occlusive disease with use of abdominal duplex scanning. *J Vasc Surg* 1993; **17**: 738–45.
19 Mallek R, Mostbeck GH, Walter RM, Stumpflen A, Helbich T, Tscholakoff D. Duplex Doppler sonography of celiac trunk and superior mesenteric artery: comparison with intra-arterial angiography. *J Ultrasound Med* 1993; **12**: 337–42.
20 Gentile AT, Moneta GL, Lee RW, Masser PA, Taylor LM, Porter JM. Usefulness of fasting and postprandial duplex ultrasound examinations for predicting high-grade superior mesenteric artery stenosis. *Am J Surg* 1995; **169**: 476–9.
21 Perko MJ, Just S, Schroeder TV. Importance of diastolic velocities in the detection of celiac and mesenteric artery disease by duplex ultrasound. *J Vasc Surg* 1997; **26**: 288–93.
22 Zwolak RM, Fillinger MF, Walsh DB, LaBombard FE, Musson A, Darling CE, Cronenwett JL. Mesenteric and celiac duplex scanning: a validation study. *J Vasc Surg* 1998; **27**: 1078–87.
23 Boijsen E. *Abram's angiography*. Boston: Little, Brown & Co., 1997, 1549–79.
24 Kadir S. *Diagnostic angiography*. Philadelphia: W. B. Saunders, 1986; 34–77.

25 Reuter S, Redman H, Cho K. *Gastrointestinal angiography*. Philadelphia: W. B. Saunders, 1986; 446–50.
26 Rubin GD, Walker PJ, Dake MD, Napel S, Jeffrey RB, McDonnell CH, *et al*. Three-dimensional spiral computed tomography angiography: an alternative imaging modality for the abdominal aorta and its branches. *J Vasc Surg* 1993; **8**: 656–65.
27 Zeman RK, Zeiberg AS, Davros WJ, Ascher SM, Cooper CJ, Weltman DI, *et al*. Routine helical CT of the abdomen: image quality considerations. *Radiology* 1993; **189**: 395–400.
28 Rubin GD, Dake MD, Semba CP. Current status of three-dimensional spiral CT scanning for imaging the vasculature. *Radiol Clin North Am* 1995; **33**: 51–70.
29 Zeman RK, Silverman PM, Vieco PT, Costello P. CT angiography. *Am J Roentgenol* 1995; **165**: 1079–88.
30 Qanadli SD, Mesurolle B, Coggia M, Barré O, Fukui S, Goeau-Brissonniére OA, *et al*. Abdominal aortic aneurysm: pretherapy assessment with dual-slice helical CT angiography. *Am J Roentgenol* 2000; **174**: 181–7.
31 Heath DG, Soyer PA, Kuszyk BS, Bliss DF, Calhoun PS, Bluemke DA, *et al*. Three-dimensional spiral CT during arterial portography: comparison of three rendering techniques. *Radiographics* 1995; **15**: 1001–11.
32 Winter TC, Nghiem HV, Schmiedl UP, Freeny PC. CT angiography of the visceral vessels. *Semin Ultrasound CT MR* 1996; **17**: 339–51.
33 Johnson PT, Heath DG, Kuszyk BS, Fishman EK. CT angiography with volume rendering: advantages and applications in splanchnic vascular imaging. *Radiology* 1996; **200**: 564–8.
34 Rubin GD, Dake MD, Napel S, Jeffrey RB Jr, McDonnell CH, Sommer FG, *et al*. Spiral CT of renal artery stenosis: comparison of three-dimensional rendering techniques. *Radiology* 1994; **190**: 181–9.
35 Brink JA, Lim JT, Wang G, Heiken JP, Deyoe LA, Vannier MW. Technical optimization of spiral CT for depiction of renal artery stenosis: in vitro analysis. *Radiology* 1995; **194**: 157–63.
36 Dix JE, Freeny PC. Corrugation artifact in spiral CT angiography: evaluation of the cause and potential for misinterpretation. Abstract 108, Radiological Society of North America, 1999.
37 Winter III TC, Freeny PC, Nghiem HV, Hommeyer SC, Barr D, Croghan AM, *et al*. Hepatic arterial anatomy in transplantation candidates: evaluation with three-dimensional CT arteriography. *Radiology* 1995; **195**: 363–70.
38 Blomley MJK, Albrecht T, Williamson RCN, Allison DJ. Three-dimensional spiral CT angiography in pancreatic surgical planning using non-tailored protocols: comparison with conventional angiography. *Br J Radiol* 1998; **71**: 268–75.
39 Gomes MN, Davros WJ, Zeman RK. Preoperative assessment of abdominal aortic aneurysm: the value of helical and three-dimensional computed tomography. *J Vasc Surg* 1994; **20**: 367–76.
40 Cikrit DF, Harris VJ, Hemmer CG, Kopecky KK, Dalsing MC, Hyre CE, *et al*. Comparison of spiral CT scan arteriography for evaluation of renal and visceral arteries. *Ann Vasc Surg* 1996; **10**: 109–16.
41 Wasser MN, Geelkerken RH, Kouwenhoven M, Hajo van Bockel J, Hermans J, Schultz Kool LJ, de Roos A. Systolically gated 3D phase contrast MRA of mesenteric arteries in suspected mesenteric ischemia. *J Comput Assist Tomogr* 1996; **20**: 262–8.
42 Meaney JFM. Non-invasive evaluation of the visceral arteries with magnetic resonance angiography. *Eur Radiol* 1991; **9**: 1267–76.
43 Miyazaki T, Yamashita Y, Shinzato J, Kojima A, Takahashi M. Two-dimensional time-of-flight magnetic resonance angiography in the coronal plane for abdominal disease: its usefulness and comparison with conventional angiography. *Br J Radiol* 1995; **68**: 351–7.
44 Ernst O, Asnar V, Sergent G, Lederman E, Nicol L, Paris J-C, L'Herminé C. Comparing contrast-enhanced breath-hold MR angiography and conventional angiography in the evaluation of mesenteric circulation. *Am J Roentgenol* 2000; **174**: 433–9.

45 Kopka L, Rodenwaldt J, Vosshenrich R, Fischer U, Renner B, Lorf T, *et al.* Hepatic blood supply: comparison of optimized dual phase contrast-enhanced three-dimensional MR angiography and digital subtraction angiography. *Radiology* 1999; **211**: 51–8.
46 Holland GA, Dougherty L, Carpenter JP, Golden MA, Gilfeather M, Slossman F, *et al.* Breath-hold ultrafast three-dimensional gadolinium-enhanced MR angiography of the aorta and the renal and other visceral abdominal arteries. *Am J Roentgenol* 1996; **166**: 971–81.
47 Prince MR, Narasimham DL, Stanley JC, Chenevert TL, Williams DM, Marx MV, Cho KJ. Breath-hold gadolinium-enhanced MR angiography of the abdominal aorta and its major branches. *Radiology* 1995; **197**: 785–92.
48 Baden JG, Racy DJ, Grist TM. Contrast-enhanced three-dimensional magnetic resonance angiography of the mesenteric vasculature. *J Magn Reson Imaging* 1999; **10**: 369–75.
49 Meaney JF, Prince MR, Nostrant TT, Stanley JC. Gadolinium-enhanced MR angiography of visceral arteries in patients with suspected chronic mesenteric ischemia. *J Magn Reson Imaging* 1997; **7**: 171–6.
50 Watanabe Y, Dohke M, Okumura A, Amoh Y, Ishimori T, Oda K, *et al.* Dynamic subtraction contrast-enhanced MR angiography: technique, clinical applications, and pitfalls. *Radiographics* 2000; **20**: 135–52.
51 Dong Q, Schoenberg S, Carlos RC, Neimatallah M, Cho KJ, Williams DM, *et al.* Diagnosis of renal vascular disease with MR angiography. *Radiographics* 1999; **19**: 1535–44.
52 Gilfeather M, Holland GA, Siegelman ES, Schnall MD, Axel L, Carpenter JP, Golden MA. Gadolinium-enhanced ultrafast three-dimensional spoiled gradient-echo MR imaging of the abdominal aorta and visceral and iliac vessels. *Radiographics* 1997; **17**: 423–32.
53 Shirkhoda A, Konez O, Shetty AN, Bis KG, Ellwood RA, Kirsch MJ. Mesenteric circulation: three dimensional MR angiography with a gadolinium-enhanced multiecho gradient-echo technique. *Radiology* 1997; **202**: 257–61.
54 Shirkhoda A, Konez O, Shetty AN, Bis KG, Ellwood RA, Kirsch MJ. Contrast-enhanced MR angiography of the mesenteric circulation: a pictorial essay. *Radiograhpics* 1998; **18**: 851–61.
55 Stafford-Johnson DB, Hamilton BH, Dong Q, Cho KJ, Turbot JG, Fontana RJ, Prince MR. Vascular complications of liver transplantation: evaluation with gadolinium-enhanced MR angiography. *Radiology* 1998; **207**: 153–60.
56 Moore JR, Finn JP, Edelman RR. Measurement of visceral blood flow with magnetic resonance imaging. *Invest Radiol* 1992; **27**: S103–8.
57 Li KCP. Magnetic resonance angiography of the visceral arteries: techniques and current applications. *Endoscopy* 1997; **29**: 496–503.
58 Hany TF, Schmidt M, Andreas RT, Schoenenberger W, Debatin JF. Contrast-enhanced three-dimensional magnetic resonance angiography of the splanchnic vasculature before and after caloric stimulation. *Invest Radiol* 1998; **33**: 682–6.

4

Clinical presentation and diagnosis of chronic visceral ischemia

NICK J.M. LONDON

INTRODUCTION

This chapter will deal with chronic mesenteric ischemia (CMI) due to stenosis or occlusion of the visceral arteries. CMI can also result from superior mesenteric venous thrombosis or celiac artery compression syndrome, but these subjects are covered in other chapters. Similarly, chronic ischemic colitis is covered in Chapter 15.

EPIDEMIOLOGY

Although CMI is uncommon, it is not rare. The frequency with which CMI is being diagnosed appears to be increasing, and it has recently been reported that it accounts for 0.1% of all hospital admissions and 1% of admissions in patients with acute abdominal pain.[1] This apparent increase may result from an increasingly elderly population,[2] a heightened awareness of the diagnosis, an increased ability of intensive care techniques to salvage critically ill patients or improved diagnostic techniques.[3]

CMI is a disease of the elderly, occurring most frequently during the sixth or seventh decade of life. The average age at diagnosis is 60 years.[4, 5] Patients with CMI are generally younger than those with acute mesenteric ischemia, and most studies have reported a predilection for the female sex, with a female-to-male ratio as high as 4 : 1.[6, 7] The reason for this sex distribution is unclear, and it is in contrast to the overall prevalence of males with peripheral vascular disease. In addition, women present at an earlier age with CMI than their male counterparts. Patients with CMI are frequently known to have vascular disease, and nearly half have undergone prior vascular surgery.[8, 9] Other affected arteries include the lower limb vessels in 40%,[10, 11] coronary arteries in 40%,[10] and renal arteries in 15%.[11] Cerebrovascular disease is also common. Risk factors for atherosclerosis are often present, including smoking in 75%,[10, 11] hypertension in 40%,[10, 12] and diabetes mellitus in 10%.[11, 12] Hypercholesterolemia is relatively uncommon because fat malabsorption due to CMI limits its development.

PATHOPHYSIOLOGY

The basic pathophysiological mechanism underlying chronic intestinal ischemia is failure to achieve normal postprandial hyperemic intestinal arterial flow. Although symptomatic intestinal ischemia is rare, atherosclerotic stenosis or occlusion of the intestinal arteries is common. A study of 120 consecutive autopsies found that there was a significant stenosis of the celiac artery in 22%, of the superior mesenteric artery (SMA) in 16%, and of the inferior mesenteric artery (IMA) in 10%.[13] The prevalence of visceral artery atherosclerosis increases with age and is higher in patients with other forms of symptomatic peripheral arterial disease. Thus in one study, 27% of patients undergoing aortography before peripheral vascular surgery were found to have an asymptomatic stenosis >50% of either the celiac artery or the SMA.[14] The rarity of CMI symptoms despite the commonness of mesenteric atherosclerosis is due to the development of collaterals. It would seem that the ability of the visceral circulation to tolerate atherosclerotic occlusions or stenoses of its major trunks reflects the slow progression of the atherosclerotic process, providing time for the gradual enlargement of collateral pathways. In the case of the patient with clinically evident intestinal ischemia, symptoms are not usually present until at least two of the three major mesenteric arteries are highly stenosed or occluded.[15, 16] Symptoms of CMI may result, however, from atherosclerotic occlusion of a single vessel, usually the SMA in patients who have undergone previous abdominal surgery that has interrupted important collateral vessels. In contrast, isolated narrowing or occlusion of the celiac artery or the IMA is well tolerated.

Intestinal arterial blood flow increases markedly after a meal, with the majority of this hyperemic flow distributed to the small intestine and pancreas, and only a relatively small proportion going to the stomach or colon. The duration of this hyperemic response is 4–6 h.[17] Mesenteric ischemic pain results from an imbalance between oxygen availability and oxygen demand. It is presumed that the metabolic demands of nutrient absorption and intestinal motility in the presence of significant arterial obstruction lead to the production of one or more anaerobic metabolites that mediate postprandial pain. An alternative hypothesis for the production of postprandial pain in patients with CMI is that the increased blood flow in the distal small-intestinal branches is achieved at the expense of flow to the stomach. This is sometimes referred to as 'gastric steal'. Reduced gastric blood flow then results in higher gastric acid production, transient peptic ulceration, and abdominal pain.[18] Although attractive, this theory is refuted by the observations that patients with CMI treated with H_2-blocking agents or proton pump inhibitors do not usually obtain symptomatic improvement.

ETIOLOGY

In more than 95% of cases, CMI is due to atherosclerotic mesenteric arterial stenosis or occlusion.[19, 20] Uncommon causes of CMI are listed in Table 4.1.[21, 22]

Table 4.1 *Causes of chronic mesenteric ischemia*

Aortic dissection
Atherosclerotic disease*
Behçets disease
Buerger's disease
Carcinoid syndrome
Cogan's syndrome
Colectomy with proximal ligation of IMA
Dermatomyositis
Drug-induced
Cocaine
Ergotamines
Essential mixed cryglobulinemia
Fibromuscular dysplasia
Giant cell arteritis
Hypertension – accelerated phase
Neurofibromatosis
Pheochromocytoma
Polyarteritis nodosa
Polycythemia rubra vera
Radiation
Rheumatoid arthritis
Sjögrens's syndrome
System lupus erythematosus
Takayasu's disease
Thrombophilic syndromes
Wegener's granulomatosis

*Atherosclerotic disease accounts for more than 95% of cases of chronic mesenteric ischemia.
IMA, inferior mesenteric artery.

CLINICAL PRESENTATION

Symptoms

The usual patient with CMI is a woman aged between 40 and 70 years who has undergone extensive investigation for long-standing abdominal pain. The period between onset of symptoms and diagnosis is frequently up to 1 year.[23] The classic triad of symptoms in patients with CMI is postprandial pain, fear of eating (sitophobia), and involuntary weight loss.[24] Abdominal pain is a cardinal feature of CMI; indeed, some series report that 100% of patients have abdominal pain.[8, 25] The pain of CMI usually begins 15–30 min after eating and peaks 2 h later. Approximately 90% of patients with pain exhibit this postprandial pattern.[26] The pain is usually epigastric or peri-umbilical and described as dull, gnawing, or aching. However, it can be crampy or colicky. The severity and duration of the pain appear to be related to the size of the meal.[27] As the severity of mesenteric ischemia increases, the pain may be induced by smaller meals and may become more severe and longer lasting until eventually it may become constant.

With increasing severity of CMI, the patient starts to lose weight. Roughly three-quarters of all patients lose weight,[4, 19] with the average weight loss being 10 kg.[11] However, weight loss can be so marked that the clinician may misdiagnose an incurable malignancy. Other symptoms include diarrhea in 35%,[28] nausea and vomiting in 25%,[29] and constipation in 20%.[27] It is said that celiac trunk occlusion tends to produce upper-gastrointestinal symptoms, such as nausea, vomiting, and bloating, whereas IMA occlusion tends to produce lower-gastrointestinal symptoms, including constipation and diarrhea. SMA occlusion does not usually produce these associated symptoms.[30]

Signs

Physical examination usually reveals signs of weight loss and malnutrition, including leg edema. Approximately two-thirds of patients have signs of peripheral vascular disease,[19] and up to 80% of patients have an abdominal bruit.[31] However, many patients without intestinal ischemia have abdominal bruits, and some authors feel that the finding of an abdominal bruit is so non-specific that it is not useful.

INVESTIGATION OF CHRONIC MESENTERIC ISCHEMIA

Laboratory studies

If CMI has led to malabsorption, then tests for malabsorption such as the D-xylose test may be abnormal. However, malabsorption may be due to myriad causes, and tests for malabsorption are therefore not recommended as part of the investigation of CMI. Similarly, other non-specific laboratory abnormalities include anemia, hypoalbuminemia, and low serum vitamin B_{12}. These abnormalities are not, however, specific for CMI. It is important to take notice of the serum creatinine because patients with CMI may have associated renovascular disease. Seventeen per cent of patients with CMI have positive fecal occult blood.[6]

Non-specific investigations

The diagnosis of CMI is usually made only after prolonged investigation of abdominal pain or weight loss. It is not unusual, therefore, for patients to undergo a number of investigations that are designed not specifically to diagnose CMI but rather to exclude other causes. Gastroduodenoscopy may demonstrate superficial gastroduodenal ischemic ulcers.[32–34] Histological examination of duodenal biopsy specimens obtained at duodenoscopy or jejunal biopsy specimens may demonstrate non-specific findings of epithelial cell flattening, villous atrophy, and chronic inflammation.[6] Although colonoscopy is important in the diagnosis of ischemic colitis, it is not usually helpful in CMI. Nonetheless, colonoscopy or barium enema examination may reveal a stricture due to CMI that is indistinguishable from the classic 'apple core' lesion of colonic carcinoma.[21] Upper-gastrointestinal investigation in the form of small-bowel follow-through or enema may show non-specific findings that are found in malabsorption, such as contrast segmentation, flocculation, bowel dilatation, hypoperistalsis, and mural thickening.[35]

The main role of abdominal computerized tomography (CT) is to help identify other causes of abdominal pain, and although there are established CT features of bowel ischemia (ascites, bowel wall edema, venous or bowel wall gas), these findings appear following the onset of infarction and are not usually present in CMI. Recently, magnetic resonance imaging (MRI) has been proposed as a non-invasive screening test for CMI. MRI is theoretically attractive because it is not impeded by the limitations of duplex scanning, such as operator dependence, bowel gas, and patient obesity. Magnetic resonance angiograms (MRA) have demonstrated postprandial retrograde flow in the SMA in a patient with an 80% SMA stenosis,[36] and postprandial hyperemia in the superior mesenteric vein has also been studied with phase-contrast cine MRI. It was found in a study comparing asymptomatic volunteers with patients suffering from CMI that postprandial augmentation of peak flow in the superior mesenteric vein was significantly less in the symptomatic patients.[37] The definitive magnetic resonance criteria of CMI are not yet established, but as MRA becomes refined further, it may supplant arteriography as the investigation of choice.

A fascinating approach to the early diagnosis of CMI is the detection of changes in the magnetic field produced by the electrical activity of the small-bowel smooth muscle to detect changes in the basic electrical rhythm (BER).[38] The BER is the slow wave of the gastrointestinal tract, which has been shown to reduce in frequency in the presence of ischemia; this decrease occurs well before the onset of irreversible ischemic damage. It is not possible to measure the BER non-invasively. However, it has been shown that a superconducting quantum interference device placed outside the abdomen can detect changes in the magnetic field from the small bowel and that these changes mirror changes in the BER.[38] Thus, it is theoretically possible to measure non-invasively small-bowel BER and thereby detect small-bowel ischemia at an early stage. Although there are significant hurdles to be overcome before this technique can be brought to clinical fruition, the basic principle is scientifically sound and the technology is currently being developed.

Angiography

Although in some vascular beds, such as the carotid arteries and lower limb arteries, it has been stated by some authors that color duplex scanning has replaced angiography as the gold standard investigation, this is not the case for the mesenteric arteries, where angiography remains the gold standard. Angiography is required to help confirm the diagnosis, to assess disease severity, and to plan revascularization. Angiography can usually be performed via a transfemoral route, but occasionally a transaxillary route is required because of iliac artery disease. Angiographic evaluation should include biplane aortography to evaluate the aorta, mesenteric vessels, and collateral vessels, with selective angiography to evaluate distal mesenteric vessels. The severity of occlusion of major mesenteric arteries is best assessed by lateral projection, whereas the presence of collaterals is best assessed by an anteroposterior projection.[19] The direction of collateral blood flow is determined by sequential radiographs. Because visceral artery lesions are usually ostial, selective catheterization of the main intestinal arteries is usually neither necessary nor advisable.

It must be stressed that visceral arteriography should be regarded as a means of demonstrating lesions rather than a means of ascribing symptoms to specific lesions. Interpretation of the significance of visceral artery collateral vessels should be cautious. Thus, poor collateralization may indicate that the visceral artery lesion is either hemodynamically insignificant or poorly compensated, whereas intensive collateralization found in association with visceral artery stenosis may reflect either an adequate collateral blood supply to the intestine or a vigorous but insufficient attempt at collateralization. Although mesenteric angiography is generally safe,[39] complications can include acute tubular necrosis or precipitation of mesenteric ischemia due to increased mesenteric arterial occlusion.[11, 26]

Angiography typically demonstrates significantly reduced flow in at least two mesenteric arteries. Such findings are generally diagnostic in symptomatic patients, but they are non-specific and non-diagnostic in asymptomatic patients. Many patients with two-vessel disease are asymptomatic because of adequate collateral perfusion. About 90% of patients with CMI have significant obstruction of both the celiac axis and the SMA, and half of these patients will also have significant obstruction of the IMA. A recent study of 16 patients with CMI showed that 24% had four-vessel disease, 44% had three-vessel disease, and 19% had two-vessel disease.[40] SMA involvement was noted in 100% of patients, celiac axis involvement in 90%, and IMA involvement in 90%. Severe occlusive disease affects both internal iliac arteries in nearly 60% of patients.

Rarely, patients develop symptoms from single-vessel occlusion (either the celiac axis or the SMA).[41] These patients have poorly developed collateral channels that cannot compensate for the single-vessel occlusion. However, it should be stressed that the diagnosis of CMI should be made carefully and infrequently when only one artery is occluded.[6]

Duplex scanning

This method of imaging is discussed in Chapter 3.

REFERENCES

1 Klein H-M, Lensing R, Klosterhalfen B, Tons C, Gunther RW. Diagnostic imaging of mesenteric infarction. *Radiology* 1995; **197**: 79–82.

2 Kaleya RN, Boley SJ. Acute mesenteric ischemia: an aggressive diagnostic and therapeutic approach. 1991 Roussel Lecture. *Can J Surg* 1992; **35**: 613–23.

3 Newman TS, Magnuson TH, Ahrendt SA, Smith-Meek MA, Bender JS. The changing face of mesenteric infarction. *Am Surg* 1998; **64**:611–16.

4 Hallett JW, Jr, James ME, Ahlquist DA, Larson MV, McAfee MK, Cherry KJ, *et al*. Recent trends in the diagnosis and management of chronic intestinal ischemia. *Ann Vasc Surg* 1990; **4**: 126–32.

5 Taylor LM Jr, Moneta GL. Intestinal ischemia. *Ann Vasc Surg* 1991; **5**: 403–6.

6 Moawad J, Gewertz BL. Chronic mesenteric ischemia: clinical presentation and diagnosis. *Surg Clin North Am* 1997; **77**: 357–69.

7 McAfee MK, Cherry KJ, Naessens JM. Influence of complete revascularisation on chronic mesenteric ischaemia. *Am J Surg* 1992; **164**: 220–4.

8 Harward TR, Brooks DL, Flynn TC, Seeger JM. Multiple organ dysfunction after mesenteric artery revascularization. *J Vasc Surg* 1993; **18**: 459–67.

9 Rapp JH, Reilly LM, Qvarfordt PG, Goldstone J, Ehrenfeld WK, Stoney RJ. Durability of endarterectomy and antegrade grafts in the treatment of chronic visceral ischemia. *J Vasc Surg* 1986; **3**: 799–806.

10 McMillan WD, McCarthy WJ, Bresticker MR, Pearce WH, Schneider JR, Golan JF, Yao JS. Mesenteric artery bypass: objective patency determination. *J Vasc Surg* 1995; **21**: 729–40.

11 Rheudasil JM, Stewart MT, Schellack JV, Smith RB, III, Salam AA, Perdue GD. Surgical treatment of chronic mesenteric arterial insufficiency. *J Vasc Surg* 1988; **8**: 495–500.

12 Calderon M, Reul GJ, Gregoric ID, Jacobs MJ, Duncan JM, Ott DA, *et al*. Long-term results of the surgical management of symptomatic chronic intestinal ischemia. *J Cardiovasc Surg* 1992; **33**: 723–8.

13 Jarvinen O, Laurikka J, Sisto T, Salenius JP, Tarkka MR. Atherosclerosis of the visceral arteries. *Vasa* 1995; **24**: 9–14.

14 Valentine RJ, Martin JD, Myers SI, Rossi MB, Clagett GP. Asymptomatic celiac and superior mesenteric artery stenoses are more prevalent among patients with unsuspected renal artery stenoses. *J Vasc Surg* 1991; **14**: 195–9.

15 Mikkelsen WP, Berne CJ. Intestinal angina: its surgical significance. *Am J Surg* 1957; **94**: 262–9.

16 Hansen HJB. Abdominal angina. *Acta Chir Scand* 1976; **142**: 319–25.

17 Moneta GL, Taylor DC, Helton WS, Mulholland MW, Strandness DE, Jr. Duplex ultrasound measurement of postprandial intestinal blood flow: effect of meal composition. *Gastroenterology* 1988; **95**: 1294–301.

18 Poole JW, Sammartano RJ, Boley SJ. Hemodynamic basis of the pain of chronic mesenteric ischemia. *Am J Surg* 1987; **153**: 171–6.

19 Cunningham CG, Reilly LM, Stoney R. Chronic visceral ischemia. *Surg Clin North Am* 1992; **72**: 231–44.

20 Rogers AI, David S. Intestinal blood flow and diseases of vascular impairment. In: Haubrich WS, Schaffner F, Berk JE (eds). *Gastroenterology*, 5th edn. Philadelphia: W. B. Saunders, 1995; 1212–34.

21 Ruotolo RA, Evans SRT. Mesenteric ischemia in the elderly. *Clin Geriatr Med* 1999; **15**: 527–57.

22 Cappell MS. Intestinal (mesenteric) vasculopathy. II. Ischemic colitis and chronic mesenteric ischemia. *Gastroenterol Clin North Am* 1998; **27**: 827–60.

23 Schneider DB, Schneider PA, Reilly LM, Ehrenfeld WK, Messina LM, Stoney RJ. Reoperation for recurrent chronic visceral ischemia. *J Vasc Surg* 1998; **27**: 276–84.

24 MacFarlane SD, Beebe HG. Progress in chronic mesenteric arterial ischemia. *J Cardiovasc Surg* 1989; **30**: 178–84.

25 McMillan WD, McCarthy WJ, Bresticker MR, Pearce WH, Schneider JR, Golan JF, Yao JS. Mesenteric artery bypass: Objective patency determination. *J Vasc Surg* 1995; **21**: 729–41.

26 Rogers DM, Thompson JE, Garrett WV, Talkington CM, Patman RD. Mesenteric vascular problems. A 26-year experience. *Ann Surg* 1982; **195**: 554–65.

27 Connelly TI, Perdue GD, Smith RB, Ansley JD, McKinnon WM. Elective mesenteric revascularization. *Am Surg* 1981; **47**: 19–25.

28 Cipolla DM, Boley SJ, Luchs S, Pasternak B, Floch C, DiCorato M, *et al.* Chronic mesenteric ischemia presenting as chronic diarrhea and weight loss with pneumatosis intestinalis. *Gastroenterologist* 1996; **4**: 134–41.

29 Crawford ES, Morris GC, Jr, Myhre HO, Roehm JO, Jr. Celiac axis, superior mesenteric artery, and inferior mesenteric artery occlusion: surgical considerations. *Surgery* 1977; **82**: 856–66.

30 Babu SC, Shah PM. Celiac territory ischemic syndrome in visceral artery occlusion. *Am J Surg* 1993; **166**: 227–30.

31 Zelenock GB, Graham LM, Whitehouse WM, Jr, Erlandson EE, Kraft RO, Lindenauer SM, *et al.* Splanchnic arteriosclerotic disease and intestinal angina. *Arch Surg* 1980; **115**: 497–501.

32 Cappell MS, Mikhail N, Gujral N. Gastrointestinal hemorrhage and intestinal ischemia associated with anticardiolipin antibodies. *Dig Dis Sci* 1994; **39**: 1359–64.

33 Casey KM, Quigley TM, Kozarek RA, Raker EJ. Lethal nature of ischemic gastropathy. *Am J Surg* 1993; **165**: 646–9.

34 Cherry RD, Jabbari M, Goresky CA, Herba M, Reich D, Blundell PE. Chronic mesenteric vascular insufficiency with gastric ulceration. *Gastroenterology* 1986; **91**: 1548–52.

35 Meyers MA, Kaplowitz N, Bloom AA. Malabsorption secondary to mesenteric ischaemia. *Am J Roentgenol* 1973; **119**: 352–8.

36 Li KCP, Whitney WS, McDonnell CH, Fredrickson JO, Pelc NJ, Dalman RL, *et al.* Chronic mesenteric ischemia: evaluation with phase-contrast cine MR imaging. *Radiology* 1994; **190**: 175–9.

37 Burkart DJ, Johnson CD, Reading CC, Ehman RL. MR measurements of mesenteric venous flow: prospective evaluation in healthy volunteers and patients with suspected chronic mesenteric ischemia. *Radiology* 1995; **194**: 801–6.

38 Seidel SA, Bradshaw LA, Ladipo JK, Wikswo JP, Jr, Richards WO. Noninvasive detection of ischemic bowel. *J Vasc Surg* 1999; **30**: 309–19.

39 Kaleya RN, Boley SJ. Acute mesenteric ischemia. *Crit Care Clin* 1995; **11**: 479–512.

40 Johnston KW, Lindsay TF, Walker PM, Kalman PG. Mesenteric arterial bypass grafts: early and late results and suggested surgical approaches for chronic and acute mesenteric ischemia. *Surgery* 1995; **118**: 1–7.

41 Roobottom CA, Dubbins PA. Significant disease of the celiac and superior mesenteric arteries in asymptomatic patients: predictive value of Doppler sonography. *Am J Roentgenol* 1993; **161**: 985–8.

5

Endovascular treatment of chronic visceral ischemia

ANNA-MARIA BELLI AND ROBERT MORGAN

INTRODUCTION

Endovascular methods of treatment for visceral ischemia consist of percutaneous transluminal angioplasty (PTA) and intra-arterial stenting. These techniques have been extrapolated from other vascular territories, such as the renal and peripheral vasculature. Endovascular methods have the advantage over surgery of being less invasive and so are particularly attractive in patients considered at high risk for operative treatment.

Only chronic visceral ischemia is considered in this chapter. The management of acute visceral ischemia and thrombolysis in mesenteric venous thrombosis are discussed in Chapters 6 and 13, respectively.

INDICATIONS FOR TREATMENT

Table 5.1 shows the indications for treatment of chronic visceral ischemia. The main indication for treatment is angiographic evidence of stenosis or occlusion of two out of the three mesenteric arteries in patients with typical symptoms of weight loss and postprandial abdominal pain (Figures 5.1 and 5.2). The first reports of successful PTA in such cases were in 1980,[1–3] which were followed by further series with encouraging results.[4–6]

As in most endovascular interventions, great care should be taken before performing an interventional procedure in an asymptomatic patient. A case can be made for prophylactic PTA in certain circumstances, e.g. before abdominal aortic aneurysm surgery, when the inferior mesenteric artery (IMA) may be ligated and there is occlusive disease of the celiac axis and superior mesenteric artery (SMA).[7, 8]

Table 5.1 *Indications for endovascular therapy in chronic visceral ischemia*

Typical symptoms with three-vessel involvement
Typical symptoms with two-vessel involvement
Atypical symptoms with two-vessel involvement
Prophylaxis, e.g. abdominal aortic aneurysm repair and celiac and superior mesenteric artery stenoses

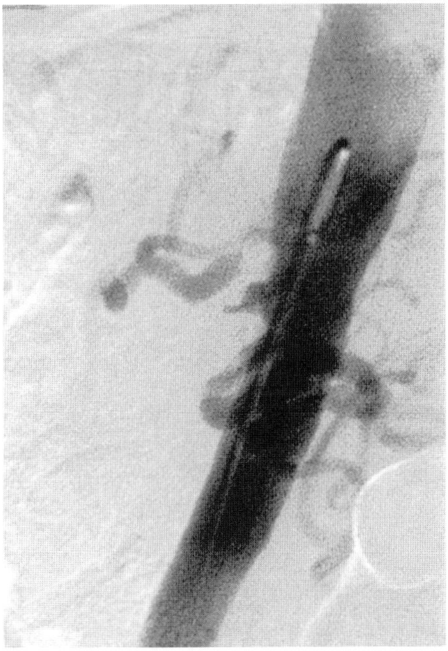

Figure 5.1 *Lateral aortogram demonstrating a stenosis at the origin of the celiac axis and a superior mesenteric artery occlusion.*

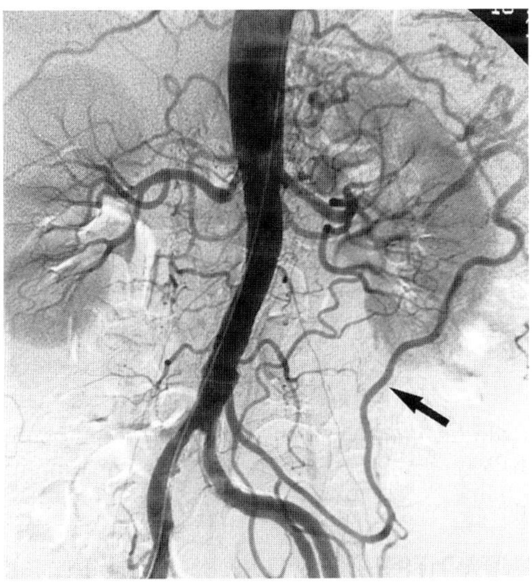

Figure 5.2 *Flush aortogram in the anterior–posterior direction showing no filling of the celiac axis or superior mesenteric artery. However, the inferior mesenteric artery is patent, and a large marginal artery is identified (arrows).*

Table 5.2 *Causes of chronic visceral ischemia*

Causes
Atherosclerosis
Fibromuscular dysplasia
Median arcuate ligament syndrome
Arteritis, e.g. Takayasu's disease
Aortic dissection

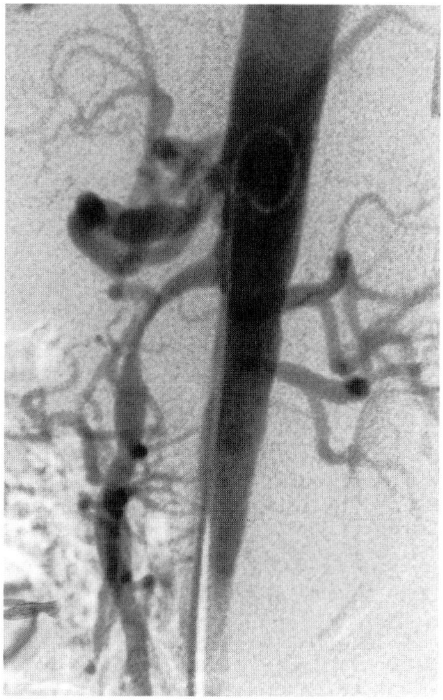

Figure 5.3 *A long non-ostial stenosis of the superior mesenteric artery in a patient with Sneddon's syndrome.*

Most of the lesions are due to atherosclerotic disease, but fibromuscular dysplasia can also be a cause of symptoms and responds well to PTA (Table 5.2).[9] Mesenteric arterial stenosis can also be part of an arteritis, e.g. Takayasu's disease,[10, 11] and is seen in some rare syndromes, e.g. pseudo-xanthoma elasticum and Sneddon's syndrome (Figure 5.3).[12, 13] Mesenteric arterial stenosis may also be associated with aortic dissection extending into the abdominal aorta.[14] In these cases, stenting is required to overcome the extrinsic compression of the true lumen by the false lumen.

There are insufficient published data to decide whether primary stenting is superior to PTA in the mesenteric arteries. It is reasonable to assume that if the lesion is ostial, then the situation is analogous to the renal artery, where primary stenting gives superior technical success rates.[15] On the other hand, stenting may be reserved for situations where there is a significant residual pressure gradient following balloon dilation, or where there is a flow-limiting dissection (Table 5.3).

CONTRAINDICATIONS TO TREATMENT

Extrinsic compression of the celiac axis by the median arcuate ligament of the diaphragm will not respond to either PTA or stenting (as the stent will be crushed or compressed by the ligament).

Table 5.3 *Indications for stenting in chronic visceral ischaemia*

Poor percutaneous transluminal angioplasy result
>30% residual stenosis
Ostial stenosis
Residual pressure gradient
Flow-limiting dissection
Aortic dissection

Other contraindications include contraindications to angioplasty, e.g. bleeding diatheses, although usually these can be controlled for the procedure. If the patient is on warfarin, this should be stopped and heparin started before the procedure.

Asymptomatic patients should not undergo any intervention unless there are overriding reasons, such as high risk of ischemia at surgery.

TECHNIQUE

Before intervention, diagnostic angiography should be performed to identify the site and morphology of the lesions. A flush aortogram in the lateral position should be performed to assess the arteries, and selective catheterization should be avoided until the time of intervention. The arteriogram helps to assess the best route of intervention, i.e. femoral or upper limb approach. It should also distinguish between stenoses and occlusions and whether the lesion is ostial or non-ostial. This information is helpful as occlusions are more difficult to cross with a guidewire and a stent may be required for an ostial lesion or occlusion.[16, 17]

The technique of PTA and stenting is standard. Generally, a femoral arterial approach is used, but if there is an acute angle of the mesenteric arteries with the aorta (as is often the case), and a good stable guidewire position cannot be achieved from this approach, then an upper-limb approach may be more successful. Either axillary or brachial artery puncture may be performed, depending on local expertise. The only disadvantage of the upper-limb approach is that the complication rate of axillary and brachial artery puncture is higher than femoral artery puncture, and hematoma around the artery can lead to neurological complications.[18–21] Once arterial access is obtained, a long sheath (approx. 35–45 cm, depending on patient length) or a guiding catheter is inserted into the abdominal aorta. This allows repeat arteriography to be performed without necessitating introduction of a separate aortic catheter or withdrawal of the wire once across the lesion. Generally, a 7F long sheath or 8F guiding catheter is advised as this allows immediate progression to stenting if required without further catheter exchange. Various stents are now available, some of which require a smaller guiding catheter, so it is advised that a sheath or catheter size compatible with the intended stent is chosen.

A selective catheter is required to catheterize the origin of the vessel. This may be a sidewinder or cobra shape, depending on the anatomy. There is a wide variety of preshaped catheters that may be used for this purpose. It is essential to be able to pass a guidewire through the selective catheter and beyond the stenosis or occlusion. Various guidewires may be used. A hydrophilic guidewire, e.g. Terumo, may be helpful for the initial crossing of the lesion, but a stiffer guidewire is usually required when passing the balloon catheter or stent, e.g. a TAD 11 wire, which has a very atraumatic platinum floppy tip and a stiff main shaft (Figure 5.4). If it is not possible to pass the guidewire through the stenosis or occlusion, the procedure cannot proceed, but once the guidewire is across, it is important to pass the catheter beyond the lesion to establish that the wire is intraluminal. At this point, heparin may be given

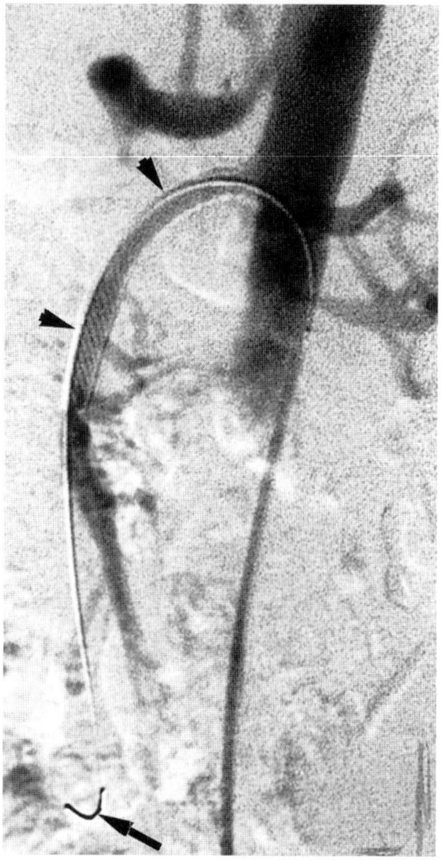

Figure 5.4 *The same patient as in Figure 5.3. A TAD 11 wire has been placed across the stenosis. The radio-opaque platinum floppy tip is advanced as far as possible (arrow), and the stiff shaft allows easy passage of a stent introduced from the femoral artery (arrowheads).*

as a bolus of 3000–5000 units, and an antispasmodic agent is helpful to prevent spasm in any of the small vessels, e.g. 150-μg aliquots of glyceryl trinitrate.

Once the guidewire is in a stable position, with its tip as far distally as is reasonably possible, a balloon catheter of appropriate dimensions is passed through the lesion and inflated. A check angiogram may then be performed though the sheath or guiding catheter to assess whether the procedure has been successful. Intra-arterial pressure measurements are helpful in assessing technical success. The pressure gradient between the aorta and beyond the original stenosis should be abolished. However, a pressure gradient of 10% or less of the systemic systolic pressure is considered acceptable in our practice. If there is a significant residual gradient following balloon dilation, this may be an indication for placing a stent.

The diameter and length of the balloon or stent should be chosen to match the lesion. The diameter of the celiac axis and SMA is usually 6–8 mm. The IMA is significantly smaller, e.g. 3 mm. Accurate assessment of size may be performed using a calibrated pigtail catheter during the initial flush aortograms, but most operators gauge the diameter from their experience in other vascular territories.

If stenting is to be performed, either a balloon expandable or self-expanding stent may be used. Balloon expandable stents of very short length (13–20 mm) can be obtained; these are ideally suited to short ostial lesions and can be placed extremely accurately to cover the stenosis with a small amount of stent (e.g. 2 mm) protruding into the aortic lumen. Self-expanding stents may also be used. When a stent is being placed, it is important to identify clearly not only the lesion but also the origin of the artery as it arises from the aorta so that the stent is positioned optimally. For the mesenteric arteries, this is usually the lateral projection, although different obliquities may have to be performed to identify the origin, particularly in elderly patients with tortuous arteries.

A 'roadmap' technique may be helpful. This is when real-time screening is superimposed on an arteriogram to allow identification of the lesion and manipulation through it. This may be degraded by breathing artifacts in the abdomen, so for accurate positioning of a stent following the initial manipulation, short check angiograms are often required. These can be performed through the guiding catheter or long sheath, and fine adjustment can be made to the position of the stent before deployment. Even though two of the three visceral arteries may be stenosed, it is not known whether treatment of all affected arteries is required. The benefit of treating more than one stenosis may be an improved rate of clinical success should one artery re-stenose or occlude. In practice, the celiac axis and SMA are the main arteries to be dilated, but the IMA may also be treated.[22, 23]

RESULTS

The published data are sparse. The results of series of PTA in visceral arteries show that technical success rates are of the order of 75–95%, even without stents, as the majority of lesions are non-ostial (i.e. >5 mm from artery origin). Clinical success rates are good, with 60–85% of patients showing immediate clinical improvement. However, with time, re-stenosis occurs and reintervention may be necessary to maintain this improvement; stenting could have a role in avoiding this (Table 5.4).

In Matsumoto's series, four out of 19 patients had extrinsic arterial compression by the median arcuate ligament, which resulted in an immediate technical failure.[6] In patients with typical symptoms of mesenteric angina, the clinical success rate was high (78%). If symptoms were atypical, the clinical success rate was 50%.

In Odurny's series, 17 out of 19 arteries were dilated successfully in 10 patients, resulting in symptomatic relief lasting for 6–24 months.[24] Symptoms recurred in five patients, and three had further relief of symptoms following balloon dilation. In this series, one technical failure was due to extrinsic compression by the median arcuate ligament, and another was due to a severe ostial stenosis that did not respond to PTA. This would now be considered an indication for a stent. In addition, the average residual pressure gradient following PTA was 40 mm Hg. This report was before the widespread use of stents, and such a high residual gradient would now also be considered an indication for stenting.

Most studies rely on angiographic criteria. A > 30% residual stenosis is considered to be technically unsuccessful in most vascular territories. However, Rose *et al.* found that this did not correlate with symptomatic relief following PTA, and even when there was a >30% residual stenosis, there was symptomatic improvement so long as there was some angiographic improvement in the lumen.[28] The recurrence rate following PTA varies from 11 to 63% (Table 5.4). Many re-stenoses are amenable to further percutaneous treatment. The majority of reported series performed repeat balloon dilation, but in more recent practice, stenting is considered appropriate for recurrent stenosis. Five out of the 12 patients treated by stenting in Sheeran's series were for post-PTA re-stenosis.[17]

The literature on stenting in mesenteric arteries is still very limited (Table 5.5). Stenting has high technical and clinical success rates (83–100%), but re-stenosis still occurs. In-stent

Table 5.4 *Outcome of mesenteric percutaneous transliminal angioplasty*

Reference	No. of patients	No. of arteries treated	Technical success	Clinical success	Re-stenosis rate	Mean follow-up(months)	Complications	30-day mortality
Matsumoto *et al.* 1995[6]	19	20	15/19 (79%)	12/19 (63%)	2/12 (17%)	25	6/19 (32%)	0/19 (0%)
Odurny *et al.* 1988[24]	10	19	8/10 (80%)	8/10 (80%)	5/8 (63%)	11	2/10 (20%)	1/10 (10%)
Maspes *et al.* 1998[25]	23	41	22/23 (96%)	17/23 (74%)	2/17 (12%)	27	2/23 (9%)	0/23 (0%)
McShane *et al.* 1992[26]	6	12	5/6 (83%)	5/6 (83%)	2/5 (40%)	18	0/6 (0%)	0/6 (0%)
Allen *et al.* 1996[27]	19	24	18/19 (95%)	15/19 (79%)	3/15 (20%)	39	2/19 (11%)	1/19 (5%)

Table 5.5 *Outcome of mesenteric stenting*

Reference	No. of patients	No. of arteries treated	Technical success	Clinical success	Re-stenosis rate	Mean follow-up(months)	Complications	30-day mortality
Sheeran *et al.* 1999[17]	12	13	11/12 (92%)	10/12 (83%)	2/11 (18%)	16	3/12 (25%)	1/12 (8%)
Liermann and Strecker 1995[29]	12	12	12/12 (100%)	12/12 (100%)	4/12 (33%)	28	1/12 (8%)	0/12 (0%)
Nyman *et al.* 1998[30]	5	9	5/5 (100%)	5/5 (100%)	3/5 (60%)	14	2/5 (40%)	0/5 (0%)

re-stenosis may be treated by PTA or further stenting.[17, 29, 30] Other therapeutic strategies for in-stent re-stenosis, such as atherectomy, brachytherapy, or cutting balloons, have not been described in mesenteric arteries.

Sheeran *et al.* stented six ostial lesions in their series. Three of these were re-stenosis following primary PTA, and the other three were primarily stented.[17] They also performed primary stenting of the three chronic occlusions, which was technically successful in all cases. There is insufficient evidence in the literature to either support or refute this approach in mesenteric arteries. On the other hand, stenting is indicated in situations where there is a significant residual pressure gradient following balloon dilation, or where there is a flow-limiting dissection (Table 5.3).

COMPLICATIONS

Matsumoto *et al.* quoted a 32% complication rate. Half of the complications in this series were minor and resolved with conservative therapy.[6] These were hematoma and cholesterol embolization. The other half were major complications, consisting of one case of hypotension and a decrease in hematocrit requiring blood transfusion but no site of hemorrhage identified, one case of occlusion of the arterial access site subsequently resulting in below-knee amputation, and in the third patient myocardial infarction, renal failure, and ipsilateral brachial nerve palsy.

There are reports of acute bowel infarction as a consequence of PTA and stenting. Allen *et al.* reported a case of SMA dissection with thrombosis and infarction, resulting in death despite emergency surgery and revascularization.[27] Flow-limiting dissection would now be considered an indication for stenting. Sheeran *et al.* also reported one death due to bowel infarction post-stenting despite an apparently technically successful procedure.[17] This was one of five patients presenting with acute worsening of chronic abdominal symptoms. All five patients developed bowel ischemia. Two had these resected before endovascular intervention and the other three after stenting as their symptoms continued. Other reported complications are related to access-site hematomas and occlusions, and are the same as any percutaneous arterial intervention. Access-site hematoma accounted for both complications reported in the series by Nyman *et al.*, which required surgical repair.[30]

CONCLUSION

Arterial insufficiency causing chronic mesenteric ischemia can be due to a variety of conditions. Atherosclerosis is the most common cause, but other causes should be remembered and are amenable to endovascular therapy. Aortic dissection compromising the mesenteric arteries requires stenting to maintain the patency of the lumen. However, extrinsic compression by the median arcuate ligament does not respond to either PTA or stenting.

The majority of the results in the literature are following PTA, with few reports of stenting. The literature suggests it is possible that the technical and clinical success rates have improved with the advent of stents, but to date the re-stenosis rate is not reduced significantly.

The low morbidity and mortality rates associated with endovascular intervention make this a desirable treatment option for those with poor operative risk. The fact that the procedure can be repeated easily also makes this a desirable treatment option, despite the recurrence rate. However, it must be remembered that all patients are at risk of bowel infarction should re-stenosis or occlusion occur.

REFERENCES

1 Furrer J, Gruntzig A, Kugelmeier J, Goebel N. Treatment of abdominal angina with percutaneous dilatation of an arteria mesenterica superior stenosis. *Cardiovasc Intervent Radiol* 1980; **3**: 43–4.

2 Uflacker R, Goldany MA, Constant S. Resolution of mesenteric angina with percutaneous transluminal angioplasty of a superior mesenteric artery stenosis using a balloon catheter. *Gastrointest Radiol* 1980; **5**: 367–9.

3 Novelline RA. Percutaneous transluminal angioplasty: newer applications. *Am J Roentgen* 1980; **135**: 983–8.

4 Golden DA, Ring EJ, McLean GK, Freiman DB. Percutaneous transluminal angioplasty in the treatment of abdominal angina. *Am J Roentgenol* 1982; **139**: 247–9.

5 Roberts L, Wertman DA, Mills SR, Moore AV, Heaston DK. Transluminal angioplasty of superior mesenteric artery: an alternative to surgical revascularization. *Am J Roentgenol* 1983; **141**: 1039–42.

6 Matsumoto AH, Tegtmeyer CJ, Fitzcharles EK, Selby JB Jr, Tribble CG, Angle JF, Kron IL. Percutaneous transluminal angioplasty of visceral arterial stenoses. Results and long-term clinical follow-up. *J Vasc Interv Radiol* 1995; **6**: 165–74.

7 Gonzalez LL, Jaffe MS. Mesenteric arterial insufficiency following abdominal aortic resection. *Arch Surg* 1966; **93**: 10 20.

8 Rogers DM, Thompson JE, Garrett WV, Talkington CM, Patman RD. Mesenteric vascular problems: a 26 year experience. *Ann Surg* 1982; **195**: 554–65.

9 Ripley HR, Levin SM. Abdominal angina associated with fibromuscular hyperplasia of the coeliac and superior mesenteric arteries. *Angiology* 1966; **17**: 297–310.

10 Ozdil E, Krajcer Z, Angelini P. Percutaneous balloon angioplasty with adjunctive stent placement in the mesenteric vessels in a patient with Takayasu's arteritis. *Circulation* 1996; **93**: 1940–41.

11 Tyagi S, Verma P, Kumar N, Arora R. Stent angioplasty for relief of chronic mesenteric ischemia in Takayasu's arteritis. *Indian Heart J* 1997; **49**: 315–18.

12 Belli A, Cawthorne S. Visceral angiographic findings in pseudoxanthoma elasticum. *Br J Radiol* 1988; **61**: 368–71.

13 Khoo LAL, Belli AM. Superior mesenteric artery stenting for mesenteric ischemia in Sneddon's syndrome. *Br J Radiol* 1999; **54**: 51–5.

14 Yamakado K, Takeda K, Nomura Y, Kato N, Hirano T, Matsumura K, *et al*. Relief of mesenteric ischemia by Z-stent placement into the superior mesenteric artery compressed by the false lumen of an aortic dissection. *Cardiovasc Intervent Radiol* 1998; **21**: 66–8.

15 Van de Ven PJG, Kaatee R, Beutler JJ, Beek FJA, Wottiez AJ, Buskens E, *et al*. Arterial stenting and balloon angioplasty in ostial atherosclerotic renovascular disease: a randomised trial. *Lancet* 1999; **353**: 282–6.

16 Maleux G, Wilms G, Stockx L, Vandeemput J, Baert AL. Percutaneous recanalization and stent placement in chronic proximal superior mesenteric artery occlusion. *Eur Radiol* 1997; **7**: 1228–30.

17 Sheeran SR, Murphy TP, Khwaya A, Sussman SK, Hallisey MJ. Stent placement for treatment of mesenteric artery stenoses or occlusions. *J Vasc Interv Radiol* 1999; **10**: 861–7.

18 Heenan SD, Grubnic S, Buckenham TM, Belli A-M. Transbrachial arteriography: indications and complications. *Clin Radiol* 1996; **51**: 205–9.

19 Baudouin CJ, Belli A-M, Peck RJ, Cumberland DC. The complications of high brachial artery puncture. *Clin Radiol* 1990; **42**: 277–80.

20 McIvor J, Rhymer JC. 245 trans axillary arteriograms in arteriopathic patients: success rate and complications. *Clin Radiol* 1992; **45**: 390–94.

21 Gagliardi JM, Batt M, Avril G, Declemy S, Hassen-Khodja R, Daune B, *et al*. Neurological complications of axillary and brachial catheter arteriography in atherosclerotic patients: predictive factors. *Ann Vasc Surg* 1990; **4**: 547–9.

22 Schneider DB, Nelken NA, Messina LM, Ehrenfeld WK. Isolated inferior mesenteric artery revascularization for chronic visceral ischemia. *J Vasc Surg* 1999; **30**: 51–8.

23 Crotch-Harvey MA, Gould DA, Green AT. Case report: percutaneous transluminal angioplasty of the inferior mesenteric artery in the treatment of chronic mesenteric ischemia. *Clin Radiol* 1992; **46**: 408–9.

24 Odurny A, Sniderman KW, Colapinto RF. Intestinal angina. Percutaneous transluminal angioplasty of the celiac and superior mesenteric arteries. *Radiology* 1988; **167**: 59–62.

25 Maspes F, Mazzetti di Pietralata G, Gahdini R, Innocenzi L, Lupatelli L, Barzi F, Simonetti G. Percutaneous transluminal angioplasty in the treatment of chronic mesenteric ischemia: results and 3 years of follow-up in 23 patients. *Abdom Imaging* 1998; **23**: 358–63.

26 McShane MD, Proctor A, Spencer P, Cumberland DC, Welsh CL. Mesenteric angioplasty for chronic intestinal ischemia. *Eur J Vasc Surg* 1992; **6**: 333–6.

27 Allen RC, Martin GH, Rees CR, Rivera FJ, Talkington CM, Garrett WV, *et al*. Mesenteric angioplasty in the treatment of chronic intestinal ischemia. *J Vasc Surg* 1996; **24**: 415–23.

28 Rose SC, Quigley TM, Raher EJ. Revascularization for chronic mesenteric ischemia; comparison of operative arterial by pass grafting and percutaneous transluminal angioplasty. *J Vasc Interv Radiol* 1995; **6**: 339–49.

29 Liermann D, Strecker EP. Tantalum stents in the treatment of stenotic and occlusive disease of abdominal vessels. In: Liermann DD (ed.). *Stents: state of the art and future developments.* Watertown: Boston Scientific Corp., 1995; 127–34.

30 Nyman U, Ivancev K, Lindh M, Uher P. Endovascular treatment of chronic mesenteric ischemia: report of five cases. *Cardiovasc Intervent Radiol* 1998; **21**: 305–13.

6

Management of acute visceral ischemia

MARIO J. PERKO, TORBEN V. SCHROEDER, AND GEORGE GEROULAKOS

INTRODUCTION

Acute visceral ischemia is defined as the sudden symptomatic reduction of blood flow, which may be the result of embolization or thrombosis occurring at a previous stenotic lesion in the proximal part of a major visceral branch of the aorta. An additional cause of acute visceral ischemia is severe and prolonged intestinal arterial spasm, a condition that is known as non-occlusive acute visceral ischemia. This may occur in patients with low cardiac output, and is frequently associated with other life-threatening disorders.

The first successful treatment of visceral ischemia reported by Elliot in 1895 included resection of necrotic bowel and creation of stomas with secondary closure.[1] This treatment method remained valid until the 1950s, when vascular reconstruction became available. However, there has been little change in the overall mortality, which is currently approximately 80%, thus it remains a lethal condition.[2–4] The prognosis is more favorable with superior mesenteric artery (SMA) embolism than that from SMA thrombosis. This occurs because more than 50% of emboli get lodged in the SMA distally to the origin of the middle colic artery. By contrast, occlusion of the SMA at its origin, as occurs commonly with thrombosis, results in extensive infarction of the bowel.[2]

The reason for the failure to improve outcome in acute visceral ischemia is due mainly to the difficulty in making a diagnosis before infarction has occurred. No simple laboratory test is available, so diagnosis is often delayed, resulting in irreversible bowel injury that requires resection. Mortality can be improved only if revascularization can be achieved before intestinal infarction. In spite of recent developments in minimally invasive catheter-based techniques, laparotomy with embolectomy, endarterectomy, and bypass remain the most commonly employed methods of revascularization on a small number of patients suitable for these procedures, and the outcome has not improved during the past decades.[4, 5]

CLINICAL FEATURES

The onset of acute visceral ischemia may be preceded for months or years by symptoms of unrecognized mesenteric angina, such as abdominal pain and weight loss. Rheudasil *et al.* reported that 25% of hospitalized patients being investigated for chronic visceral ischemia developed acute symptoms.[6] By contrast, if the cause of the acute symptoms is embolic, then associated cardiac diseases can be identified, including atrial fibrillation, history of myocardial infarction, mitral valve disease, and left ventricular aneurysm.

The key feature of acute intestinal ischemia is abdominal pain, which is present in all patients with this condition. In the early phases of the disease, the patient is in great pain, which is out of proportion to the physical signs. This pain is due to an intense muscle spasm, which is the immediate consequence of withdrawal of blood supply and is associated with hyperactive bowel sounds and only slight abdominal tenderness. In the later phases, there is paralytic ileus and the bowel sounds disappear. Rebound tenderness and guarding are present. The location and nature of the pain are variable. In a large series of 103 patients, in addition to pain, 56% gave a history of vomiting, 29% had diarrhea, and 25% had occult blood in the gastric, rectal contents or both. In the same report, the duration of the symptoms did not correlate with the reversibility of the intestinal ischemia with revascularization.[2]

LABORATORY INVESTIGATIONS

There is no biochemical or hematological marker in any combination with sufficient sensitivity and specificity to diagnose acute visceral ischemia, therefore blood tests have limited value. Jamieson *et al.* suggested that acidosis, leukocytosis, and hyperphosphatemia constitute a useful diagnostic triad.[7] However, this has not been confirmed on other reports. Mamode presented eight patients with acute mesenteric ischemia who had serum phosphate concentration measured; only one had raised phosphate.[4] Furthermore, in another large series, 10% of the patients had a white blood count of less than 10 000 mm^3.[2] Serum amylase levels are mildly elevated in approximately 50% of the patients.[2, 4] There has been some interest in the use of intestinal mucosa enzymes as early markers, since the mucosa is the most sensitive layer of the bowel wall in acute ischemia. In an experimental model of intestinal ischemia in dogs, serum levels of seromuscular enzymes (creatinine phosphokinase, lactic dehydrogenase, serum glutamic oxaloacetic transaminase) were compared with mucosal enzymes (diamine oxidase, alkaline phosphatase). The seromuscular enzymes were more elevated in this study during intestinal ischemia.[8]

PREOPERATIVE IMAGING

Abdominal X-rays are non-diagnostic but may show edematous bowel wall that demonstrates the thumbprinting appearance. In advanced stages, plain films may show gas in the bowel wall and the portal vein. In a series of 77 patients with acute mesenteric ischemia, more than half had a normal appearing plain film, while signs of ileus were a common finding.[9] Although a contrast computed tomography (CT) scan rarely provides a definite diagnosis, it may show occlusions of the main visceral arteries. This finding strongly suggests the presence of acute ischemia in a patient with relevant history and clinical picture. Ultrasonography is most valuable in chronic ischemia because the patients are thin and bowel gas is seldom a problem. Patients suspected of acute ischemia may not be examined readily because the excess abdominal gas interferes with the procedure.

Arteriography should be considered in patients presenting with pain alone if acute intestinal ischemia is suspected. In these patients, diagnosis is required because pain alone is not an indication for laparotomy. In addition, it could help in the differentiation between thrombosis and embolism, which is useful in the planning of the surgical approach.[9] This approach may spare selected patients with non-occlusive ischemia from inappropriate dissection of the mesenteric vessels.[10] However, in patients who present with a typical history and signs of peritonitis or a rigid abdomen, preoperative angiography results in unnecessary delay and should not be performed.

PREOPERATIVE MANAGEMENT

Regardless of whether the diagnosis has been established, patients suspected of acute mesenteric ischemia should be monitored thoroughly. Intravenous lines are used for fluid management, pain relief, heparinization, antibiotics, and correction of acidosis. A bladder catheter is inserted for monitoring urine output. The patient's customary medication should be continued until operation. This is valid especially for drugs for controlling hypertension, heart failure, and ischemic heart disease. It is difficult, if not impossible, to predict the extent of operation before the abdomen is opened, therefore sedation has to be profound to allow for abdominal aortic surgery with intestinal resection. The most common sedation is oral benzodiazepine (diazepam 0.15 mg/kg or an equivalent dose of lorazepam or triazolam) given 90 minutes preoperatively with an intramuscular opiate (morphine 0.1 mg/kg) given 30–60 min before the operation. However, in patients with coexisting pulmonary or cardiac diseases with reduced cardiac output, opiates and other sedatives (with or without direct cardiovascular effects) can induce respiratory depression, hypoxemia, and hypotension.[11, 12] Therefore, in these patients premedication should be administered in reduced doses.

SURGICAL OR ENDOVASCULAR APPROACH?

Patients undergoing arteriography for investigation of symptoms and signs of acute intestinal ischemia may be considered for thrombolytic therapy, angioplasty, or both if an embolism or stenosis is found in the mesenteric vessels. However, in spite of the wide use of endovascular techniques for the management of peripheral arterial occlusions, their role in the management of acute visceral ischemia is limited and confined to a number of case reports and small series. Leduc *et al.* reported on a patient who had a diagnosis of acutely ischemic but non-necrotic bowel made at diagnostic laparoscopy. Angiography showed an occluded SMA and a 95% stenosis of the origin of the celiac axis and of the inferior mesenteric artery. Angioplasty of the stenotic vessels was followed by complete resolution of his symptoms. A second-look laparoscopy 24 h later showed no evidence of bowel infarction.[13] Vandeinse reported another patient with acute intestinal ischemia who was treated successfully with angioplasty of the SMA.[14] In a selected series of ten patients with no peritoneal signs on clinical examination, catheter-directed thrombolytic treatment for the management of SMA embolism achieved good clinical and radiological results in seven patients. In this small series, the resolution of the abdominal pain within the first hour of fibrinolytic treatment was the best marker of clinical success.[15] It has been suggested that in a stable patient with no signs of peritonitis, laparoscopy should be performed 24 h following the completion of the endovascular revascularization. Such an approach offers the diagnostic advantages of a laparotomy but avoids the morbidity in patients in whom no further procedure is indicated.[13, 16]

The high mortality associated with the standard operative approach means that exploration of the potential of endovascular techniques in reducing the mortality and morbidity of this condition is reasonable.

EXPLORATORY LAPAROTOMY

The goals of the operative treatment are the resection of necrotic intestine and the re-establishment of arterial flow to ischemic but potentially viable intestine. When the viability of resected bowel edges is equivocal, double stomata should be performed and the anastomosis delayed for a second-look procedure or later. A second-look procedure is indicated when the viability of the bowel at the first operation is questionable.[9]

Owing to a risk of mesenteric vasoconstriction aggravating ischemia, the operating room should be warmed and humidified. The skin of the groin and the upper thigh should be exposed in case the saphenous vein is needed for a bypass or venous patches. A midline incision from the xyphoid process to below the umbilicus is preferred because it provides wide exposure of all abdominal organs. The omentum with transverse colon is lifted upwards, and examination of SMA blood flow/pulsation in the root of the mesentery is performed. This is followed by general inspection of the bowel to evaluate the degree of ischemia and the presence of perforations.

Necrosis of the entire small bowel, regardless of the cause, is not treatable at present. Perforation with partial intestinal necrosis is not a contraindication for further treatment, and resection of the ischemic bowel should be considered.

Restoration of adequate flow in the SMA can be achieved either by mesenteric embolectomy or with an SMA bypass. The choice of appropriate technique depends on the cause of the ischemia.

EMBOLECTOMY/THROMBECTOMY

The ligament of Treitz is divided completely to gain access to the small-bowel mesentery directly over the SMA, which is exposed as closely to the aorta as possible, avoiding all side branches (Figure 6.1).[17–19] The SMA is opened through a transverse arteriotomy and, if possible, proximal to the middle colic artery. The transverse opening, as opposed to the longitudinal, diminishes the risk of narrowing the lumen of small-calibrated splanchnic vessels. Embolectomy using a 3 Fr. catheter is carried out proximally towards the aorta as well as distally. The procedure should be carried out carefully, as the SMA is fragile, and should be repeated until all thrombotic material is removed and a pulsatile inflow achieved. At the end of the procedure, the artery is irrigated with heparin saline and papaverine hydrochloride. Closure of the arteriotomy is accomplished with a 6–0 or 7–0 monofilament continuous or interrupted suture. If a longitudinal incision is used, then the arterial lumen is restored preferably using a vein patch.

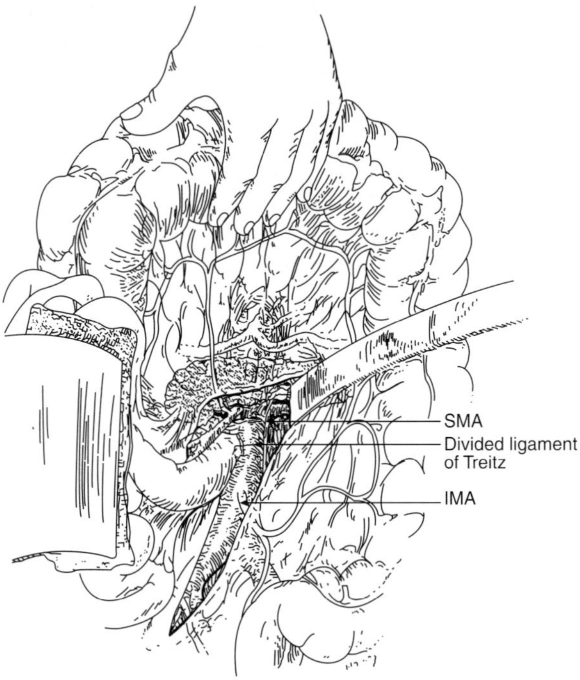

Figure 6.1 *Superior mesenteric artery (SMA) embolectomy. Exposure of infrarenal aorta and proximal SMA after division of ligament of Treitz. Transverse arteriotomy is followed by balloon catheter extraction and arteriotomy closure. IMA, inferior mesentric artery.*

BYPASS OR RESECTION AND RE-ANASTOMOSIS

If a preoperative diagnosis of thrombosis of the SMA has been established, or if the embolectomy fails to restore pulsatile flow as a result of a tight stenosis at the origin of the SMA, the patient is better treated with an aortomesenteric bypass. When the entire intestine appears salvageable before revascularization, a prosthetic graft is often preferred, but prosthetic material should be avoided if some of the intestine is obviously necrotic, requiring resection. In these cases, an autologous saphenous vein should be used. Alternatively, the part of the SMA distal to the occlusion may be transected and reimplanted into the aorta a few centimeters lower down. The surgical techniques do not differ from those used during surgery for chronic ischemia (Chapters 8 and 9).

INTESTINAL VIABILITY

Upon completion of revascularization, the need for intestinal resection must be determined.[17–19] Typically, the bowel contains obvious viable areas, obvious necrotic areas, and areas of questionable viability. A wide variety of techniques have been reported to achieve accurate intraoperative assessment of intestinal viability. The intraoperative assessment of small-intestinal viability includes inspection of the return of the color, the presence of visible arterial pulsation in the mesenteric arcade vessels, the amount of bleeding from the cut edges of the bowel, and the presence of visible peristalsis. Reperfusion should be allowed for at least 20–30 min after SMA flow has been established before bowel resection is considered. In a series of 28 consecutive patients operated for acute intestinal ischemic disease, standard clinical judgment had an overall accuracy of 89% in determining viability of the small intestine.[20] Doppler examination has also been used to determine the precise limits of the resection. The disappearance of the Doppler signal is monitored along the mesentery, mesenteric border of the bowel, and the antimesenteric bowel surface, always moving from normal tissue towards the ischemic margins. In an operative canine model of intestinal ischemia, the sensitivity and specificity of this technique in estimating intestinal viability were 88% and 66%, respectively.[21] With the intravenous fluorescein method, the operative room is darkened and the bowel is illuminated with a standard, hand-held, long-wave (3600 Å) ultraviolet light. A complete absence of fluorescence is diagnostic of non-viability. In a prospective, randomized, controlled study, Bulkley demonstrated an improved accuracy than what was achieved with standard clinical judgment.[20] The requirements of special equipment that will be used only rarely in view of the low incidence of this catastrophic surgical emergency is a significant disadvantage to the clinical use of this method. We and others rely on a combination of clinical judgment and Doppler ultrasound to determine intestinal viability. Since every method has the potential of false negative testing, a second-look laparotomy should be performed in patients with equivocal viability of the bowel after revascularization and resection 18–36 h later.

POSTOPERATIVE MANAGEMENT

Extubation immediately after operation is not recommended as circulatory, respiratory, and renal insufficiency are to be expected. Moreover, hemorrhage, metabolic instability, and the need for high doses of opioids are not uncommon. If a second-look operation is planned for the following day, then extubation should be delayed until the day after.

Intensive care with control of hemodynamic status, renal function, fluid balance, electrolyte balance, and acid–base balance is the same as that following any major vascular surgery. Therapy involves the same basic principles used in the management of patients with cardiovascular instability and/or shock. Nasogastric suction should continue until effective peristalsis returns, and the patient should be given broad-spectrum antibiotics and heparin sulfate.

One of the most important, but often ignored, principles on postoperative care units is that of normothermia. Hypothermia is common after major abdominal surgery, but it is especially dangerous for patients with arteriosclerosis and after revascularization of ischemic bowel. Its main adverse effect is local vasoconstriction, increasing the risk of sustained ischemia, necrosis of anastomoses, and reduction of cardiac output and regional blood flow, which further intensifies ischemia. Treatment includes external warming with thermal blankets or ceiling lights, warm intravenous fluids and blood, and supplemental oxygen.

If the mesenteric embolus arose because of atrial fibrillation, anticoagulants should be started as soon as the risk of postoperative hemorrhage is eliminated. Digitalis and beta-blockers should be avoided if possible.

NON-OCCLUSIVE ACUTE INTESTINAL ISCHEMIA

The management of non-occlusive ischemia involves first of all treatment of causative factors that contributed to the circulatory insufficiency, e.g. myocardial infarction or sepsis. Pain relief and heparinization should be initiated early. It is critical to recognize whether the regional circulatory insufficiency is associated with a low or high cardiac output. In the first case, the preload can be increased easily by intravenous fluids until the mean right atrial pressure is about

15 mm Hg. Above this limit, there is a risk of overhydration and a shift towards the descending part of the Starling curve, with a resultant fall in cardiac output. If the patient is monitored with a Swann–Ganz catheter, the wedge pressure can be raised to 20 mm Hg. However, the risk of pulmonary edema is increased at this level. Further treatment of normovolemic patients with cardiac-failure-induced non-occlusive mesenteric ischemia includes catecholamines. Confusion in the literature regarding the value of this treatment originates from inappropriate extrapolation of animal data to human physiology. In dogs, dopamine reduces intestinal blood flow by 10–60%,[22] while in the cat it increases it by 25–50%.[23] We know that dopamine at low doses (2–4 μg/kg/min) increases total splanchnic and selective renal blood flow with only minimal effects on systemic vascular resistance.[24] Initial infusion of dopamine at 2.5 μg/kg/min can be increased advantageously to 10 μg/kg/min. However, above this limit further circulatory response cannot be expected. In some cases, dobutamine can be more effective than dopamine, owing to greater augmentation of myocardial blood flow. Dobutamine can be added to dopamine gradually in similar doses.

The use of digitalis in patients with acute non-occlusive mesenteric ischemia should be avoided.[25]

When the central hemodynamics are adjusted adequately, further treatment can be focused on the mesenteric vessels. If mesenteric vasospasm is suggested by ultrasound scanning or angiography,[26] then infusion of papaverine at 30–60 mg/h selectively to the SMA for at least 24 h can be helpful. During infusion, careful monitoring of blood pressure and control of appropriate placement of the catheter in the mesenteric artery are warranted. Intravenous infusion of glucagon (0.5–1 μg/kg/min) increases total splanchnic and selective mesenteric blood flow; it also increases cardiac output, thus its proven beneficial effect in an animal model can also be expected in humans.[27] Theoretical consideration also indicates an advantageous effect of calcium-channel blockers, potent vasodilators used in myocardial ischemia and cardiac surgery to protect against spasms of arterial grafts.[28, 29] However, clinical data validating this theory are absent.

Laparotomy should be considered when vasodilators fail to resolve vasoconstriction and the abdominal pain continues or when rebound tenderness is present.

ACKNOWLEDGMENTS

The authors express their appreciation to Mr Adrian Marston for his useful comments.

REFERENCES

1 Elliot JW. The operative relief of gangrene of the intestine due to occlusion of the mesenteric vessels. *Ann Surg* 1895; **21**: 9.
2 Ottinger LW. The surgical management of acute occlusion of the superior mesenteric artery. *Ann Surg* 1978; **188**: 721–1.
3 Sachs SM, Morton JH, Schwartz SI. Acute mesentric ischaemia. *Surgery* 1982; **92**: 646–53.
4 Mamode N, Pickford I, Leiberman P. Failure to improve outcome in acute mesenteric ischaemia: seven-year review. *Eur J Surg* 1999; **165**: 203–8.
5 Klempnauer J, Grothues F, Bektas H, Pichlmayr R. Long-term results after surgery for acute mesenteric ischaemia. *Surgery* 1997; **121**: 239–43.
6 Rheudasil JM, Stewart MT, Shellack JV, *et al.* Surgical treatment of chronic mesenteric artery insufficiency. *J Vasc Surg* 1988; **8**: 495 500.

7 Jamieson WG, Marchuk S, Rowsom J, Durand D. The early diagnosis of massive acute intestinal ischaemia. *Br J Surg* 1982; **69** (suppl): S52–3.
8 Thompson JS, Bragg LE, West WW. Serum enzyme levels during intestinal ischaemia. *Ann Surg* 1990; **211**: 369–73.
9 Levy PJ, Krausz MM, Manny J. Acute mesenteric ischaemia: Improved results: a retrospective analysis of ninety-two patients. *Surgery* 1990; **107**: 372–80.
10 Boley SJ, Sprayregan S, Siegelman SS, Veith EY. Initial results from an aggressive roentgenological and surgical approach to acute mesenteric ischaemia. *Surgery* 1977; **82**: 848–55.
11 Stone JG, Foex P, Sear JW, Johnson LL, Khambatta HJ, Triner L. Myocardial ischemia in untreated hypertensive patients: effect of a single small oral dose of a beta-adrenergic blocking agent. *Anesthesiology* 1988; **68**: 495–500.
12 Stanley TH, Lathrop GD. Urinary excretion of morphine during and after valvular and coronary-artery surgery. *Anesthesiology* 1977; **46**: 166–9.
13 Leduc FJP, Pestieau SR, Detry O, Hamoir E, Honore P, Trotteur G, Jacquet N. Acute mesenteric ischaemia: minimally invasive management by combined laparoscopy and percutaneous transluminal angioplasty. *Eur J Surg* 2000; **166**: 345–7.
14 Vandeinse WH, Zawacki JK, Phillips D. Treatment of acute mesenteric ischaemia by percutaneous transluminal angioplasty. *Gastroenterology* 1986; **91**: 475–8.
15 Simo G, Echengusia AJ, Camunez F, Turegano F, Cabrera A, Urbano J. Superior mesenteric arterial embolism: local fibrinolytic treatment with urokinase. *Radiology* 1997; **204**: 775–9.
16 MacSweeney STR, Postlethwaite JC. Second-look laparoscopy in the management of acute mesenteric ischaemia. *Br J Surg* 1994; **81**: 90.
17 Christoffersen JK. Mesenteric ischemia. In: Chant ADB, D'Sa AABB (eds). *Emergency vascular practice*. London: Arnold, 1997; 105–18.
18 Kazmers A. Operative management of acute mesenteric ischemia. *Ann Vasc Surg* 1998; **12**: 187–97.
19 Vicente DC, Kazmers A. Acute mesenteric ischemia. *Curr Opin Cardiol* 1999; **14**: 453–8.
20 Bulkley GB, Zuidema GD, Hamilton SR, O'Mara CS, Klacsmann PG, Horn SD. Intraoperative determination of small intestinal viability following ischaemic injury. A prospective controlled trial of two adjuvant methods (Doppler and fluorescein) compared with standard clinical judgement. *Ann Surg* 1981; **193**: 628–37.
21 Pearce WH, Jones DN, Warren GH, Bartle EJ, Whitehill TA, Rutherford RB. The use of infrared photoplethysmography in identifying early intestinal ischaemia. *Arch Surg* 1987; **122**: 308–10.
22 Pawlik W, Shepherd AP, Mailman D, Shanbour LL, Jacobsen ED. Effects of dopamine and epinephrine on intestinal blood flow and oxygen uptake. *Adv Exp Med Biol* 1976; **75**: 511–16.
23 Ross G, Brown AW. Cardiovascular effects of dopamine in the anesthetized cat. *Am J Physiol* 1967; **212**: 823–8.
24 Hollenberg NK, Adams DF, Mendell P, Abrams HL, Merrill JP. Renal vascular responses to dopamine: haemodynamic and angiographic observations in normal man. *Clin Sci Mol Med* 1973; **45**: 733–42.
25 Bynum TE, Hanley HG. Effect of digitalis on estimated splanchnic blood flow. *J Lab Clin Med* 1982; **99**: 84–91.
26 Siegelman SS, Sprayregen S, Boley SJ. Angiographic diagnosis of mesenteric arterial vasoconstriction. *Radiology* 1974; **112**: 533–42.
27 Kazmers A, Zwolak R, Appelman HD, Whitehouse WM Jr, Wu SC, Zelenock GB, *et al*. Pharmacologic interventions in acute mesenteric ischemia: improved survival with intravenous glucagon, methylprednisolone, and prostacyclin. *J Vasc Surg* 1984; **1**: 472–81.
28 Resnekov L. Silent myocardial ischemia: therapeutic implications. *Am J Med* 1985; **79**(3A): 30–34.
29 Hayes EC, L'Ecuyer KM. A standard of care for radial artery grafting. *Am J Crit Care* 1998; **7**: 429–35.

7

Small-bowel transplantation

NADEY S. HAKIM

INTRODUCTION

Many patients die each year lacking only a functional small bowel to survive. The minimum amount of small-intestinal absorptive surface required to sustain life varies from patient to patient. Prolonged survival with oral alimentation alone has been reported in a few patients with an intact duodenum and as little as 15–45 cm of residual jejunum. However, without long-term total parenteral nutrition (TPN), prolonged patient survival is the exception rather than the rule.[1]

Chronic parenteral nutrition is associated with complications, including sepsis, venous thrombosis, metabolic disorders, and liver dysfunction.[2] From studies of patients currently on long-term TPN, it would appear that two to three patients per million of population per year develop irreversible small-bowel failure.[3] It is estimated that 20 new patients per year in the UK receiving home TPN would be potential candidates for small-bowel transplantation.[1]

INDICATIONS

Short bowel syndrome comprises a spectrum of conditions resulting from loss of absolute bowel surface to loss of appropriate function. It occurs when the patient is unable to absorb enough nutrients from the intestinal tract to maintain adequate nutrition, health, and a feeling of wellbeing.[4] This may be due to a short bowel following, for example, extensive resection secondary to acute mesenteric ischemia,[5] extensive inflammatory or radiation-induced bowel disease, congenital disease affecting the microvilli of the mucosa, or motility disorders. The common causes of short bowel syndrome in adults and children are given in Table 7.1.

There are two categories of candidates for small-bowel transplantation: patients with small-bowel failure alone, and patients with advanced liver disease associated with small bowel failure. Small-bowel transplantation is currently indicated only for patients with

Table 7.1 *Causes of intestinal failure*

Adults	Children
Crohn's disease	Intestinal atresia
Superior mesenteric artery thrombosis	Gastroschisis
Superior mesenteric vein thrombosis	Microvillous involution disease
Trauma	Necrotizing enterocolitis
Desmoid tumor	Mid-gut volvulus
Volvulus	Chronic intestinal pseudo-obstruction
Radiation enteritis	Crohn's disease
Pseudo-obstruction	

intestinal failure who are failing on TPN[7] and who cannot be improved with alternative surgical procedures (e.g. gut lengthening or reversed bowel loops).[8]

Pediatric patients with congenital abnormalities of the gastrointestinal tract form the majority of potential candidates for small-bowel transplantation.[6] Some of those pediatric patients will have other congenital abnormalities that require careful evaluation to determine whether they are suitable for small-bowel transplantation. Congenital immunodeficiency syndromes appear to be a basic contraindication because of unrestrained graft-versus-host disease (GVHD)[9] mediated by donor lymphocytes from the graft lymphoid organs that are transplanted with the intestine. There is no upper age limit for patients that can undergo intestinal transplantation, although elderly patients are more likely to have other adverse conditions. Likewise, there is no clear lower age limit for pediatric patients.

SMALL-BOWEL PROCUREMENT AND PRESERVATION

The small-bowel graft should be retrieved from a stable cadaveric heart-beating donor who is approximately 20% smaller than the recipient. If the donor is too large, it may be impossible to transplant the small bowel in the (usually) contracted peritoneal cavity of the recipient. The intestinal mucosa is extremely sensitive to ischemia and hypoxia secondary to cardiac or respiratory instability of the donor. The donor should undergo selective bowel decontamination with antibiotic and antifungal agents given by nasogastric tube. For patients undergoing combined liver and small-bowel transplantation, good liver function in the donor is extremely important.

Antilymphocyte preparations[10] to prevent GVHD have been used,[11] but some authors consider them unnecessary.[12] For the retrieval operation, the donor is heparinized and the supraceliac aorta is cross-clamped. In adults, 1 l of University of Wisconsin (UW) solution is infused into the aorta and 1 l of UW into the portal vein, venting the vena cava either into the chest or through a cannula into an iliac vein. Excessive perfusion may damage the intestinal graft. Appropriately smaller volumes of perfusate should be administered to pediatric donors. The abdominal cavity is packed with ice slush to assist cooling. If both the small bowel and the liver are to be used for separate isolated transplants, the liver and the intestine can be retrieved en bloc and separated on the back table.[12]

RECIPIENT OPERATION

The allograft arteries are anastomosed directly onto the descending aorta with a Carrel patch. The venous drainage of clusters that include the liver flows into the recipient's vena cava via a direct anastomosis or patch of the donor's vena cava. When the liver is not included, the preferred venous drainage is onto the recipient's portal vein. Any residual recipient viscera are drained into either the portal vein or the vena cava.

A gastrostomy or jejunostomy is performed to permit continuous enteral feeding. A Bishop–Koop ileostomy is also performed to permit frequent endoscopic and histologic monitoring. The primary technical difference is related to the number of organs replaced at the time of the transplant. Early evidence showed that the liver may protect other organs from rejection, and many patients received composite liver–intestine or multivisceral allografts. This complex operation often requires replacement of the entire stomach, pancreas, duodenum, liver, and small intestine, and sometimes includes the kidneys. These are technically challenging transplants. Alternatively, isolated small-bowel transplant can be performed in patients with short gut syndrome but a near-normally functioning liver. It is important to recognize patients developing liver dysfunction while on TPN early in their course of treatment. If there is no severe fibrosis or cirrhosis, the isolated small-bowel transplant may return the liver to normal functioning after discontinuation of TPN.[13]

POSTOPERATIVE MANAGEMENT

Intensive care unit

Patients who are well nourished may be extubated 24–48 h after surgery. Malnourished patients often require a longer period of ventilation. Fluid losses from the stoma, feces, jejunal and gastric tubes, and abdominal drains must be closely monitored and replaced. The small-bowel graft may become edematous during the early postoperative period, with fluid sequestration into the tissues that is later released back into the circulation. Early graft viability can be assessed by observing the color of the ileostomy; if necessary, an ileoscopy can be performed.

Immunosuppression

FK506 is the drug of choice, at least in the early postoperative period. An initial intravenous dose of 0.15 mg/kg per day as a continuous infusion is used. During the first 2 weeks, FK506 is indicated to maintain constant blood levels at the upper limit of the therapeutic range (about 10–15 ng/ml). FK506 may be combined in a triple-therapy regimen with azathioprine 2 mg/kg per day, titrated against the white cell count, together with prednisolone or hydrocortisone. When the small bowel starts to function, the patient can be changed to enteral FK506 in a twice-daily dosage, aiming initially for whole blood levels of 10 ng/ml. When the patient has been stabilized on the oral regimen, and provided that there is no evidence of rejection, a careful serial reduction in the dose of the triple immunosuppressive agents is indicated.

Recent evidence from Mount Sinai,[13] the University of Miami, and the University of Pittsburgh has demonstrated what appears to be a decrease in very severe, life-threatening, early, accelerated rejections by using some of the humanized monoclonal anti-CD3 antibodies, such as basiliximab and daclizumab. This is a great step forward in reducing morbidity and mortality related to intestinal transplantation. Other post-transplant management strategies are on the horizon. Co-stimulatory blockade has been studied extensively in other organs and is now being applied to intestinal transplant models in animals with some measure of success. Totally novel treatment strategies specific to the immunology of the mucosa of the gastrointestinal transplant should be considered. There is a high production of tumor necrosis factor (TNF) and interferon alfa in the intraepithelial component of the small bowel. These types of cytokine expression patterns are similar to those seen in inflammatory bowel disease, and certain modifications of immunosuppression along such lines would be important in the near future. In fact, we are currently studying some of these in our laboratory.[13]

Monitoring of rejection

Clinical observations, endoscopic examination, and bowel histologic findings are used to monitor small-bowel rejection. Clinical features that correlate well with histologic rejection are sudden onset of fever, abdominal distension, ileus, increased ileostomy output, hematemesis, stomal bleeding, and discoloration of stoma. Septicemia due to gut-derived organisms correlates strongly with intestinal rejection. Endoscopy is performed once or twice weekly for the first postoperative month.[14] Over the next 5 months, the frequency of endoscopy is reduced to once every 3–4 weeks, or more frequently if clinically indicated. The earliest endoscopic signs of rejection are edema, erythema, and aperistalsis. More advanced rejection is characterized by aphthous ulcers. Severe rejection is marked by broad-based ulcers or completely denuded mucosa, sometimes covered by pseudomembrane. Acute cellular rejection is marked by the presence of activated lymphocytes attacking crypt epithelium to produce cryptitis.[15] Apoptosis of crypt cells then occurs. Untreated rejection leads to complete endothelial sloughing, which results in pseudomembranous enteritis. A wasting syndrome of chronic rejection has been identified in one adult but not in children. In this syndrome, patchy villous atrophy, smooth muscle hypertrophy, and obliterative vasculitis occur.

Infections

After small-bowel transplantation, bacterial, fungal, and viral infections are very common.[16] In the Pittsburgh series, more than 90% of patients experienced infectious complications, often with multiple episodes.[17] Technical complications (anastomotic leak, infected fluid collection, wound infection) may result in bacterial infection in the early postoperative period. Intravenous lines are particularly susceptible to infection in immunosuppressed patients. Infection is more common during rejection episodes because of the need to increase immunosuppression and because of bacterial translocation associated with loss of mucosal integrity. Viral infections are especially common in pediatric patients.[18] Respiratory viruses such as parainfluenza may be fatal. Cytomegalovirus (CMV) infection is common in both adults and children. Lymphoproliferative disease associated with the Epstein–Barr virus is proving to be a major problem, particularly in children.[19] Lymphoma is more likely to occur in patients who have required high doses of immunosuppression to control multiple rejection episodes.

Graft-versus-host disease

Donor lymphocytes in mesenteric lymph nodes, Peyer's patches, and the lamina propria are transplanted with the small bowel and can be activated against the recipient. Clinical symptoms of GVHD include fever, malaise, anorexia, anemia, and rash.[20] If the recipient bowel mucosa is involved, bacterial translocation and septicemia may occur. GVHD has rarely been recorded following small-bowel transplantation in humans, and the only recorded fatality occurred in a child with an immunodeficiency syndrome.[9]

INTESTINAL FUNCTION

Small-bowel motility usually recovers within 24–48 h post-transplant. Small-bowel function is gauged by measuring the volume and composition of the stomal effluent. Glucose is introduced to the diet, and reducing substances in the stomal effluent are measured as an indicator of sugar absorption. Alternatively, glucose or xylose tolerance tests can be performed.[6] Once carbohydrate absorption has been established, peptide solutions can be introduced and the nitrogen balance determined. Finally, fat in the form of medium-chain triglycerides is added to the diet and the stomal fat content measured.

The incidence of diarrhea is high following small-intestinal transplantation. This may be due to rapid intestinal transit associated with lack of extrinsic neuronal control, to bacterial overgrowth, to malabsorption, or to the loss of the ileocecal valve. It is becoming clear that in addition to transplanting the small bowel, the ileocecal valve and right colon should be transplanted in continuity with the small bowel.[21] The donor colon can be brought out as a colostomy or anastomosed to the recipient colon. An ileostomy is fashioned to allow ileal biopsies.

RESULTS OF SMALL-BOWEL TRANSPLANTATION

The first successful recipient of a combined small-bowel–liver transplantation remains well more than 7 years after the transplant.[22] The most extensive experience has been gained at the University of Pittsburgh, where 59 intestinal transplants were reported at the third international symposium on small-bowel transplantation as isolated transplants (34%), combined intestine and liver grafts (54%), or multivisceral transplants (12%).[23] The 1-year patient survival rate was 82% and the graft survival rate was 73%. Graft and patient survival rates were slightly better in adults than in pediatric patients. Most of the postoperative deaths were due to sepsis, either in association with acute rejection or as a result of technical failure, line infection, or chest infection. Lymphoproliferative disease is a major cause of late postoperative mortality, and its incidence is likely to increase with extended follow-up of these patients. The only patients to have survived graft loss have been those with isolated small-bowel transplants. Over 50% of the patients that were transplanted at the University of Pittsburgh were free of TPN and on an unrestricted oral diet.

Currently, patient survival at 2 years is approximately 70% after intestinal transplantation. Graft survival is slightly less, but long-term success remains in the 55–60% range.[13]

The field of small-bowel transplantation is moving rapidly and has become a realistic alternative for patients with short bowel syndrome maintained on long-term TPN.

REFERENCES

1 Sarr MG, Hakim NS (eds). *Enteric physiology of the transplanted intestine.* Austin, TX: R. G. Landes, 1994; 3–7.

2 Quigley EM, Marsh MN, Shaffer JL, Markin RS. Hepatobiliary complications of total parenteral nutrition. *Gastroenterology* 1993; **104**: 286–301.

3 Mughal M, Irving M. Home parenteral nutrition in the United Kingdom and Ireland. *Lancet* 1986; **2**: 383–7.

4 Lennard Jones JE. Indications and need for long-term parenteral nutrition: implications for intestinal transplantation. *Transplant Proc* 1990; **22**: 2427–9.

5 Hancock BJ, Wiseman NE. Lethal short-bowel syndrome. J Pediatr Surg 1990; 25: 1131–4.

6 Reyes J, Tzakis AG, Todo S, Nour B, Casavilla A, Abu-Elmagd K, *et al.* Nutritional management of intestinal transplant recipients. *Transplant Proc* 1993; **24**: 1200–1.

7 Clark CI. Recent progress in intestinal transplantation. *Arch Dis Child* 1992; **67**: 976–9.

8 Thompson JS. Recent advances in the surgical treatment of the short-bowel syndrome. *Surg Annu* 1990; **22**: 107–27.

9 Tzakis AG, Todo S, Reyes J, Abu-Elmagd K, Casavilla A, Ohya T, *et al.* Clinical intestinal transplantation: focus on complications. *Transplant Proc* 1992; **24**: 1238–40.

10 Diflo T, Shaffer D, Maki T, Monaco AP. The effects of antilymphocyte serum and cyclosporine on orthotopic small bowel allografts in the rat. *Transplant Proc* 1988; **20** (Suppl 1): 200–2.

11 Grant D, Garcia B, Wall W, Oheme-Fianko D, Zhong R, Mimeault R, *et al.* Graft-versus-host disease after clinical small bowel/liver transplantation. *Transplant Proc* 1990; **22**: 2464.

12 Casavilla A, Selby R, Abu-Elmagd K, Tazakis A, Todo S, Reyes J, *et al.* Logistics and technique for combined hepatic-intestinal retrieval. *Ann Surg* 1992; **216**: 605–9.

13 Fishbein TM. Intestinal transplantation: the past, the present and the future. *Physician's Forum on Transplantation*, American Society of Transplantation, Chicago, October 1999.

14 Garau P, Orenstein SR, Neigut DA, Putnam PE, Reyes J, Tzakis AG, Kocoshis SA. Role of endoscopy following small intestinal transplantation in children. *Transplant Proc* 1994; **26**: 136–7.

15 Abu-Elmagd KM, Tzakis A, Todo S, Reyes J, Fung J, Nakamura K, *et al.* Monitoring and treatment of intestinal allograft rejection in humans. *Transplant Proc* 1993; **25** (1 Pt 2): 1202–3.

16 Reyes J, Abu-Elmagd K, Tzakis A, Nour B, Casavilla A, Kisne S, *et al.* Infectious complications after human small bowel transplantation. *Transplant Proc* 1992; **24**: 1249–50.

17 Kusne S, Manez R, Bonet H, Abu-Elmagd K, Furukawa H, Irish W, *et al.* Infectious complications after small bowel transplantation in adults. *Transplant Proc* 1994; **26**: 1682–3.

18 Green M. *Third International Symposium on Small Bowel Transplantation*, Paris 1993.

19 Reyes J, Tzakis A, Green M, Nour B, Nalesnik M, Van Thiel D, *et al.* Posttransplant lymphoproliferative disorders occurring under primary FK 506 immunosuppression. *Transplant Proc* 1991; **23**: 3044–6.

20 Slavin RE, Santos GW. The graft versus host reaction in man after bone marrow transplantation: pathology, pathogenesis, clinical features, and implication. *Clin Immunol Immunopathol* 1973; **1**: 472–98.

21 Tzakis AG, Todo S, Reyes J, Nour B, Fung JJ, Starzl TE. Piggyback orthotopic intestinal transplantation. *Surg Gynecol Obstet* 1993; **176**: 297–8.

22 McAlister V, Wall W, Ghent C, Zhong R, Duff J, Grant D. Successful small intestine transplantation. *Transplant Proc* 1992; **24**: 1236–7.

23 Abu-Elmagd K, Todo S, Tzakis A, Reyes J, Nour B, Furukawa H, *et al.* Three years clinical experience with intestinal transplantation. *J Am Coll Surg* 1994; **179**: 385–400.

8

Visceral revascularization for chronic visceral ischemia: transabdominal approach

KENNETH J. CHERRY, JR

INTRODUCTION

Reconstruction of the mesenteric vessels seldom requires a formal thoracic component of exposure in the modern era. Classically, endarterectomy of the celiac and superior mesenteric arteries, as described initially by Wylie and co-workers, was performed through a low thoracoabdominal, or 'thoracoretroperitoneal', incision, the so-called 'Pillsbury incision', a modification of the full thoracoabdominal incision used for thoracoabdominal aortic aneurysm repair.[1, 2] Bypass grafting was also performed through this approach if the abdomen was hostile because of radiation changes, perforated bowel, previous vascular reconstructions, or previous general surgical operations, especially upon the esophagus or stomach. This approach, used mainly for endarterectomy of the visceral vessels, either alone or in combination with infrarenal grafting, remained the standard from the time of its description until vascular surgeons began to utilize medial visceral rotation through a purely transabdominal approach for exposure of the paravisceral and supraceliac abdominal aorta. That exposure was first described by trauma surgeons as an approach to trauma of the upper abdominal aorta and its branches as well as the viscera.[3–6] The surgeons at the University of California – San Francisco some years later recognized its utility for elective vascular exposures that had previously required thoracoabdominal incisions and approaches.[7, 8] In a retrospective review, they found that patients requiring thoracotomy and incision of the diaphragm had a higher morbidity, primarily pulmonary.[7] Therefore, approaching these nutritionally depleted patients without entering the thorax and incising the diaphragm appeared to be beneficial, and thus they initiated the use of medial visceral rotation for exposure of the upper abdominal aorta, i.e. the paravisceral aorta.[8] It is hard, of couse, to determine what changes over the course of several decades have helped improve results, as the variables are many: changes in incisions, approaches, number of arteries reconstructed, routes and configurations of grafting, source of inflow, and in anesthetic techniques are but the most salient of the modifications that have evolved in the past 20–25 years. Chronic mesenteric ischemia is rare enough that controlled trials are unlikely, and there are simply not enough patients to allow trials aimed at delineating any one single variable. Surgeons therefore must rely on well-done retrospective reviews from institutions with large referral practices of patients with chronic mesenteric ischemia. The group from San Francisco, while championing medial visceral rotation, nonetheless pointed out the problems they had encountered with this extensive intra-abdominal mobilization: bowel infarction, pancreatitis, and splenic injury.[8] The necessity of gentle mobilization and retraction of the viscera should be apparent to all surgeons performing medial visceral rotation.

Stoney and colleagues feel that no patient needs a thoracic component for this exposure currently.[8] On the other hand, we feel that there are still patients who require thoracoabdominal approaches to the paravisceral aorta and the mesenteric vessels. These are usually patients undergoing endarterectomy whose body habitus mitigates against a purely abdominal approach. Patients with narrow flaring of their costal margins, and whose rib cages approach within 3 inches or less of the iliac crest, are better served by a low-lying thoracoabdominal approach, as the surgeon's angle of approach to the mesenteric vessels themselves is suboptimal, despite the fact that the paravisceral aorta may be exposed by medial visceral rotation. Working over a 'ridge' of ribs with restricted access to the orifice of the mesenteric vessels and poor orientation for extending the endarterectomy along the axis of the artery mitigates against a technically satisfactory operation.

The other body type for whom purely abdominal approaches are less than optimal is that of the very large or obese. Fortunately, people in the latter group, with rare exceptions, seldom need mesenteric reconstruction. If a thoracoabdominal incision is needed, the diaphragm may, for the most part, be spared. It may be incised laterally in a very limited manner to allow spread of the ribs, and vertical incision may be made in the para-aortic diaphragm to allow appropriate clamping and exposure, but the great expanse of the diaphragm may, in most cases, be left intact.

BYPASS GRAFTING

Supraceliac grafting of the mesenteric vessels is performed through an abdominal approach. An upper midline incision, extending well up along the xyphoid, is made. It usually does not need to extend below the umbilicus. The left lobe of the liver is mobilized by incision of its superior attachment, the left triangular ligament, and folded back under itself. The lesser omentum is entered. The esophagus is palpated (with a nasogastric tube in place) and retracted to the patient's left. Gentle retraction by hand using a laparotomy pad for appropriate friction is used to bring the stomach caudad. The surgeon is then looking at only the diaphragmatic crura overlying the aorta. These are incised. With sharp dissection, the supraceliac aorta is resected free. This exposure of the aorta is done to an extent that is sufficient to allow placement of a supraceliac cross-clamp and a more distal clamp. The dissection is carried down onto the celiac and superior mesenteric arteries (Figure 8.1). The celiac is exposed as far as necessary, usually to the origins of the hepatic and splenic arteries. We have found that the left gastric is most often quite small and have not hesitated to divide it to allow a better conformation of the celiac limb of the graft and an easier route for the superior mesenteric artery limb.

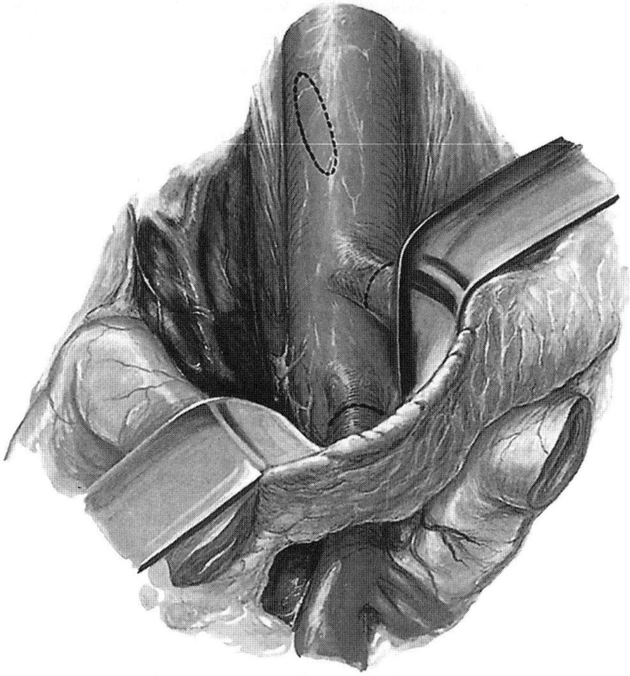

Figure 8.1 *Exposure of the supraceliac and paravisceral aorta. Note the site of proposed aortotomy and the transection sites of the celiac and superior mesenteric artery before the bypass.*

The superior mesenteric artery may be exposed either cephalad or caudad to the pancreas. If it is to be exposed cephalad, the adventitial tissue along the superior margin of the pancreas is incised to allow mobilization of the pancreas. An additional 2–4 cm of exposure of the superior mesenteric artery may be gained thus. If the superior mesenteric artery lesion extends too far distally to allow for an anastomosis proximal to the pancreatic margin, a retropancreatic tunnel may be created by gentle finger dissection. The superior mesenteric artery is then exposed beyond the ligament of Treitz in the small-bowel mesentery. This is performed in a manner similar to the transperitoneal approach to the infrarenal aorta, with the mesentery placed on a stretch by the first assistant.

The supraceliac aorta is clamped. In most cases, patients are heparinized. Additionally, they are given low-dose dopamine, frusemide, and mannitol. Partial occlusion of the aorta in general works less well than cross-clamping in this situation, resulting at times in technical compromise of the proximal anastomosis; we use it only rarely. Cross-clamping of the aorta is our preferred technique at the Mayo Clinic. We have found the Cherry Supraceliac Clamp® and the Wylie Hypogastric Clamp® to work very well in this situation. The angulation of the limbs and the handle of the supraceliac clamp keep the clamp out of the surgeon's way. The angle of the hypogastric clamp allows not only distal clamping of the aorta but occlusion of the posterior intercostal or lumbar vessels, thus allowing a dry field. A vertical or slightly oblique aortotomy is made. A small (usually 12×7 cm), knitted polyester graft is sutured into place using 3-0 or 4-0 permanent vascular suture. Clamp time should rarely need to exceed 20 min, and is most often in the range of 12–15 min.

The celiac anastomosis is performed end to end with a tongue out onto one of the two major branch arteries if necessary (usually the hepatic artery).

If the superior mesenteric artery anastomosis is cephalad to the pancreas, an end-to-end anastomosis is performed. If it is to the distal superior mesenteric artery beyond the ligament of Treitz in the small bowel mesentery, then an end-to-side anastomosis is performed more easily.

Supraceliac grafting may be a poor choice in patients with severely compromised cardiac or pulmonary function. In addition, some patients, especially those older than 70 years, have extensive atherosclerosis or circumferential calcification of the supraceliac aorta, making both clamping and anastomosis in that area unattractive options. In such instances, we prefer infrarenal sources of inflow, from either the aorta or the common iliac arteries. When performing infrarenal reconstruction, more times than not we reconstruct only the superior mesenteric artery. The proximal, or inflow, anastomosis is to the anterolateral aorta. Control of the aorta may be obtained with either cross- or partial-occlusion clamps, depending on aortic size, atherosclerosis, and calcification.

We have seen an interesting subset of patients with extensive circumferential calcification of the supraceliac and infrarenal aorta in whom the common iliac arteries are soft and provide good donor vessels. These patients undergo cross-clamping of the common iliac artery. In some cases, the left common iliac artery provides a better donor source than the right because of its spatial relationship to the superior mesenteric artery with less angulation.

The superior mesenteric artery is dissected distal to the ligament of Treitz. The distal anastomosis is performed end to side. While setting up the field, retraction on the bowel and mesentery is relaxed to allow the superior mesenteric artery to assume its anatomic location more nearly, thereby reducing the chances of graft elongation, angulation, and kinking. Interestingly, in a retrospective review, we have found these retrograde grafts to work as well as the antegrade grafts. The patients with retrograde grafts were older, however, and had reduced lifespans.[9]

ENDARTERECTOMY

Endarterectomy of the mesenteric vessels may be preferable to prosthetic bypass because of bacterial contamination, perforated bowel, radiation changes, or other hostile conditions in the abdomen. Extensive atherosclerosis of the supraceliac, paravisceral, and infrarenal aorta may make this preferable to bypass grafting because an acceptable graft donor site, either supraceliac or infrarenal, is simply not available. This may be done alone or in combination with infrarenal aortic reconstruction. When a transabdominal rather than a thoracoabdominal route is chosen for endarterectomy, the paravisceral aorta and the mesenteric vessels may be approached through either a midline or an oblique abdominal incision. In most cases, we prefer the midline approach. This incision is usually full length. Medial rotation of the viscera is effected. To facilitate exposure of the superior mesenteric artery, the left kidney is left in place and the dissection is carried out anterior to it (Figure 8.2). The superior mesenteric artery, if diseased extensively, may be dissected out via this approach to a surprising degree. The patient is heparinized and treated with appropriate vasoactive and diuretic drugs. The supraceliac aorta and infrarenal aorta are clamped. Aortotomy through either a straight incision or a trapdoor incision is made. Endarterectomy of the paravisceral aorta, the celiac artery, and the superior mesenteric artery is performed (Figures 8.3 and 8.4). The endarterectomy is ended above the renal artery orifices in most cases. If renal or infrarenal aortic reconstruction is indicated, these may be performed in conjunction with the mesenteric reconstruction. The endarterectomy will have a good endpoint at the bifurcation of the celiac artery into the common hepatic and splenic arteries. Often, the superior mesenteric artery disease will extend beyond that which ostial endarterectomy techniques can address. In this case, the aortotomy

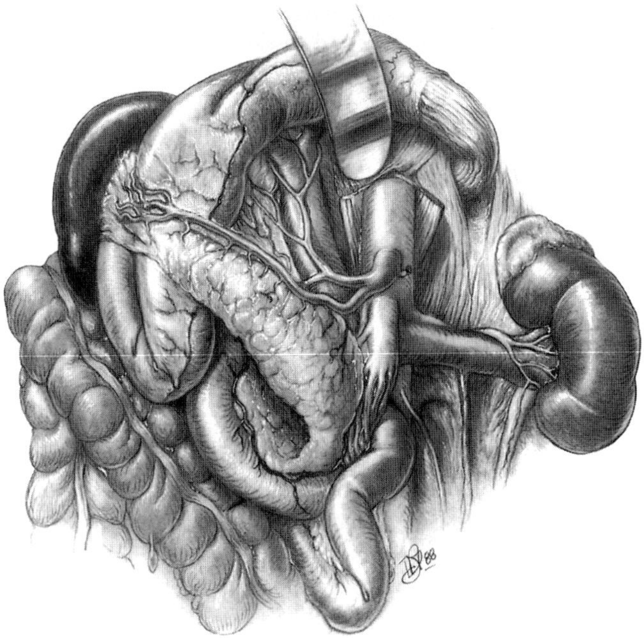

Figure 8.2 *The wide exposure of the thoracoabdominal aorta provided by left medial visceral rotation, anterior to the left kidney.*

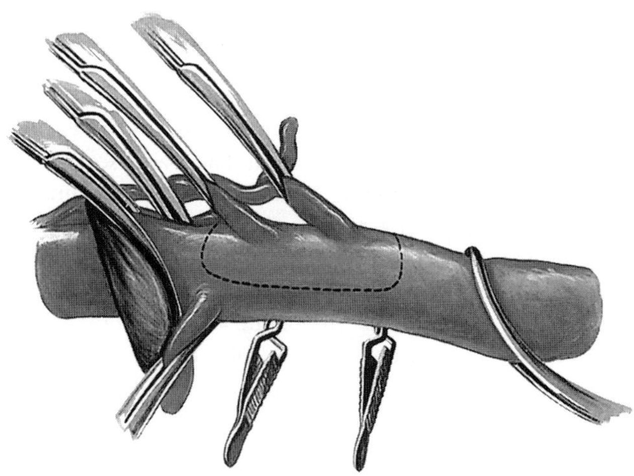

Figure 8.3 *The paravisceral aorta and its major branches are controlled with clamps. The intercostals posteriorly are occluded temporarily with bulldog clamps. The trapdoor aortotomy surrounding the orifices of the celiac axis and superior mesenteric artery is outlined.*

is closed with permanent suture and flow is restored to the distal aorta and the celiac artery. Tacking sutures at the distal endarterectomy site are placed to prevent distal aortic dissection. The superior mesenteric artery is then clamped separately, and a more distal completion endarterectomy is performed without further aortic and renal ischemia time. That incision

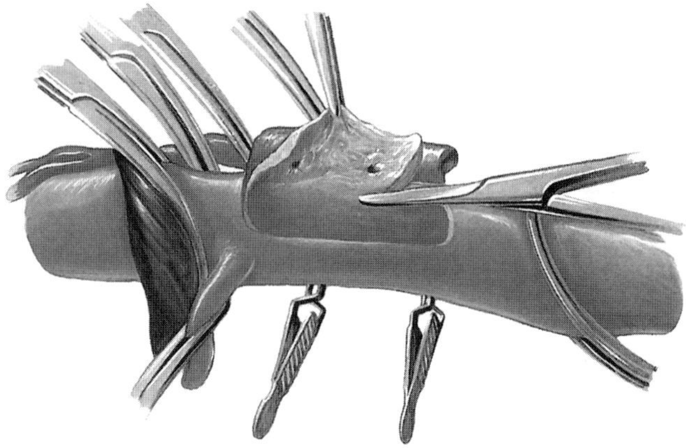

Figure 8.4 *The ventral atheroma in the aorta is being divided at the hinge of the trapdoor with scissors.*

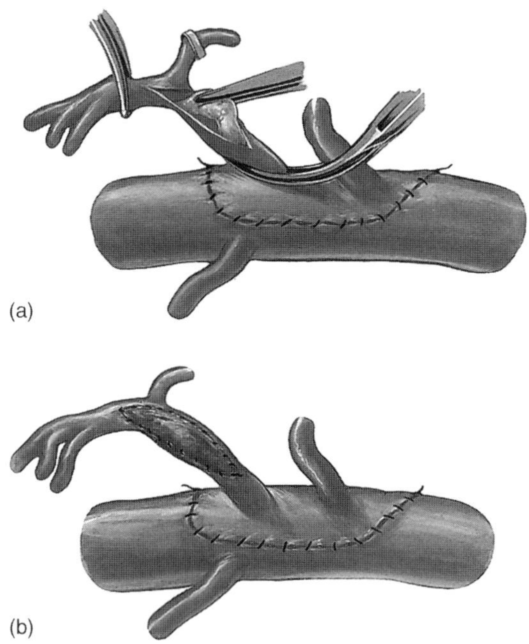

Figure 8.5 *(a) The aortotomy has been closed with a continuous suture, and flow has been restored to the paravisceral aorta and celiac axis. A distal atheroma with thrombus associated with the superior mesenteric artery occlusion was present. A distal superior mesenteric arteriotomy allows precise removal of this lesion, ensuring a satisfactory endpoint. (b) The completed visceral revascularization is shown. A saphenous vein patch applied to the superior mesenteric arteriotomy prevents narrowing.*

may be transverse or longitudinal. If the disease by palpation extends into the branch vessels of the superior mesenteric artery, we prefer a longitudinal incision to allow use of appropriate techniques at these branch sites. The longitudinal arteriotomy is patched with vein or prosthetic, depending on clinical indications (Figure 8.5).

Intraoperative assessment is via duplex scanning and/or measurement of blood flow. We have found flows of ≥600 mm/min to be a predictor of long-term patency.[9] The abdominal incisions are closed in the standard manner.

REFERENCES

1 Stoney RJ, Wylie EJ. Surgical management of arterial lesions of the thoracoabdominal aorta. *Am J Surg* 1973; **126**: 157–64.
2 Stoney RJ, Ehrenfeld WK, Wylie EJ. Revascularization methods in chronic visceral ischemia caused by atherosclerosis. *Ann Surg* 1977; **186**: 468–76.
3 Shirkey AL, Quast DC, Jordan GL, Jr. Superior mesenteric artery division and intestinal function. *J Trauma* 1967; **7**: 7–24.
4 Buscaglia LC, Blaisdell FW, Lim RC, Jr. Penetrating abdominal vascular injuries. *Arch Surg* 1969; **99**: 764–9.
5 Elkins RC, DeMeester TR, Brawley RK. Surgical exposure of the upper abdominal aorta and its branches. *Surgery* 1971; **70**: 622–7.
6 Mattox KL, McCollum WB, Jordan GL, Jr, Beall AC, Jr, DeBakey ME. Management of upper abdominal vascular trauma. *Am J Surg* 1974; **128**: 823–8.
7 Cunningham CG, Reilly LM, Rapp JH, Schneider PA, Stoney RJ. Chronic visceral ischemia. *Ann Surg* 1991; **214**: 276–88.
8 Reilly LM, Ramos TK, Murray SP, Cheng SWK, Stoney RJ. Optimal exposure of the proximal abdominal aorta: a critical appraisal of transabdominal medial visceral rotation. *J Vasc Surg* 1994; **19**: 375–90.
9 Park WM, Cherry KJ, Jr, Chua HK, Clark RC, Noel AA, Panneton JM, *et al.* Current results of open revascularization for chronic mesenteric ischemia: a standard for comparison. Presented at the 25th annual meeting of the Midwest Vascular Surgical Society, Chicago, USA; 21–22 September 2001.

9

Antegrade visceral revascularization via a thoracoabdominal approach for chronic mesenteric ischemia

GEORGE GEROULAKOS, GIANNI BELCARO, AND WILLIAM L. SMEAD

INTRODUCTION

Since the first successful thromboendarterectomy of the superior mesenteric artery (SMA) was first performed by Shaw and Maynard in 1958,[1] a variety of techniques have been developed to correct atherosclerotic obstruction of the major visceral branches of the aorta. Techniques of revascularization include reimplantation, bypass grafting, endarterectomy, and balloon angioplasty with or without stent placement. There is no consensus regarding the best surgical approach for the treatment of chronic visceral ischemia. This condition is encountered infrequently, and it is unlikely that a single center can treat enough patients and accumulate sufficient experience to develop principles of treatment by demonstrating significant differences between the various mesenteric revascularization strategies. However, the majority of reports on the surgical management of chronic visceral ischemia advocate revascularization via an abdominal approach, and only a few recommend a thoracoabdominal approach.[2,3] It has been suggested that patients with chronic visceral ischemia are elderly and emaciated, hence they may not tolerate the considerable morbidity caused by antegrade visceral revascularization via a thoracoabdominal approach.[4,5] The purpose of this study is to assess the efficacy of the thoracoabdominal approach for the treatment of chronic mesenteric ischemia. This approach was selected because it provides an excellent exposure for mesenteric revascularization plus the option of originating the graft from several areas of the distal thoracic and upper-abdominal aorta. Our preliminary experiences with this technique have already been published,[6] and this chapter provides further information on the incidence of complications with this approach on a larger number of patients.

MATERIAL AND METHODS

The medical records including follow-up mesenteric duplex scans, angiograms, and office records of all patients undergoing bypass grafting for symptomatic chronic visceral ischemia at the Ohio State University Hospital between 1988 and 1999 were examined. Nineteen patients were admitted with symptoms of chronic visceral ischemia and eventually underwent antegrade visceral revascularization via a thoracoabdominal approach. This patient population consisted of 18 women and one man, with a mean age of 66.5 years (range 50–83 years). The cause of the chronic visceral ischemia was atherosclerosis in all cases. The median follow-up was 28 months (range 3–93 months).

All patients complained of postprandial abdominal pain and weight loss. Other gastrointestinal complaints included diarrhea in nine patients, nausea or vomiting in eight patients, and constipation in three patients. Physical examination confirmed substantial weight loss in all patients. The median weight loss was 13.6 kg. Each of the patients in this series had one or more associated risk factors (Table 9.1). The patients in these series had undergone extensive investigations before being referred to the vascular service. Contrast studies, endoscopy, and computed tomography (CT) were not essential to the diagnosis but proved important in eliminating other sources of abdominal discomfort. In one patient, however, the endoscopic finding of mucosal ulceration in the ascending colon contributed to a successful diagnosis. Three patients had cholecystectomy a few months before the vascular referral because of findings of gallstones, apparently with no improvement of their symptoms. In all instances, lateral views of biplane aortography demonstrated visceral occlusive lesions compatible with the diagnosis. No patient in this series had single-vessel disease. The combination of visceral artery involvement is shown in Table 9.2.

Surgical management

Visceral revascularization was performed through a transperitoneal thoracoabdominal approach. The thoracic portion of the exposure was developed through an ninth interspace incision, which is extended obliquely across the abdomen. The patients underwent complete diaphragm division through both the muscular and central tendinous portion. The left crus

Table 9.1 *Patient profile (some patients had two or more risk factors)*

Risk factor	Patients (n)
Coronary artery disease	8
Congestive cardiac failure	2
Chronic obstructive airways disease	4
Renovascular disease	5
Diabetes mellitus	5
Cerebrovascular accident	2
Peripheral vascular disease	8

Table 9.2 *Combination of visceral arteries involved*

Visceral arteries involved	Patients (*n*)
CA, SMA, IMA	5
CA, SMA	12
SMA, IMA	2

CA, celiac artery; IMA, inferior mesenteric artery; SMA, superior mesenteric artery.

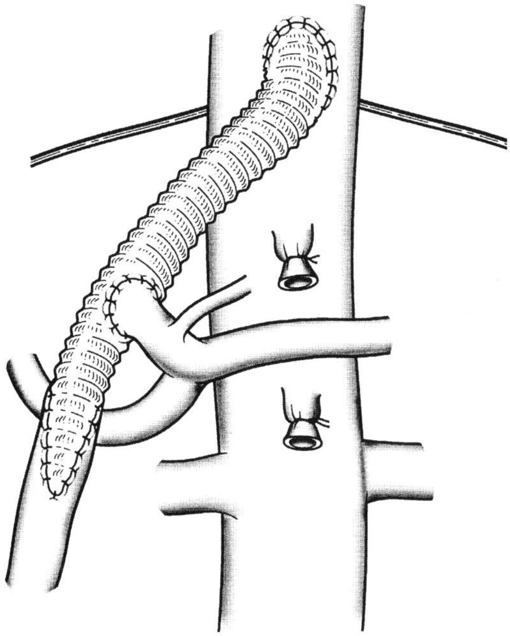

Figure 9.1 *Diagrammatic demonstration of the antegrade bypass to the superior mesenteric artery. The celiac axis is transected and implanted end to side on the graft.*

was divided, and medial visceral rotation was performed. This technique provides excellent exposure of the lower thoracic and the complete length of the infradiaphragmatic aorta.

The conduit of choice in our practice for the antegrade visceral revascularization is an 8-mm knitted Dacron. The proximal anastomosis is sutured end to side on the anterolateral surface of the distal thoracic aorta, while the distal anastomosis is sutured end to end beyond the site of the atherosclerotic involvement on the SMA. As a variation of the usually described techniques, after completing the distal end-to-end anastomosis, the celiac artery (CA) was transected and implanted end to side on the graft in 15 patients (Figure 9.1). One patient in addition had revascularization of the inferior mesenteric artery. Two others had revascularization of the left renal artery with a 6-mm knitted Dacron graft.

Duplex examination

Ten patients underwent duplex scanning of the visceral arteries, performed by an experienced vascular technologist during their last visit in the clinic. Examination was performed with a color duplex with a 3.5-MHz sector transducer. The bypass, CA, and SMA were imaged after an overnight fast in a supine position with the head of the bed elevated at 30 degrees (Plate 14).

Diagnostic criteria for the quantification of visceral artery stenosis were used as validated previously.[7]

RESULTS

There were no postoperative deaths in this series. Early postoperative complications occurred in five patients. Three developed pulmonary infections and two required reintubation. Three of these patients had a history of chronic obstructive pulmonary disease (COPD) and one of congestive cardiac failure (CCF). The mean length of stay in the hospital following surgery was 36 days. By contrast, the 14 patients with no pulmonary complications were discharged after a mean of 10 days (range 7–23 days). At a mean follow-up of 25 months (range 3–96 months), ten patients underwent mesenteric duplex that revealed that all grafts were patent. One patient developed a high-grade graft–SMA anastomotic stenosis, which was followed by recurrence of the symptoms (Figure 9.2a). This was dilated by balloon angioplasty on three occasions within a period of 17 months. On the last occasion, a stent was inserted, and since then the patient remains asymptomatic (Figure 9.2b). Visceral revascularization provided complete symptomatic relief for the remaining patients. One patient died 2 months following his operation from CCF. There were no other deaths during the follow-up. The median weight gain was 4.5 kg.

DISCUSSION

Chronic visceral ischemia is uncommon, and its precise incidence is unknown. As a result of an abundant network of collateral vessels, clinical symptoms present when at least two of the three visceral arteries have significant disease. This study confirms previous reports of a female patient distribution of this condition.[2, 8] The reason for this peculiar sex distribution remains undetermined. However, it has been suggested that it could be the result of the inclusion of cases of Takayasu's aortitis in reports of atherosclerotic chronic visceral ischemia.[9] Takayasu's aortitis closely mimics atherosclerosis of the abdominal aorta, and it has a marked female predominance. In three patients, the discovery of gallstones (which in retrospect were incidental findings) led to the performance of cholecystectomy with no symptomatic relief. Cholecystectomy performed in error in this group of patients has also been reported by others.[10] There were no perioperative deaths in this series, which compares well with an operative mortality for symptomatic chronic mesenteric insufficiency of 5–17% reported by others.[4]

Bypass grafting is the most common type of visceral revascularization performed today. It provides immediate restoration of flow and may originate from several different locations, including the supraceliac aorta, the infrarenal aorta, and the common iliac arteries. Regardless of the bypass technique used, the status of the donor artery is critical to success.[7] The distal thoracic aorta is usually free of atherosclerotic disease and is an excellent origin of a short antegrade bypass to the SMA. The bypass is placed in the direction of normal blood flow (antegrade orientation), thus reducing anastomotic turbulence. In addition, this design eliminates the possibility of kinking and thrombosis by compression or traction from the overlying intestinal mesentery, which may be observed with retrograde grafts.[11] The distal portion of the thoracic aorta may be approached from the abdomen through division of the crura. However, a thoracoabdominal approach gives better exposure and provides the surgeon with more options on the rare occasions when circumferential atherosclerosis precludes the use of the distal thoracic aorta for the origin of the bypass (Figure 9.3). It is worth noting that in two of our patients, the proximal anastomosis was performed in the mid-thoracic aorta as a result of atherosclerotic involvement of the distal thoracic aorta. In addition, this exposure provides optimal exposure of the visceral vessels for the performance of the distal anastomosis. It has been hypothesized that the morbidity rate of the thoracoabdominal approach may be more than patients with chronic visceral ischemia can tolerate.[4, 5] However, no data were provided to support this statement. Our experience demonstrates that this procedure can be tolerated with an acceptable rate of perioperative morbidity.

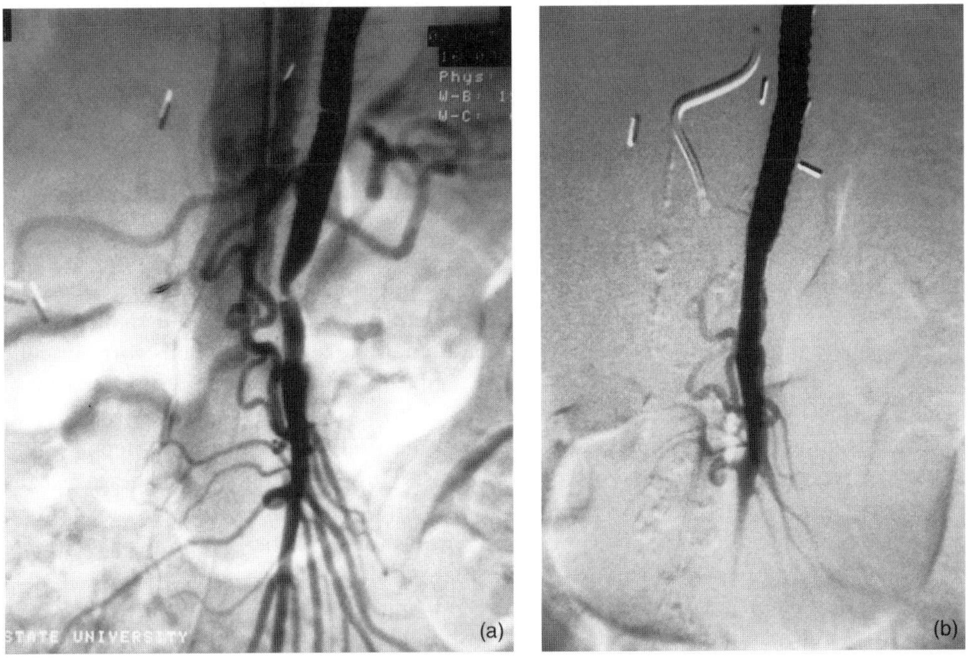

Figure 9.2 *(a) Stenosis at the distal anastomosis of an aortosuperior mesenteric bypass. (b) This was treated successfully with stenting.*

The prevalence of symptomatic renovascular disease in patients with chronic visceral ischemia is not known. Two of our patients with symptomatic left renal artery stenosis underwent renal artery bypass with good functional results. The thoracoabdominal approach provides an excellent approach for left renal artery reconstruction. However, the approach to the right renal artery is technically demanding. Under this setting, an endarterectomy should be considered. Transaortic mesenteric endarterectomy can be used effectively in association

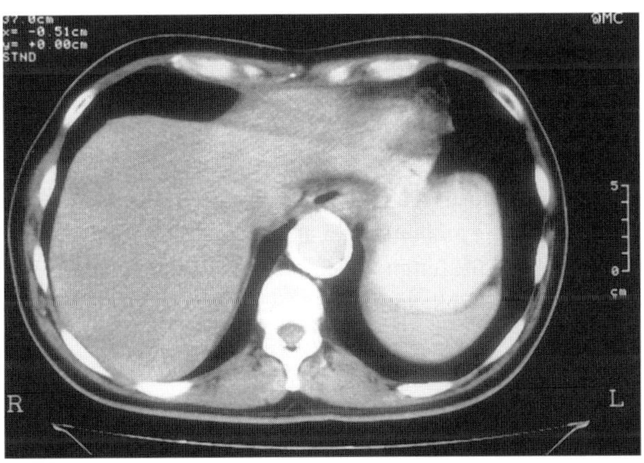

Figure 9.3 *Computerized tomography scanning showing circumferential atherosclerosis and calcification of the supraceliac aorta. An abdominal approach in this case may not provide a good origin for an antegrade bypass.*

with renal artery endarterectomy when clinically significant ostial atherosclerosis is present on both renal arteries.[11, 12]

All the substantial postoperative morbidity (26%) in this series relates to pulmonary complications. A similar high incidence of pulmonary complications (34%) was also reported postoperatively in a series of 85 patients who had elective treatment for chronic mesenteric occlusive disease with a variety of surgical techniques at the Cleveland Clinic.[13] Three of our patients developed pulmonary infections and two required reintubation, which significantly prolonged their hospitalization. Since we do not have a control group of patients treated with a single-cavity procedure, we cannot comment on whether the incidence of pulmonary complications would have been different had we used an abdominal approach. However, we would now advocate more selective use of the thoracoabdominal approach, excluding patients with a history and signs of COPD. In addition, preservation of the central tendinous portion of the diaphragm is an independent factor associated with early ventilator weaning and may be of benefit for these patients.[14]

There is no uniform agreement about the graft material of choice. In early reports, vein grafts had patency rates inferior to synthetic grafts.[15, 16] More recent reports described the use of either autogenous veins or prosthetic grafts with excellent long-term function and no difference in patency rates.[11, 17] Our own preference is for synthetic Dacron bypass because it is always available, spares the patient from the morbidity of one more incisions for the harvesting of the vein, and provides good early and long-term results.

Aorto-SMA bypass alone is usually sufficient to provide good symptomatic relief as a result of the extensive collateral circulation, even when all three visceral arteries are occluded. Hollier has shown that complete revascularization in multi-vessel disease resulted in a late recurrence of 11%, while when one of three stenotic vessels was revascularized the recurrence rate was 50%.[18] He concluded that it is preferable to revascularize as many vessels as possible to provide the best chance of long-term relief. The only recurrence of symptoms in this series was noted in a patient with three-vessel involvement who was treated with aorto-SMA bypass. This was due to a stenosis in the distal graft–SMA anastomosis. This patient was treated with endovascular techniques. Others have also used endovascular techniques for the management of recurrent symptoms.[13] The durability of these techniques is not known, but eventually they may become a reasonable option for some patients who otherwise require reoperation.

REFERENCES

1 Shaw RS, Maynard EP. Acute and chronic thrombosis of mesenteric arteries associated with malabsorption: a report of two cases successfully treated by thromboendarterectomy. *N Engl J Med* 1958; **258**: 874–8.

2 Zelenock G, Graham LM, Whitehouse WM, Erlandson EE, Kraft RO, Lindenauer SM, Stanley JC. Splanchnic arteriosclerotic disease and intestinal angina. *Arch Surg* 1990; **115**: 497–501.

3 McCready RA, McIntyre AB, Hyde GL, Giffen WO. Recurrent visceral ischemia: revascularisation by thoracic aorta to splenic artery bypass: a case report. *Contemp Surg* 1985; **27**: 49–56.

4 Harward TRS, Brooks DL, Flynn TC, Seeger JM. Multiple organ dysfunction after mesenteric artery revascularization. *J Vasc Surg* 1993; **18**: 459–69.

5 Beebe HG, MacFarlane S, Raker EJ. Supraceliac aortomesenteric bypass for intestinal ischemia. *J Vasc Surg* 1987; **5**: 749–54.

6 Geroulakos G, Tober JC, Anderson L, Smead WL. Antegrade visceral revascularisation via a thoracoabdominal approach for chronic mesenteric ischemia. *Eur J Vasc Endovasc Surg* 1999; **17**: 56–9.

7 Moneta GL, Lee RW, Yeager RA, Taylor LM, Porter JM. Mesenteric duplex scanning: a blinded prospective study. *J Vasc Surg* 1993; **17**: 779–86.

8 Geelkerken RH, van Bockel JH, De Ross WK, Hermans J, Terpstra JL. Chronic mesenteric vascular syndrome. Results of reconstructive surgery. *Arch Surg* 1991; **126**: 1101–6.

9 Lande A. Abdominal Takayasu's aortitis, the middle aortic syndrome and atherosclerosis. *Int Angiol* 1998; **17**: 1–9.

10 Babu SC, Shah PM. Celiac territory ischemic syndrome in visceral artery occlusion. *Am J Surg* 1993; **166**: 227–30.

11 Cunningham CG, Reilly LM, Rapp JM, Schneider PA, Stoney RJ. Chronic visceral ischemia. *Ann Surg* 1991; **214**: 276–87.

12 Hansen KJ, Deitch JS. Transaortic mesenteric endarterectomy. *Surg Clin North Am* 1997; **77**: 397–407.

13 Mateo RB, O'Hara P, Hertzer NR, Mascha EJ, Beven EG, Krajewski LP. Elective surgical treatment of symptomatic chronic occlusive disease: early results and late outcomes. *J Vasc Surg* 1999; **29**: 821–32.

14 Engle J, Safi H, Miller C, Campbell MP, Harlin SA, Letsou GV, *et al*. The impact of diaphragm management on prolonged ventilator support after thoracoabdominal aortic repair. *J Vasc Surg* 1999; **29**: 150–6.

15 Rheudasil JM, Stewart MT, Schellack JV, Smith RB, Salam AA, Perdue GD. Surgical treatment of chronic mesenteric arterial insufficiency. *J Vasc Surg* 1988; **8**: 495–500.

16 Rob C. Surgical diseases of the celiac and mesenteric arteries. *Arch Surg* 1966; **93**: 21–30.

17 Stoney RJ, Ehrenfeld WK, Wylie EJ. Revascularization methods in chronic visceral ischemia caused by atherosclerosis. *Ann Surg* 1977; **186**: 468–76.

18 Hollier LH, Bernatz PE, Pairolero PC, Spencer Payne W, Osmundon PJ. Surgical management of chronic intestinal ischemia. A reappraisal. *Surgery* 1981; **90**: 940–6.

10

Endovascular versus surgical reconstruction for the management of chronic visceral ischemia: a comparative analysis

PETER ROBLESS, ANNA-MARIA BELLI, AND GEORGE GEROULAKOS

INTRODUCTION

The diagnosis and management of symptomatic chronic mesenteric ischemia (CMI) poses a difficult challenge for several reasons. Although the reported incidence has increased, it still remains an uncommon and under-recognized condition [1]. In several large series of patients, the average time interval between the onset of symptoms and the diagnosis being made was 12 months.[2–4] Patients with CMI are often elderly and malnourished and have significant co-morbidity, which increases the risk of major surgical intervention. Several aspects of the management remain controversial, such as choice of endovascular or surgical approach, method of revascularization, number and priority of vessels that necessitate reconstruction, configuration of the bypass, and choice of graft material. There are many series reporting endovascular or surgical revascularization for CMI. These are generally small studies, and long-term follow-up is often reported only as the incidence of symptomatic recurrence. It is unlikely that a single center can treat enough patients and accumulate sufficient experience to develop principles of treatment by demonstrating differences between various revascularization techniques. It seems reasonable to review the literature on the early and long-term results of endovascular and surgical management to provide a framework for the appropriate use of these techniques.

SURGICAL MANAGEMENT OF CHRONIC MESENTERIC ISCHEMIA

The aims of revascularization in patients with symptomatic CMI are to relieve symptoms and prevent intestinal infarction. A transperitoneal approach provides the opportunity to exclude non-vascular causes for the patient's symptoms before revascularization. Historically, endarterectomy was the first method used in visceral arterial revascularization.[5] The drawback in this technique is the inability to identify adequately the endpoint of the endarterectomy on the visceral vessel, risking the inadvertent creation of an intimal flap and early thrombosis of the reconstruction. However, the use of intraoperative duplex ultrasound may reduce this risk significantly.[6]

Bypass procedures have gained widespread acceptance in visceral arterial reconstruction. There are three decisions to make: the inflow site, the outflow site, and the type of bypass graft material. The infrarenal aorta, the supraceliac aorta, and the iliac artery are all satisfactory choices as inflow sites. Further details on the merits of the different surgical approaches are presented in Chapters 8 and 9.

In this chapter, we provide a comparative analysis of the merits of surgical versus endovascular management of CMI. A standard search strategy was used to identify all relevant published series of surgical revascularization for CMI using a combination of Medical subheadings and free text searches from the National Library of Medicine's MEDLINE and EMBASE databases from January 1980 to January 2002. To avoid publication bias of reports with favorable outcomes, smaller series of fewer than ten patients or case reports were not included. In studies involving patients with acute and chronic mesenteric ischaemia, only patients with CMI were included in the analysis.

During the past two decades, several studies have reported the merits of different methods of revascularization for CMI.[2, 3, 6–27] Controversy surrounds two issues: the merits of antegrade versus retrograde bypass grafting in providing optimal revascularization, and the need for complete revascularization to reduce the risk of recurrence. A summary of the data published over the past two decades for the surgical management of CMI is given in Table 10.1. Twenty-two studies involving 732 patients undergoing surgery for CMI over the last 20 years were included.[2–4, 6–28] A total of 1312 vessels were revascularized by a variety of surgical approaches. The mean number of patients in each study was 33, with a mean follow-up of 57 months. Most patients had two-vessel revascularization, with good technical and clinical success rates. The mean 30-day mortality was 5.0%, with a 30% mean complication rate. The mean 5-year survival was 69%, with 87% having long-term pain relief.

In 1981, Hollier *et al.* advocated complete revascularization, proposing that re-establishing blood flow to all diseased vessels reduces the risk of recurrent symptoms and associated complications related to graft thrombosis or stenosis.[2] The recurrence rate was reduced from 50% to 17% using this technique. McAfee *et al.* showed a similar trend in a series of 58 patients.[15] At 5 years, clinical graft patency in patients who underwent three-vessel revascularization was 90% compared with 54% for two-vessel revascularization and 0% for single-vessel reconstruction.[15] These reports have to be contrasted with those of Foley and colleagues, who subsequently showed that isolated retrograde bypass grafts to the superior mesenteric artery (SMA) provide comparable results and long-term prevention of recurrence with bypass grafts to multiple visceral vessels.[25, 26] The authors reported 4-year primary patency and patient survival rates of 89% and 82%, respectively, and symptomatic improvement was noted in all patients available for follow-up.[26] Other proponents of single-vessel revascularization have reported similar long-term results.[4, 7, 10, 11] Interestingly, Schneider *et al.* demonstrated that in selected cases where antegrade aortovisceral bypass grafting or transaortic endarterectomy is not possible due to technical or

Table 10.1 *Review of published series for the surgical management of patients with chronic mesenteric ischemia*

References	Patients (*n*)	Arteries (*n*)	Technical success (%)	Clinical success (%)	Patency (%)	Mean-follow up (months)	Complications (%)	30-day mortality	5-year survival (%)	Long-term pain relief (%)
1	56	103	95	96	NA	38	NA	8.9	NA	73
6	17	20	100	100	100	61	NA	12	71	100
7	10	20	90	100	90	43	NA	0	NA	80
8	41	62	98	98	87	42	54	5	78	79
9	60	69	100	NA	75	102	NA	3.5	NA	NA
10	32	90	100	NA	91	69	NA	9	NA	NA
5, 11, 12	74	194	100	100	NA	71	17	12	66	86
13	14	23	90	93	90	141	14	7	79	100
14	58	119	100	NA	90	60	41	10	NA	90
15	20	36	100	100	NA	36	20	0	NA	100
16	34	50	94	59	80	36	20	3	NA	NA
17	53	121	100	NA	NA	55	NA	0	66	66
18	16	24	100	100	89	35	12	6	NA	97
19	21	43	100	NA	86	120	19	0	NA	NA
20	24	38	100	100	78	60	45	4	71	78
21	10	18	100	100	66	40	20	0	NA	90
22	10	14	100	100	100	28	50	0	NA	90
3	42	66	98	98	66	33	30	10	70	86
23	11	15	73	100	79	44	63	3	50	NA
2, 27	85	130	100	100	76	57.6	33	8	64	87
24, 25	26	26	96	96	79	44	30	3	61	77
26	18	31	100	100	100	35	17	6	76	100
Mean	33.27	59.64	97.00	96.47	84.56	56.85	30.31	5.02	68.60	87.00
SD	22.32	47.88	6.23	9.87	10.44	29.36	15.83	4.14	8.80	10.42
Total	732	1312								

patient-related factors, isolated inferior mesenteric artery (IMA) revascularization may be sufficient.[24] In our view, a weakness of the studies favoring single-vessel revascularization is that no comparison is made with a group of patients treated during the same period of time with complete revascularization. In addition, thrombosis of the single bypass in five patients in the Foley series resulted in death in three patients, with a fourth requiring colectomy.

Stoney *et al.* at the University of California, San Francisco advocate transaortic endarterectomy or antegrade bypass grafting from the supraceliac aorta with medial visceral rotation.[6, 12, 13] Using this approach on 74 patients over 30 years, they have achieved 1- and 5-year patency rates of 97% and 86%, respectively, with a higher complication rate for the transaortic endarterectomy group (17% *v.* 0%, $P = 0.03$).

Foley *et al.* advocate retrograde bypass graft from the infrarenal aorta because of its familiar exposure and reduced risks with clamping and dissection. They report an operative mortality rate of 3%, and have achieved a 5-year patency rate of 79% verified with duplex surveillance.[26] Johnston *et al.* treated 21 patients with antegrade or retrograde bypass grafts. They reported 1- and 5-year survival rates of 100% and 86%, respectively.[20] Of note, 19% of these procedures were performed through a thoracoabdominal incision. More recently, Geroulakos *et al.* reported an antegrade revascularization via a transabdominal thoracoabdominal approach with low morbidity and no mortality.[22] However, long-term results are awaited. To date, no method has been proved to be superior because most studies have a limited cohort size as a result of the low prevalence of this disease. Furthermore, earlier reports lack objective follow-up data and have based their life-table patency solely on clinical observations, which may underestimate the incidence of graft thrombosis and stenosis.[2, 6, 15, 17, 18] McMillan and colleagues demonstrated primary patency rates of 89% at 72 months' follow-up using mesenteric duplex ultrasound or arteriography for symptomatic recurrences.[19] They found that symptomatic follow-up had a sensitivity of only 33% compared with objective measurement. They found no difference in objective patency rates at 36 months between antegrade versus retrograde bypass or saphenous vein graft versus prosthetic graft. Kihara *et al.* reported a large series of 42 patients with 66 visceral artery bypass grafts with an overall 3-year objective patency rate of 65%. Using a multivariate analysis they demonstrated that their patency rates were independent of conduit type, but male patients and short preoperative duration of symptoms (1 month) were associated with a greater risk for graft thrombosis.[4] With improvements in duplex scanning, recent studies have had more objective measures of long-term graft patency.[3, 4, 21, 24, 26]

ENDOVASCULAR INTERVENTION

Until recently, surgical revascularization was the only treatment available for patients with CMI. The complex anatomy, technically demanding surgical exposure, and morbidity often associated with surgical treatment in these patients have initiated the development of endovascular options. These consist of percutaneous transluminal angioplasty (PTA) and intra-arterial stenting using techniques that have been extrapolated from other vascular territories, such as the renal and peripheral vasculature. Endovascular methods have the advantage over surgery of being less invasive and are particularly attractive in patients considered to be at high risk for operative treatment. Details on the indications and the techniques for endovascular intervention are provided in Chapter 5.

Results of endovascular treatment of chronic mesenteric ischemia

The published data are sparse. A summary of the data published for the endovascular management of CMI with PTA is given in Table 10.2.[28, 29–34] Six studies were included, involving 93 patients and 141 revascularized vessels. The mean number of patients in each study was 16, with a mean follow-up of 25 months. Two-thirds of patients had two-vessel revascularization with reasonable technical and clinical success rates. The mean 30-day mortality was 3.5%, with a mean complication rate of 13%. Although 5-year survival is not available, 70% of patients had pain relief for the duration of follow-up.

The results of PTA in visceral arteries show that technical success rates are of the order of 75–96%, even without stents, as the majority of lesions are non-ostial (i.e. >5 mm from artery origin). Clinical success rates are good, with 60–85% of patients showing immediate clinical improvement. However, with time, re-stenosis occurs, and reintervention may be necessary to maintain this improvement; stenting could have a role in avoiding this.

In Matsumoto's series, four of 19 patients had extrinsic arterial compression by the median arcuate ligament that resulted in immediate technical failure.[29] In patients with typical symptoms of mesenteric angina, the clinical success rate was high (78%). If symptoms were atypical, then the clinical success rate was 50%.[29]

In Odurney's series, 17 of 19 arteries were dilated successfully in ten patients, resulting in symptomatic relief lasting for 6–24 months.[30] Symptoms recurred in five patients, and three had further relief of symptoms following balloon dilation. In this series, one technical failure was due to extrinsic compression by the median arcuate ligament, and the second was due to a severe ostial stenosis that did not respond to PTA. This would now be considered an indication for a stent. In addition, the average residual pressure gradient following PTA was 40 mm Hg. This report was before the widespread use of stents, and such a high residual gradient would now also be considered an indication for stenting.

Allen *et al.* report a series of 19 patients with CMI who were considered poor surgical candidates.[33] PTA was technically successful in 18 of 19 patients and 23 of 24 stenoses. The lone technical failure resulted in SMA dissection with thrombosis and bowel infarction; the patient died despite emergency laparotomy and surgical revascularization. Complete symptomatic relief was attained in 15 (79%) patients, with follow-up showing continued relief of symptoms for a mean of 39 months. Recurrent symptoms developed in three (20%) patients at a mean interval of 28 months. Repeat PTA performed in two patients provided good technical results and relief from clinical symptoms.

Most studies rely on angiographic criteria. A >30% residual stenosis is considered to be technically unsuccessful in most vascular territories. However, Rose *et al.* found that this did not correlate with symptomatic relief following PTA, and even when there was a >30% residual stenosis, there was symptomatic improvement so long as there was some angiographic improvement in the lumen.[35]

The recurrence rate following PTA varies from 21 to 48% (Table 10.2). Most re-stenoses are amenable to further percutaneous treatment. The majority of reported series performed repeat balloon dilation, but in more recent practice, stenting is considered appropriate for recurrent stenosis. Five of the 12 patients treated by stenting in Sheeran's series were for post-PTA re-stenosis.[36]

The literature on stenting in mesenteric arteries is still very limited (Table 10.3).[28, 36–38] A summary of the data published for the endovascular management of CMI with angioplasty and stenting consists of four studies involving 57 patients and 66 revascularized vessels. The mean number of patients in each study was 14, with a mean follow-up of 25 months. The mean number of vessels revascularized per patient was 1.2. The mean 30-day mortality was 4.8%, with a mean complication rate of 22%. Although 5-year survival is not available, 74%

Table 10.2 *Review of published series for the management of patients with chronic mesenteric ischemia with percutaneous transluminal angioplasty*

Reference	Patients (*n*)	Arteries (*n*)	Technical success (%)	Patency (%)	Mean follow-up (months)	Complications (%)	30-day mortality	Long-term pain relief (%)
40	10	19	80	37	11	20	10	NA
41	6	12	83	60	18	0	0	NA
42	16	25	84	75	28	6	6	75
33	19	20	79	83	25	32	0	52
43	19	24	95	80	39	11	5	79
44	23	41	96	88	27	9	0	75
Mean	15.50	23.50	86.17	70.50	24.67	13.00	3.50	70.25
SD	6.35	9.73	7.47	19.00	9.52	11.38	4.18	12.31
Total	93	141						

Table 10.3 *Review of published series for the management of patients with chronic mesenteric ischaemia with angioplasty and stenting*

Reference	Patients (*n*)	Arteries (*n*)	Technical success (%)	Patency (%)	Mean follow-up (months)	Complications (%)	30-day mortality (%)	Long-term pain relief (%)
27	28	32	100	73	36	18	11	66
46	12	13	92	74	16	NA	8	75
47	5	9	100	40	21	40	0	80
48	12	12	100	67	28	8	0	NA
Mean	14.25	16.50	98.00	63.50	25.25	22.00	4.75	73.67
SD	9.74	10.47	4.00	15.97	8.69	16.37	5.62	7.09
Total	57	66						

of patients had pain relief for the duration of follow-up. Stenting has high technical and clinical success rates (92–100%), but re-stenosis still occurs. In-stent re-stenosis may be treated by PTA or further stenting.[36–38] Other therapeutic strategies for in-stent re-stenosis, such as atherectomy, brachytherapy, and cutting balloons, have not been described in mesenteric arteries.

There are not enough published data to decide whether primary stenting is superior to PTA in the mesenteric arteries. Sheeran *et al.* stented six ostial lesions in their series.[36] Three of these were re-stenoses following primary PTA, and the other three were primarily stented.[36] They also performed primary stenting of the three chronic occlusions, which was technically successful in all cases.

The complications that have been described with endovascular revascularization include minor complications, such as access-site hematoma and cholesterol embolization.[29] Major complications include major blood loss, femoral artery occlusion, myocardial infarction, renal failure, ipsilateral brachial nerve palsy, and acute bowel infarction.[33, 36]

MANAGEMENT OF RECURRENT CHRONIC MESENTERIC ISCHEMIA

Recurrent visceral ischemia is not uncommon after primary visceral revascularization. It has been reported in up to one-third of patients who undergo visceral revascularization.[2, 8, 18, 39] Schneider and co-workers found a 17% recurrence rate of visceral ischemia and report a series of 33 patients with recurrent visceral ischemia (26 chronic visceral ischemia, seven acute visceral ischemia).[39] Thirty visceral reoperations were performed for recurrent visceral ischemia in 24 patients. Of seven patients who had recurrent acute visceral ischemia, six died of bowel infarction. Recurrent ischemia was associated with younger age, greater weight loss, and modification of surgical technique at the time of initial operation. The life-table symptom-free survival rate after reoperation was 77.3% and 62.8% at 1 and 5 years, respectively. The life-table overall survival rate after reoperation was 74.6% at 5 years. Their results show that reoperation for recurrent CMI can be performed safely and effectively with established revascularization techniques. The authors conclude that relief of symptoms and life expectancy rates after repeat visceral revascularization are comparable with those of patients who undergo primary visceral revascularization.[39]

Some series have reported successful endovascular treatment for recurrent visceral ischemia following primary surgical revascularization.[3, 4, 22] Geroulakos *et al.* report one symptomatic recurrence in a series of ten patients.[22] This was confirmed on mesenteric duplex to be a high-grade graft–SMA stenosis. The patient underwent angioplasty and stenting of the stenosis and remained symptom free thereafter. Mateo *et al.* report successful endovascular intervention in the management of three patients who developed recurrent symptoms following successful revascularization.[3] Kihara *et al.* report that seven patients in their series underwent further intervention for symptomatic recurrence.[4] Six patients were treated with surgery and one had successful endovascular stenting. A recent report by Phipp and colleagues describes a single case of recurrent CMI 3 years after antegrade mesenteric bypass with a prosthetic graft.[40] Graft thrombectomy was performed with limited success. The residual graft stenosis was treated with a Palmaz stent, which has resulted in a 2-year pain-free follow-up. The durability of these techniques is not known, but the techniques may be a reasonable option for some patients with a treatable lesion who may otherwise require an operation.

SURGERY VERSUS PERCUTANEOUS TRANSLUMINAL ANGIOPLASTY/ STENTING FOR CHRONIC MESENTERIC ISCHEMIA

In 1995, Rose and co-workers published a retrospective comparison between surgery and endovascular treatment for CMI in a small series of patients.[35] Their initial technical success rate for PTA was 30% (using a residual stenosis <30% as the endpoint) compared with 100% in the surgery group. If a residual stenosis of <50% was taken as the endpoint, they found their primary technical success rate increased to 88%. The reported mortality rate (PTA 13% *v.* surgery 11%) and the incidence of non-fatal major complications (PTA 25% *v.* surgery 33%) were similar in both groups. Patients who underwent surgery had more vessels revascularized compared with the angioplasty group. Long-term pain relief was 67% at 9 months in the PTA group compared with 88% at 35 months in the surgery group. The authors concluded that long-term pain relief was better in the surgery group compared with the PTA group, which may be due to the number of vessels revascularized.

There were similar observations in a recent retrospective study comparing the safety and efficacy of open surgery versus percutaneous angioplasty and stenting in patients with CMI in a single center.[28] Over a 3.5-year period, 28 patients (32 vessels) underwent endovascular revascularization for CMI. These patients were compared with a series of 85 patients (130 vessels) treated with surgery. Both groups were similar with respect to baseline co-morbidities, duration of symptoms, and the number of vessels involved. However, the patients who underwent endovascular treatment were significantly older. Patients who underwent mesenteric angioplasty had fewer vessels revascularized (1.1 ± 0.4) compared with the surgery group (1.5 ± 0.6, $P = 0.001$). Occlusions were more likely to be treated with open bypass grafting compared with angioplasty (4% PTA *v.* 51% surgery). No difference was noted in the perioperative complication or mortality rates. Although the 3-year cumulative recurrent stenosis and mortality rates were similar, the endovascular group had a higher recurrence rate (34% *v.* 13%, $P = 0.001$).[28]

Tables 10.1–10.3 demonstrate the results of published series for surgery, PTA and PTA/stenting for the management of CMI. Overall, more vessels per patient were revascularized in the surgery group (mean 1.8 vessels per patient) compared with the PTA/stenting group (mean 1.4 vessels per patient). This difference may be due to patient selection or from preferentially treating stenosis rather than occlusions with angioplasty. Endovascular recanalization of occlusions is usually more difficult with a higher incidence of complications and is associated with lower long-term patency. Early pain relief was more likely in the surgery group (96%) compared with the PTA group (77%). Long-term pain relief was noted in 87% of patients in the surgery group during a mean follow-up of 57 months compared with 72% of patients in the PTA group over a 25-month mean follow-up period. Overall, complication rates were higher in the surgery group (30%) compared with the PTA group (16%). There was no difference in 30-day mortality between the two groups (surgery 5.0% *v.* PTA 4.0%). Owing to variable reporting standards, firm conclusions cannot be made from this retrospective review of published studies. However, using early and long-term pain relief as fairly reliable indicators of technical success and durability of treatment, most published surgical series for CMI give consistently better results compared with the PTA series. One of the possible factors for this is the number of vessels revascularized, which has been reported to influence long-term outcome.[3, 15]

CONCLUSION

In this review, 22 studies involving open surgical revascularization and ten studies of endovascular revascularization were included. Patients treated with endovascular techniques had fewer vessels per patient revascularized, a lower number of totally occluded vessels, and a higher symptomatic recurrence rate. However, most were treated with balloon angioplasty alone, before stents were readily available. Endovascular techniques have the advantage of being easily repeatable and performed under local anesthesia. Increased experience and longer follow-up with arterial stenting are required to compare results with surgical series. Percutaneous angioplasty and stenting is an attractive option for surgical recurrences and higher-risk patients. Surgical revascularization has been shown to be effective and durable, but it also has a higher complication rate. Although more data are required, and the availability of local expertise should be considered, based on current evidence surgical revascularization should remain the first option in patients with CMI. Mesenteric angioplasty and stenting is a valuable option in patients who are at high risk from surgery.

REFERENCES

1 Bradbury A, Brittenden J, McBride K, Ruckley C. Mesenteric ischaemia: a multidisciplinary approach. Br J Surg 1995; **82**: 1446–9.

2 Hollier L, Bernatz P, Pairolero P, Payne W, Osmundson P. Surgical management of chronic intestinal ischemia: a reappraisal. *Surgery* 1981; **90**: 940–6.

3 Mateo R, O'Hara P, Hertzer N, Mascha E, Beven E, Krajewski L. Elective surgical treatment of symptomatic chronic mesenteric occlusive disease: early results and late outcomes. *J Vasc Surg* 1999; **29**: 821–31.

4 Kihara T, Blebea J, Anderson K, Friedman D, Atnip R. Risk factors and outcomes following revascularization for chronic mesenteric ischemia. *Ann Vasc Surg* 1999; **13**: 37–44.

5 Shaw R, Maynard E. Acute and chronic thrombosis of the mesenteric arteries associated with malabsorption: a report of two cases treated by thromboendarterectomy. *N Engl J Med* 1958; **258**: 974.

6 Cunningham C, Reilly L, Rapp J, Schneider P, Stoney R. Chronic visceral ischemia. Three decades of progress. *Ann Surg* 1991; **214**: 276–88.

7 Stanton P, Hollier P, Seidel T, Rosenthal D, Clark M, Lamis P. Chronic intestinal ischemia: diagnosis and therapy. *J Vasc Surg* 1986; **4**: 338–44.

8 Beebe H, MacFarlane S, Raker E. Supraceliac aortamesenteric bypass for intestinal ischemia. *J Vasc Surg* 1987; **5**: 749–54.

9 Rheudasil J, Stewart M, Schellack J, Smith RI, Salam A, Perdue G. Surgical treatment of chronic mesenteric arterial insufficiency. *J Vasc Surg* 1988; **8**: 495–500.

10 Kieny R, Batellier J, Kregz J. Aortic reimplantation of the superior mesenteric artery for atherosclerotic lesions of the visceral arteries: sixty cases. *Ann Vasc Surg* 1990; **4**: 122–5.

11 Cormier J, Fichelle J, Vennin J, Laurian C, Gigou F. Atherosclerotic occlusive disease of the superior mesenteric artery: late results and reconstructive surgery. *Ann Vasc Surg* 1991; **5**: 510–18.

12 Stoney R, Ehrenfeld W, Wylie E. Revascularization methods in chronic visceral ischemia caused by atherosclerosis. *Ann Surg* 1977; **186**: 468–76.

13 Rapp J, Reilly L, Qvarfordt P, Goldstone J, Ehrenfeld W, Stoney R. Durability of endarterectomy and antegrade grafts in the treatment of chronic visceral ischemia. *J Vasc Surg* 1986; **3**: 799–806.

14 Geelkerken R, van Boeckel H, de Roos E, Hermans J, Terpstra J. Chronic mesenteric vascular syndrome: results of reconstructive surgery. *Arch Surg* 1991; **126**: 1101–6.

15 McAfee M, Cherry K, Naessens J, Pairolero P, Hallett J, Gloviczki P, *et al.* Influence of complete revascularization on chronic mesenteric ischemia. *Am J Surg* 1992; **164**: 220–4.

16 Calderon M, Reul G, Gregoric I, Jacobs M, Duncan J, Ott D, *et al.* Long-term results of the surgical management of symptomatic chronic intestinal ischemia. *J Cardiovasc Surg* 1992; **33**: 723–8.

17 Sandmann W, Bohner H, Kniemeyer H, Schramm J. Chronic mesenteric ischemia. *Dtsch Med Wochenschr* 1994; **119**: 979–84.

18 Christensen M, Lorentzen J, Schroeder T. Revascularization of atherosclerotic mesenteric arteries: experience in 90 consecutive patients. *Eur J Vasc Surg* 1994; **8**: 297–302.

19 McMillan W, McCarty W, Bresticker M, Pearce W, Schneider J, Golan J, *et al.* Mesenteric artery bypass: objective patency determination. *J Vasc Surg* 1995; **21**: 729–40.

20 Johnston K, Lindsay T, Walker P, Kalman P. Mesenteric arterial bypass grafts: early and late results and suggested surgical approach for chronic and acute mesenteric ischemia. *Surgery* 1995; **118**: 1–7.

21 Moawad J, Mckinsey J, Wyble C, Bassiouny H, Schwartz LB, Gewertz GB. Current results of surgical therapy for chronic mesenteric ischemia. *Arch Surg* 1997; **132**: 613–19.

22 Geroulakos G, Tober J, Anderson L, Smead W. Antegrade visceral revascularisation via a thoracoabdominal approach for chronic mesenteric ischemia. *Eur J Vasc Endovasc Surg* 1999; **17**: 56–9.

23 Wolf Y, Verstandig A, Sasson T, Eidelman L, Anner H, Berlatzky Y. Mesenteric bypass for chronic mesenteric ischemia. *Cardiovasc Surg* 1998; **6**: 34–41.

24 Schneider D, Nelken N, Messina L, Ehrenfeld W. Isolated inferior mesenteric artery revascularization for chronic visceral ischemia. *J Vasc Surg* 1999; **30**: 51–8.

25 Gentile A, Moneta G, Taylor L, Park T, McConell D, Porter J. Isolated bypass to the superior mesenteric artery for intestinal ischemia. *Arch Surg* 1994; **129**: 926–32.

26 Foley M, Moneta G, Abou-Zamzam A, Edwards J, Taylor L, Yeager R, *et al.* Revascularisation of the superior mesenteric artery alone for treatment of intestinal ischemia. *J Vasc Surg* 2000; **32**: 37–47.

27 Farber M, Carlin R, Marston W, Owens L, Burnham S, Keagy B. Distal thoracic aorta as inflow for the treatment of chronic mesenteric ischemia. *J Vasc Surg* 2001; **33**: 281–7.

28 Kasirajan K, O'Hara P, Gray B, Hertzer N, Clair D, Greenberg R, *et al.* Chronic mesenteric ischemia: open surgery versus percutaneous angioplasty and stenting. *J Vasc Surg* 2001; **33**: 63–71.

29 Matsumoto A, Tegtmeyer C, Fitzcharles E, Selby J, Tribble C, Angle J, *et al.* Percutaneous transluminal angioplasty of visceral arterial stenoses. Results and long-term clinical follow-up. *J Vasc Interv Radiol* 1995; **6**: 165–74.

30 Odurney A, Sniderman K, Colapinto R. Intestinal angina. Percutaneous transluminal angioplasty of the celiac and superior mesenteric arteries. *Radiology* 1988; **167**: 59–62.

31 McShane M, Proctor A, Spencer P, Cumberland D, Welsh C. Mesenteric angioplasty for chronic intestinal ischemia. *Eur J Vasc Surg* 1992; **6**: 333–6.

32 Hallisey M, Deschaine J, Illescas F, Sussman S, Vine H, Ohki S, *et al.* Angioplasty for the treatment of visceral ischemia. *J Vasc Interv Radiol* 1995; **6**: 785–91.

33 Allen R, Martin G, Rees C, Rivera F, Talkington C, Garrett W, *et al.* Mesenteric angioplasty in the treatment of chronic intestinal ischemia. *J Vasc Surg* 1996; **24**: 415–23.

34 Maspes F, Mazzetti di Pietralata G, Gahdini R, Innocenzi L, Lupatelli L, Barzi F, *et al.* Percutaneous transluminal angioplasty in the treatment of chronic mesenteric ischemia: results and 3 years of follow-up in 23 patients. *Abdom Imaging* 1998; **23**: 358–63.

35 Rose S, Quigley T, Raher E. Revascularization for chronic mesenteric ischemia: comparison of operative arterial bypass grafting and percutaneous transluminal angioplasty. *J Vasc Interv Radiol* 1995; **6**: 339–49.

36 Sheeran S, Murphy T, Khwaya A, Sussman S, Hallisey M. Stent placement for treatment of mesenteric artery stenoses or occlusions. *J Vasc Interv Radiol* 1999; **10**: 861–7.

37 Nyman U, Ivancev K, Lindh M, Uher P. Endovascular treatment of chronic mesenteric ischemia: report of five cases. *Cardiovasc Intervent Radiol* 1998; **21**: 305–13.

38 Liermann D, Strecker E. Tantalum stents in the treatment of stenotic and occlusive disease of abdominal vessels. In: Liermann D (ed.). *Stents: state of the art and future developments.* Watertown: Boston Scientific, 1995; 127–34.
39 Schneider D, Peter A. Schneider P, Reilly L, Ehrenfeld W, Messina L, Stoney R. Reoperation for recurrent chronic visceral ischemia. *J Vasc Surg* 1988; **27**: 276–86.
40 Phipp L, Scott D, Kessel D, Robertson I. Stent implantation for mesenteric bypass graft stenosis. *J Endovasc Ther* 2000; **7**: 1320–3.

11

Celiac artery compression syndrome

DAVID BERGQVIST

INTRODUCTION

Celiac or arcuate ligament compression is the compression of the celiac trunk, axis, or artery by fibronervous structures and/or diaphragmatic muscle bands/ligaments, seen especially during expiration. A clinical compression syndrome with this name has been suggested in patients with epigastric pain (often severe) and weight loss, in combination with the external compression of the celiac artery. Harjola in 1963 described a 57-year-old male patient with a 'rare obstruction of the celiac artery'.[1] He had postprandial pain for several years, a rough systolic bruit in the epigastrium, and an angiographically shown narrowing of the celiac axis. He was relieved after treatment, which consisted of surgical release of the artery from a fibrotic celiac ganglion. Harjola's interpretation of the cause was retroperitoneal fibrosis. Rob and co-workers later suggested an anatomic variation of the celiac ganglion to be the cause of the arterial compression.[2]. Since then, the existence of the syndrome and the clinical relevance of the morphologic findings of external celiac compression have been debated.

There are two pathophysiological problems that have to be dealt with in the discussion of celiac compression syndrome. The first is the great collateral potential within the visceral circulation. As discussed in other chapters, ischemic symptoms usually mean severe stenotic or occlusive lesions of at least two of the three main visceral arteries, and isolated stenosis or occlusion of the celiac artery does not give symptoms. The second problem is that about one-third of people undergoing autopsy have a median arcuate ligament compression of the celiac artery.[3] In some people, this external compression or constriction may lead to intimal fibrosis with stenosis, and perhaps in the long run to development of arteriosclerosis. Again, one vessel lesion is not enough to compromise the visceral perfusion and cause symptoms of bowel ischemia.

Celiac compression syndrome is also called celiac band syndrome, median arcuate ligament syndrome (to lead the thinking away from a circulatory cause), and Dunbar's syndrome.[4] In German-language literature, it has been called a neurovascular compression syndrome of the aortic hiatus (*Neurovaskuläres Kompressionssyndrom des Hiatus aorticus*).[5, 6]

EPIDEMIOLOGICAL ASPECTS

The prevalence of celiac compression syndrome is not known. One reason for this lack of knowledge is the controversial view on the disease entity, which does not lend itself to systematic research. Another reason is that if it does exists, then the syndrome is very rare indeed: only small case series without a defined population basis have been published. Less than 500 patients have been reported in the literature. Over a 15-year period, seven patients were treated at the Texas Heart Institute; during the same time, 12 000 patients were treated for other peripheral vascular disorders.[7] During a 20-year period, only one celiac artery compression syndrome was seen when more than 85 operations were performed for symptomatic atherosclerotic occlusive disease of the bowel arteries.[8] Except for various opinions on the syndrome and thereby the lack of standards and criteria for classifying it, there are many inconsistencies in reporting routines, which makes it difficult to analyze the literature in a systematic way.

There are no randomized studies on treatment alternatives because of the rarity of the syndrome. The largest series (47 patients,[9] 31 patients,[10] and 51 patients[11]) are too small for meaningful statistical analysis.

The anatomical prerequisite for the syndrome, the external compression of the celiac axis or artery, has, however, been investigated in some radiological studies. The studies are not population based, but they at least give an impression of how often compression is seen in patients undergoing angiography. The indication for angiography varied widely.

In a series of 152 lateral aortograms for various indications, Colapinto and co-workers analyzed the proximal part of the celiac artery.[12] The stenoses found were classified into four types. In type I, the ostium was normal with a smooth, asymmetric narrowing, the artery being displaced in the inferior direction. This finding was considered to be associated with or caused by external compression. The other three types of stenosis were considered to be arteriosclerotic in origin. Of the 152 interpretable angiograms, 74 (49%) showed a celiac artery stenosis, 47 of which were classified as type I (31% of the total series). In 17 of the patients, the indication for angiography was abdominal pain. The conclusion, therefore, was that typical morphology of celiac compression can be found in asymptomatic patients. Similar rates of isolated celiac artery stenosis in asymptomatic patients have been reported by others.[13–16] Szilagyi *et al.* found a similar frequency of celiac narrowing in three groups of patients with a total of 157 angiographies: suspected intestinal angina (44%), gastrointestinal disease not characterized primarily by pain (53%), and miscellaneous non-alimentary diagnostic problems (50%).[17] Dunbar and colleagues performed abdominal aortography in 27 patients with epigastric bruit and abdominal pain unexplained by other diagnostic procedures.[4] They found four arteriosclerotic occlusive lesions, two visceral aneurysms, six normal investigations, and in 15 (56%) patients typical external celiac compression.

Lindner and Kemprud made a thorough autopsy study on 75 fresh postmortem specimens.[3] They found that both the celiac artery and the median arcuate ligament of the diaphragm were extremely variant in location. In patients with the syndrome, an abnormally high origin of the celiac artery has been described. They also found that the artery could be compressed both by the ligament and by the celiac plexus and its supplemental fibers. They thereby verified the findings of Harjola and Lahtiharju, who described three different mechanisms of stenoses: abnormally thick and fibrotic celiac ganglion, fibrotic margin of the aortic hiatus with distal location, and a combination of the two.[18] There were anatomical prerequisites for a stenosis in about half of the autopsy cases. Microscopic studies also showed that the arcuate ligament had autonomic nerve fibers within or in close relation to it, which was also shown by Mulder *et al.*[19] This autopsy study confirmed the observation by Lipshutz in 1977 that the celiac artery is not 'infrequently covered at its origin by the diaphragm'.[20] It was not until 1955, however, that Michels suggested a crural constriction of the artery.[21]

For reasons already stated, the natural history of the lesion is unknown. Spontaneous regression has been reported in some publications.[9, 14, 22, 23] Whether these patients suffer from celiac compression syndrome is a matter of opinion and definition.

PATHOPHYSIOLOGY

One problem with the syndrome is the lack of a distinct pathophysiological understanding of symptoms, especially pain. The very rich collateral visceral circulation makes ischemia less probable. In patients with classical abdominal angina, at least two of the visceral arteries must be severely compromised. There is considerable evidence that clinical and experimental interruption of the celiac artery is well tolerated in a situation where there is a normal collateral network.[24] However, in a recent case report, gastric tonometry showed a mucosal pH indicating ischemia (7.29).[25] Gastroscopy showed thickened gastric folds. After skeletonizing the celiac artery, gastric mucosal pH was normalized. Jungo suggested an effort-induced lactic acidosis as a metabolic consequence of the syndrome.[26] Several authors have measured a pressure gradient across the stenosis, but this gives no information on the flow situation distal to the first important collateral.[18, 27–30] In 11 patients, Geelkerken *et al.* found a pressure gradient of 15–40 mm Hg.[31] After division of the constricting structures, a significant increase in flow volume in the hepatic and splenic arteries has been shown (from 240 to 590 ml/min and from 180 to 450 ml/min, respectively).[32]

Stanley and Fry suggested a steal phenomenon from the superior mesenteric artery to the celiac vasculature to be responsible for the rather non-specific symptoms of the compression syndrome.[33] The possibility exists that, at least in some patients, an inadequacy of the collateral circulation is more important than the number of stenosed main vessels to induce a component of ischemia. When the superior mesenteric artery is open, there is no steal phenomenon.[24] Considering steal and retrograde flow in the common hepatic artery, Lampert *et al.* suggested proximal ligation of the splenic artery in combination with division of the arcuate ligament.[34] This, however, seems illogical as decompression in itself should normalize the flow direction. In a case report, a male patient was shown to have a median arcuate-ligament-induced occlusion of the celiac artery in combination with a tight stenosis of the superior mesenteric artery.[35] There was no evidence of arteriosclerosis during surgical exploration. Unfortunately, hemodynamic evaluation was not made preoperatively, but the patient's symptoms were relieved by dividing the left diaphragmatic crus. His abdominal pain had been intermittent and relieved by bending forward, and thus resembled pain described in patients with celiac compression syndrome. There are a few more patients reported where the compression has involved both the celiac and superior mesenteric arteries.[36, 37] This is not surprising knowing that the origin of the arteries is sometimes very close (1 cm). Median arcuate ligament compression of both arteries has been described in monozygotic twins, indicating that the anatomical prerequisite for the syndrome, at least sometimes, may be congenital.[38] An extremely rare anatomic anomaly was described by Mankowitz, with a common origin for the celiac axis, superior mesenteric artery, and both renal arteries.[39] A compression in such a case obviously has the potential of inducing more dramatic and more severe symptomatology.

Occlusive thrombosis may develop in the stenotic celiac artery, but infarction of the viscera has never been described in a patient with the syndrome.[40] Microembolization into pancreas vessels causing pancreatitis has been suggested to explain the pain.[41]

An alternative explanation is that the pain is caused by nervous entrapment of the celiac plexus, which could explain why division of the structure around the artery, including nerves, could give pain relief.[42] It has been suggested that there is chronic irritation of the celiac plexus by arterial pulsative motion with secondary vasoconstriction. Microscopy often shows a combination of muscular and nervous structures surrounding the artery.

There are other rare causes of compression, such as sarcoidosis,[43] and it has also been described in association with pancreatic cancer.[44] In a few cases, association of pancreatico-duodenal artery aneurysm and median arcuate ligament compression of the celiac has been reported.[45–47]

A very special situation in which the celiac compression may be deleterious is in recipients of a liver transplant. In these patients, the only source of arterial inflow to the transplanted liver is through the celiac artery, and it is therefore important to remove any obstruction of the celiac artery to prevent potential graft loss because of arterial thrombosis. This has been found in up to 10% of patients undergoing orthotopic liver transplantation.[48–50] Another group of patients where celiac decompression is important is those undergoing pancreatico-duodenectomy and at the same time having a celiac compression syndrome; life-threatening bowel ischemia may follow.[51, 52] Sometimes, the pathology is known preoperatively, but the knowledge of this possibility is important when ischemic signs develop intraoperatively during a pancreaticoduodenectomy, where immediate decompression is necessary.[52]

The clinical importance of a celiac compression has also been demonstrated in a patient where the superior mesenteric artery perfusion was compromised through a dissection, probably caused by a seat belt injury.[53]

Sometimes there is an intimal thickening, with smooth-muscle cell proliferation and an increase in extracellular matrix, probably as a result of repetitive trauma.[38, 42]

SYMPTOMS

Symptomatology is rather vague and usually of long duration, often several years. One patient has been reported with more of an emergency history of abdominal symptoms.[54] The most consistent symptom leading a patient to see a doctor is severe but rather unspecific epigastric pain, often cramping, grinding, or stabbing,[55] or even described as bizarre.[56] Certain movements or body positions may precipitate or accentuate the pain.[57] It is often postprandial but not as regular as in patients with abdominal angina. Mihas and co-workers pointed out the relatively common occurrence of pain at night.[23] Some patients report that the pain is alleviated when they are sitting or half-lying with the arms on the table or when they are assuming the knee-to-chest position. The pain is usually refractory to various forms of pharmacological treatment. The pain is very variable in location, intensity, and frequency of attacks. Many patients have been investigated extensively because of the pain, without positive findings and perhaps with conflicting opinions on the diagnosis, and some of them have also undergone a negative laparotomy. In many patients, there may be associated pathological findings, which could explain the symptoms and make the diagnostic situation even more difficult, such as pancreatitis, cholecystitis, peptic ulcer, constipation, and adhesions.[17, 58, 59] Dunbar *et al.* noticed that patients with compression of the celiac trunk had 'obscure abdominal symptoms not too dissimilar from abdominal angina'.[4]

Other unspecific complaints that are often reported are a feeling of epigastric fullness or distension, nausea and vomiting, and/or (less often) diarrhea.

Because of the symptoms, especially fear of food because of pain and nausea, many patients lose weight, but they are rarely as emaciated as patients with abdominal angina. The weight loss may be in the range of 5–10 kg.[31, 40] A classical 'small-meal syndrome' is less obvious than in patients with atherosclerotic occlusive disease.

SIGNS

The female sex dominates (about two-thirds of the patients are females). The majority of patients are aged 30–50 years, the youngest being 13 years and the oldest 85 years in published series. A clear difference from classical abdominal angina is in regard to age and sex. The body habitus is often leptosomic or asthenic.[30, 42, 60] Except for the patient being underweight, not much is to be found on a clinical examination. The weight loss usually is between 4 and 6 kg, and only exceptionally above 10 kg. In about one-third of the patients, weight loss is mentioned as an important complaint. Mild diffuse or localized epigastric tenderness is frequently found.[31, 43, 61] There is often an epigastric systolic bruit that is altered by body position and respiration; this phenomenon being an important part of the differential diagnosis was first pointed out by Edwards.[28] The bruit is often accentuated at expiration, but it may also disappear if the celiac artery becomes occluded at expiration.[62] In patients undergoing laparotomy for other reasons, Harjola and Lahtiharju found a clear thrill when palpating the celiac axis.[18] In a few cases, an epigastric thrill can be felt on abdominal palpation.[61] An epigastric bruit is, however, also found in healthy normal volunteers with a frequency of as high as 30%.[60, 63, 64]

DIAGNOSTIC TESTS

Unfortunately, there are no specific diagnostic tests that will establish a diagnosis with certainty, nor are there any laboratory findings that are specific for the celiac compression syndrome. The diagnosis is often one of exclusion.

When there is a high degree of clinical suspicion, the next diagnostic step is a contrast angiography with lateral views to view adequately the outflow of the celiac artery from the aorta. The typical picture is a smooth asymmetric narrowing of an external compression and without arteriosclerosis in the ostium.[4, 12] In the majority of patients (76% in the series of 51 patients reported by Reilly *et al.*[11]), there is either post-stenotic dilation of the celiac artery or evidence of increased collateral flow, or both. In elderly patients, there may be atheromatous changes in the aorta but not in the celiac artery.[55] Provocation or accentuation of the findings may be obtained at expiration. The background for this finding is the caudal movement of the celiac artery and the aorta during inspiration at the same time as the median arcuate ligament moves ventrally, and reversed movements during expiration. This dynamic situation is also the explanation why a bruit is often provoked during expiration. Similar information can be obtained with spiral computerized tomography (CT) angiography, where the compression mechanism from the diaphragm is seen more distinctly.[65] The ligamentum compression can also be identified on plain CT.[66]

The stenosis may also be identified with a duplex investigation.[25, 51, 67, 68] An increase in peak systolic velocity at expiration is suggestive but unfortunately not specific, and a decrease has also been described.[40] The important observation is probably that the velocity undergoes changes with the respiratory cycle and sometimes with body position.[68] A reversed flow in the hepatic artery may be seen. Whether a negative duplex investigation excludes the syndrome is not known at present. Vang and colleagues suggested that pain provocation by the injection of bradykinin was fairly specific for the syndrome, but this observation has not been evaluated further and is not used as a diagnostic test.[69]

The impairment of D-xylose absorption test has been suggested on the basis of one patient where the test normalized after division of the median arcuate. [70] Recently, a case report indicated gastrointestinal mucosal ischemia (pH 7.29, as measured with gastric tonometry) that was normalized after celiac trunk decompression (pH 7.48 24 h postoperatively).[25] Whether this is a regular finding has to be explored further. It is important to exclude other common diseases causing upper-abdominal pain, such as gallbladder disease, peptic ulceration, gastroesophageal reflux, and chronic pancreatitis.

TREATMENT

When Harjola reported his patient with external compression of the celiac axis by a fibrotic celiac ganglion, he suggested division of the constrictive structures as a way of treatment.[1] This was followed by several small case series claiming a beneficial effect of division of the median arcuate ligament and/or fibrotic ganglion structures.[2, 4, 18, 27, 71–73] Other authors, however, have reported less impressive results. Because of problems with the diagnosis and frequent concomitant intra-abdominal disorders, it is important to make a thorough exploration of the abdominal cavity. Thereafter, the constricting and compressing structures are divided to free the entire celiac artery to its origin from the aorta. Often, the tissue is very firm, and in some patients it is necessary to perform a ganglionectomy.[2, 11]. If, after surgical decompression, a stenosis is still present in the artery, some authors advocate a vascular reconstruction,[7, 11, 40] but it is not known whether this is necessary. One way to get further insight into the syndrome would be to perform a bypass without decompression. Pain relief in such a situation would speak in favor of a vascular genesis of the pain. Mihas *et al.* reported four patients in whom they reimplanted the post-stenotic celiac artery into the aorta (two patients) or made a short bypass (two patients) without local decompression.[23] In one patient there was no effect at all, and in the others pain recurred. In two patients, angiography was again performed showing no stenosis.

The role of percutaneous angioplasty has yet to be determined, but there are a few unsuccessful attempts reported and most authors are fairly pessimistic because of the firm elastic external compression where balloon deflation leads immediately to re-stenosis.[74] Balloon dilation after external decompression has been proposed in certain patient groups with a remaining pressure gradient.[11, 55, 74]. Over the years, there have also been critical voices claiming that the syndrome does not exist and that consequently surgical treatment is not indicated.[17, 23, 60, 75]

Indications for surgery should be considered in patients with the type of abdominal pain described previously together with weight loss in combination with an at least 75% angiographic stenosis of the celiac artery. Often, the indication is made on the basis of excluding other causes of symptomatology.

RESULTS

There are only a few case series with large enough numbers of patients and with long enough follow-up times to be able to give a true picture of the results. Postoperative mortality and morbidity are very low (near to zero). Follow-up times in publications are usually a couple of years but often much shorter. A follow-up time of more than 4 years is exceptional, used in only seven studies out of 69 in which follow-up time is mentioned.[7, 9, 11, 31, 32, 76, 77] Reilly *et al.* found a correlation between postoperative symptom relief of postprandial pain after celiac decompression if the patient was female, the pain was postprandial, weight loss was more than 20 lbs, age was 40–60 years, and angiography showed the presence of collaterals and a post-stenotic dilation.[11] There should also be an absence of psychiatric disorder or history of drug abuse. Postoperative weight gain is an indication of a successful operation. A regression analysis to identify factors predicting a favorable outcome was undertaken by Williams *et al.*[78] Patients likely to benefit had epigastric pain related to food intake and without other unrelated and diffuse symptoms.

Huber and Schwarz stated that about 70% of patients undergoing division of the arcuate ligament became pain free.[79] A similar effect was suggested by Takach and co-workers, and was somewhat higher if the decompression was combined with a revascularization.[7] Reilly *et al.* reported a series of 51 patients who were treated with celiac decompression ($n = 16$), celiac decompression in combination with graduated celiac dilation ($n = 17$), or celiac decompression and celiac reconstruction ($n = 18$).[11] At a mean follow-up of 9 years 53, 79, and 73%, respectively, were asymptomatic.

Contradictory results may have several reasons, such as patient selection, poor follow-up, and various types of surgical techniques. Incorrect diagnosis may be one reason for poor results and must be considered. The best long-term results seem to be obtained when combining division of the median arcuate ligament and excision of the celiac plexus.[73, 76, 78]

Recurrent symptoms are not unusual but frequently occur several years after decompression (in the study by Geelkerken *et al.*, after a mean of 6.8 years[31]). Evans found 83% of patients to be asymptomatic 6 months postoperatively, but only 41% remained asymptomatic 3–13 years later.[9] Thus, with a long enough follow-up period, the results seem to be less good. Nonetheless, in many patients a reasonably long symptom-free period may exist. Some critics and skeptics have explained the symptom relief as a placebo effect of the operation. Such an effect may well exist, but the longer the symptom-free interval, the less probable is the explanation. Recurrence after a pain-free interval of several years does not exclude the existence of the syndrome. One reason for recurrent symptoms could be re-stenosis because of scar tissue formation; another possible reason is re-innervation. In a few patients, disabling symptoms persist after the decompression. In this group, laparotomies have usually been performed with various types of interventions, such as cholecystectomy, gynaecological resections, division of adhesions, and appendectomy.[11, 31] A chronic pain syndrome may develop with the need for pharmacological and/or psychological treatment and support. Plate and colleagues noted that at long-term follow-up, only eight of 15 patients considered themselves improved, but six of these had developed other unspecific abdominal symptoms of chronic character.[80] Angiography has been performed in very few patients with symptomatic recurrence, so we do not know whether the recurrence also means a residual stenosis of the celiac artery. Several authors have reported recurrent pain problems without an angiographically proven re-stenosis. There are definitely patients with long-term relief of symptoms, i.e. several years.[76]

CRITICAL POINTS

Ever since the syndrome was first described, there have been critical voices, some even claiming that there is no such disease as the celiac compression syndrome. Factors speaking against the existence of a specific celiac compression syndrome can be summarized as follows:

- Celiac compression syndrome is not a well-defined clinical disease entity.
- One stenotic or occluded visceral artery is not enough to cause ischemia, which speaks against ischemia playing an important role.

- Angiographic celiac compression is not seen infrequently in normal people or in patients without abdominal pain.
- The symptomatology is fairly vague and inconsistent.
- Apart from the angiographic findings, there is a lack of objective diagnostic tests.
- There is a lack of a plausible etiological theories.
- Patients with celiac compression often have other abdominal explanations for the symptoms.

SUMMARY

Almost since the first publication on celiac compression syndrome, the existence of the syndrome has been controversial. Despite the large number of publications, this controversy still exists. The anatomical entity is well described but its clinical significance is debated. In any case, the syndrome is infrequent and any form of randomized treatment studies would be virtually impossible to perform, particularly on a multicenter basis. With the high frequency of asymptomatic celiac stenosis, we must heed great caution before linking abdominal pain to arterial stenosis. However, there are patients with distressing abdominal pain and a morphological celiac extraluminal compression and without other explanation for the pain who are helped by surgical decompression of the celiac artery also on a long-term basis.

KEY REFERENCES

Harjola P-T. A rare obstruction of the coeliac artery. Report of a case. *Ann Chir Gynaecol Fenn* 1963; **2**: 47–50.

The first report of the syndrome.

Dunbar J, Molnar W, Beman F, Marable S. Compression of the celiac trunk and abdominal angina. *Am J Roentgenol* 1965; **95**: 731–44.

A paper that gave the celiac compression syndrome the eponym 'Dunbar's syndrome'.

Reilly L, Ammar A, Stoney R, Ehrenfeld W. Late results following operative repair for celiac artery compression syndrome. *J Vasc Surg* 1985; **2**: 79–91.

The largest series analyzing various combinations of treatment with long-term follow-up.

Szilagyi D, Rian R, Elliot J, Smith R. The celiac artery compression syndrome: does it exist? *Surgery* 1972; **72**: 849–63.

A critical article pointing to several factors speaking against the syndrome as a clinical entity.

Bech F. Celiac artery compression syndromes. *Surg Clin North Am* 1997; **77**: 409–24.

A comprehensive review.

REFERENCES

1 Harjola P-T. A rare obstruction of the coeliac artery. Report of a case. *Ann Chir Gynaecol Fenn* 1963; **2**: 547–50.

2 Snyder M, Mahoney E, Rob C. Symptomatic celiac artery stenosis due to constriction by the neurofibrous tissue of the celiac ganglion. *Surgery* 1967; **61**: 372–6.

3 Lindner H, Kemprud E. A clinicoanatomical study of the arcuate ligament of the diaphragm. *Arch Surg* 1971; **103**: 600–605.

4 Dunbar J, Molnar W, Beman F, Marable S. Compression of the celiac trunk and abdominal angina. *Am J Roentgenol* 1965; **95**: 731–44.

5 Schwilden E. Zur Problematik des Zöliakakompressionssyndroms. *Angiologie* 1985; **2**: 90.

6 Feindt P, Walter P, Omlor G. Das Kompressionssyndrom des Truncus coeliacus: Ein seltenes Kranheitsbild. *Vasa* 1992; **21**: 307–9.

7 Takach T, Livesy J, Reul G, Cooley D. Celiac compression syndrome: tailored therapy based on intraoperative findings. *J Am Coll Surg* 1996; **183**: 606–10.

8 Moneta G. Diagnosis of intestinal ischemia. In: Rutherford R (ed.). *Vascular surgery*, 5th edn. Philadelphia: W. B. Saunders, 2000; 1501–11.

9 Evans W. Long-term evaluation of the celiac band syndrome. *Surgery* 1974; **76**: 867–71.

10 Beger H, Apitzsch D, Meves M. Die Arteria coelica-Kompression. *Langenbecks Arch Chir* 1976; **341**: 253–69.

11 Reilly L, Ammar A, Stoney R, Ehrenfeld W. Late results following operative repair for celiac artery compression syndrome. *J Vasc Surg* 1985; **2**: 79–91.

12 Colapinto R, McLoughlin M, Weisbrod G. The routine lateral aortogram and the celiac compression syndrome. *Radiology* 1972; **103**: 557–63.

13 Meaney T, Kistner RL. Evaluation of intra-abdominal disease of obscure cause. *Arch Surg* 1967; **94**: 811–16.

14 Bron K, Redman H. Splanchnic artery stenosis and occlusion. Incidence, arteriographic and clinical manifestations. *Radiology* 1969; **92**: 323–8.

15 Cornell S. Severe stenosis of the celiac artery. *Radiology* 1971; **99**: 311–16.

16 Koikkalainen K, Köhler R. Stenosis and occlusion in the coelic and mesenteric arteries. A comparative and angiographic and clinical study. *Ann Chir Gynaecol Fenn* 1971; **60**: 9–24.

17 Szilagyi D, Rian R, Elliot J, Smith R. The celiac artery compression syndrome: does it exist? *Surgery* 1972; **72**: 849–63.

18 Harjola P-T, Lahtiharju A. Celiac axis syndrome. Abdominal angina caused by external compression of the celiac artery. *Am J Surg* 1968; **115**: 864–9.

19 Mulder D, Rubush J, Lawrence M, Ehrenhaft J. Celiac axis compression syndrome. *Can J Surg* 1971; **14**: 122–6.

20 Lipshutz B. A composite study of the caelic axis artery. *Ann Surg* 1977; **65**: 159–69.

21 Michels N. *Blood supply and anatomy of the upper abdominal organs*. Philadelphia: Lippincott, 1955; 140.

22 Drapanas T, Bron K. Stenosis of the celiac artery. *Ann Surg* 1966; **164**: 1085–8.

23 Mihas A, Laws H, Jander H. Surgical treatment of the celiac axis compression syndrome. *Am J Surg* 1977; **133**: 688–91.

24 Brandt L, Boley S. Celiac axis compression syndrome. A critical review. *Dig Dis* 1978; **23**: 633–40.

25 Faries P, Narula A, Veith F, Pomposelli F, Marsan B, LoGerfo F. The use of gastric tonometry in the assessment of celiac artery compression. *Ann Vasc Surg* 2000; **14**: 20–23.

26 Jungo O. Anstrengungsinduzierte Lactazidose bei Coeliacastenose. *Schweiz Med Wochenschr* 1990; **120**: 349–53.

27 Lord R, Stoney R, Wylie E. Coeliac axis compression. *Lancet* 1963; **2**: 795–8.

28 Edwards A. Coeliac axis compression syndrome. *Proc R Soc Med* 1969; **62**: 488–90.

29 Meves vM, Beger H. Das coeliaca-Kompressions-Syndrom. Ein kasuistischer Beitrag. *Fortschr Geb Rontgenstr Nuklearmed* 1973; **118**: 451–4.

30 Lord R, Tracy G. Coeliac artery compression. *Br J Surg* 1980; **67**: 590–93.

31 Geelkerken R, van Bockel J, de Roos W, Hermans J. Coeliac artery compression syndrome: the effect of decompression. *Br J Surg* 1990; **77**: 807–9.

32 Kernhohan R, Barros D'Sa A, Cranley B, Johnston H. Further evidence supporting the existence of the celiac artery compression syndrome. *Arch Surg* 1985; **120**: 1072–6.

33 Stanley J, Fry W. Median arcuate ligament syndrome. *Arch Surg* 1971; **103**: 252–8.

34 Lampert V, Holz D, Jäger K, Landmann J. Nouvelle approache thérapeutique dans le syndrome de compression du tronc celiaque. *Helv Chir Acta* 1991; **58**: 169–72.
35 Lawson J, Ochsner L. Median arcuate ligament syndrome with severe two-vessel involvement. *Arch Surg* 1984; **119**: 226–7.
36 Curl J. Median arcuate ligament compression of the celiac and superior mesenteric arteries. *Ann Surg* 1971; **173**: 314–20.
37 Daily P, Fogarty T. Simplified revascularization of the celiac and superior mesenteric arteries. *Am Surg* 1976; **131**: 762–5.
38 Bech F, Loesberg A, Rosenblum J, Glagov S, Gewertz B. Median arcuate ligament compression syndrome in monozygotic twins. *J Vasc Surg* 1994; **19**: 934–8.
39 Mankowitz B. A celiac artery anomaly associated with the celiac artery compression syndrome. *Angiology* 1974; **25**: 419–21.
40 Loffeld R, Overtoom H, Rauwerda J. The celiac axis compression syndrome. *Digestion* 1995; **56**: 534–7.
41 Cleator I, Mcpherson A, Fraser G. Celiac artery compression. *J R Coll Surg Edinb* 1971; **16**: 96–9.
42 Bech F. Celiac artery compression syndromes. *Surg Clin North Am* 1997; **77**: 409–24.
43 Stamler J, Whittemore A, Loscalzo J. Celiac axis compression syndrome caused by sarcoidosis: an acquired form of the syndrome. *Am J Med* 1989; **86**: 225–7.
44 Fortner J, Watson R. Median arcuate ligament obstruction of celiac axis and pancreatic cancer. *Ann Surg* 1981; **194**: 698–700.
45 Suzuki K, Kashimura H, Sato M, Hassan M, Yokota H, Nakahara A, *et al.* Pancreaticoduodenal artery aneurysms associated with celiac axis stenosis due to compression by median arcuate ligament and celiac plexus. *J Gastroenterol* 1998; **33**: 434–8.
46 Sutton D, Lawton G. Coeliac stenosis or occlusion with aneurysm of the collateral supply. *Clin Radiol* 1973; **24**: 49–53.
47 Proud G. Aneurysm formation on the small pancreatic arteries in association with coeliac axis compression. *Ann R Coll Surg Engl* 1978; **60**: 294–7.
48 Jurim O, Shaked A, Kiai K, Mills M, Colquhoun S, Busuttil R. Celiac compression syndrome and liver transplantation. *Ann Surg* 1993; **218**: 10–12.
49 Fukuzawa K, Schwartz M, Katz E. The arcuate ligament syndrome in liver transplantation. *Transplantation* 1993; **56**: 223–4.
50 Ducerf C, Rode A, De la Roche E, Beck F, Adham M, Berthoux N, Baulieux J. Compression du tronc celiaque par le ligament argué du diaphragme au cours de la chirurgie de l'étage sus-mésocolique. *Ann Chir* 1998; **52**: 495–502.
51 Kohler T, Debas H, Crames M, Strandness D. Pancreaticoduodenectomy and the celiac artery compression syndrome. *Ann Vasc Surg* 1990; **4**: 77–80.
52 Bull D, Hunter G, Crabtree T, Bernhard V, Putnam C. Hepatic ischemia, caused by celiac axis compression, complicating pancreaticoduodenectomy. *Ann Surg* 1993; **217**: 244–7.
53 Solis M, Ranval T, McFarland D, Eidt J. Surgical treatment of superior mesenteric artery dissecting aneurysm and simultaneous celiac artery compression. *Ann Vasc Surg* 1993; **7**: 457–62.
54 Edhag O, Hogstedt C, Kövamees A, Werner B. A case of severe compression of the coeliac artery. *Acta Med Scand* 1977; **201**: 589–91.
55 Watson W, Sadikali F. Celiac axis compression. *Ann Intern Med* 1977; **86**: 278–84.
56 McSherry JA. Coeliac artery compression syndrome: can we believe our ears? *J R Coll Gen Pract* 1977; **27**: 684–6.
57 Barakat M, Mahmoud J, Bentif P. Celiac-axis compression syndrome. A report of 3 cases. *Dig Dis* 1972; **17**: 373–7.
58 Beger HG, Meves M, Apitzsch D, Kraas E, Bittner R. Diagnose und operative Behandlung bei Arteria-coeliaca-Kompression. *Dtsch Med Wochenschr* 1975; **100**: 464–71.

59 Bretzke VG. Das Dunbar-Syndrom in der Differentialdiagnose von abdominellen Beswerden. *Z Ärtzl Forbild* 1980; **74**: 1000–1002.

60 Edwards A, Hamilton J, Nichol W, Taylor G, Dawson A. Experience with coeliac axis compression syndrome. *BMJ* 1970; **1**: 342–5.

61 Ghosn P, Rabbat A, Trudel J, D'Amico P, Lecours R, Trudel J. Celiac compression syndrome. *Can J Surg* 1982; **25**: 377–9.

62 Reuter S. Accentuation of celiac compression by the median arcuate ligament of the diaphragm during deep expiration. *Radiology* 1971; **98**: 561–4.

63 Julius S, Stewart B. Diagnostic significance of abdominal murmurs. *N Engl J Med* 1972; **276**: 1175–8.

64 Watson W, Williams P, Duffy G. Epigastric bruits in patients with and without celiac axis compression. *Ann Intern Med* 1973; **79**: 211–15.

65 Kopecky K, Stine S, Dalsing M, Gottlieb K. Median arcuate ligament syndrome with multivessel involvement: diagnosis with spiral CT angiography. *Abdom Imaging* 1997; **22**: 318–20.

66 Patten R, Coldwell D, Ben-Menachem Y. Ligamentous compression of the celiac axis: findings in five patients. *Am J Roentgenol* 1990; **156**: 1101–3.

67 Erden A, Yurdakul M, Cumhur T. Marked increase in flow velocities during deep expiration: a duplex Doppler sign of celiac artery compression syndrome. *Cardiovasc Intervent Radiol* 1999; **22**: 331–2.

68 De Pauw M, Voet D, Kunnen M, Barbier F, Afschrift M. Arcuate ligament syndrome mimicking celiac occlusion. *Am J Gastroenterol* 1992; **87**: 1482–4.

69 Vang J, Hegedüs V, Jóhannesson E. Arteria-coeliaca-kompressionssyndromet. *Nord Med* 1971; **85**: 124–5.

70 Garvin P, Sawyer O, Cabel E, Kaminski D, Codd J. Malabsorption and abdominal pain secondary to celiac artery entrapment. *Arch Surg* 1977; **112**: 655–7.

71 Marable S, Molnar M, Beman F. Abdominal pain secondary to celiac axis compression. *Am J Surg* 1966; **111**: 493–5.

72 Marable S, Kaplan M, Beman F, Molnar W. Celiac compression syndrome. *Am J Surg* 1968; **115**: 97–102.

73 Carey J, Stemmer E, Conolly J, Calif I. Median arcuate ligament syndrome. *Arch Surg* 1969; **99**: 441–6.

74 Choghari C, Bosschaert T, Barroy J. Dunbar's syndrome. *Acta Chir Belg* 1994; **94**: 207–9.

75 Marston A. Celiac-axis compression. *Lancet* 1977; **2**: 32.

76 Holland A, Ibach E. Long-term review of coeliac axis compression syndrome. *Ann R Coll Surg Engl* 1996; **78**: 470–2.

77 Glover J. Late results following operative repair for celiac compression syndrome. *J Vasc Surg* 1985; **2**:90.

78 Williams S, Gillespie P, Little J. Celiac axis compression syndrome: factors predicting a favourable outcome. *Surgery* 1985; **98**: 879–87.

79 Huber F, Schwarz H. Angina intestinalis und Truncus-coeliacus-Kompressions-Syndrom. *Helv Chir Acta* 1974; **41**: 55–63.

80 Plate G, Eklöf B, Vang J. The celiac compression syndrome: myth or reality? *Acta Chir Scand* 1980; **147**: 201–3.

12

Visceral artery aneurysms

SANDRA C. CARR AND WILLIAM H. PEARCE

INTRODUCTION

Aneurysms of the visceral arteries have been described for over 100 years, mostly in autopsy series. In 1770, Beuassier first reported a splenic artery aneurysm discovered during injection of the aorta for anatomic demonstration in a 60-year-old female cadaver.[1] An important event in American history was in 1881 when President James A. Garfield died from a ruptured splenic artery aneurysm 2 months after being shot in the abdomen by an assassin.[2] The earliest treatment regimens for visceral artery aneurysms (VAA) consisted of rest, hot baths, emollients, and the administration of morphine. External compression was also suggested for some early aneurysms. During the late 1800s and early 1900s, wiring of the aneurysm sac was used to induce thrombosis. Sometimes, elective current was applied in an attempt to produce thrombosis. In 1895, Stevenson published a case report of visceral aneurysm treated by introducing 2 yards of steel coil into the sac in a patient who eventually died of hemorrhage.[3] In 1910, Lumsden and colleagues reported on a celiac artery aneurysm that was treated by the introduction of a total of 150 inches of guilded wire into the lesion.[4] The aneurysmal pulsation became less pronounced and the palpable abdominal mass was firm at 7 weeks' follow-up. Kehr, who treated a common hepatic artery aneurysm with ligation, reported the first successful surgical therapy in 1903.[5] In 1932, Lindboe electively operated on a splenic artery aneurysm with a successful outcome.[6] The first successful repair of a superior mesenteric artery (SMA) aneurysm was published in 1953 by DeBakey and Cooley, who reported on a patient with a mycotic aneurysm of the SMA who was treated with ligation of the vessel without revascularization.[7] Over the decades, there has been tremendous change in the pattern of disease as well as in the diagnosis and treatment of VAA.

VAAs are rare but potentially lethal vascular lesions. The incidence of these lesions in routine autopsy series is only 0.01–0.2%.[8] The most commonly involved visceral vessels include the splenic, hepatic, superior mesenteric, and celiac arteries. The gastroduodenal, pancreatoduodenal-pancreatic, gastric-gastroepiploic, jejunal-ileal-colic, and inferior mesenteric arteries are involved less often (Figure 12.1). Today, with the increased use of computed tomography (CT)

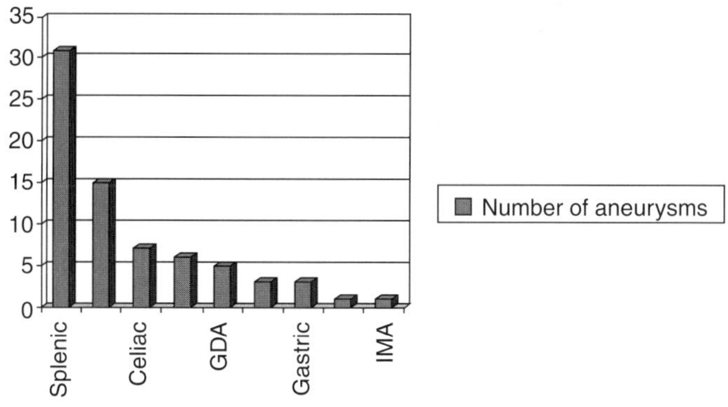

Figure 12.1 *Distribution of visceral artery aneurysms. These data represent combined retrospective series from Northwestern University, Chicago and University of Wisconsin, Madison. GDA, gastroduodenal artery; IMA, inferior mesenteric artery.*

and angiography, VAAs are being diagnosed with increasing frequency, often in the asymptomatic state. Despite advances in diagnostic technology, the majority of VAAs still present with rupture. Although the natural history and risk of rupture for most of these rare aneurysms are not well documented, the morbidity and mortality from rupture is significant. Nearly 22% of all reported VAAs present as clinical emergencies, including 8.5% that result in death.[9]

ETIOLOGY AND ASSOCIATED CONDITIONS

The precise cause of many VAAs is not known, but an etiological risk factor is often identified. There are also several conditions known to be associated with visceral aneurysms. Aneurysms of the visceral arteries may occur as pseudoaneurysms or as true aneurysms. Many present as pseudoaneurysms caused by inflammation or trauma to the visceral vessel. For example, vessel inflammation caused by pancreatitis and pseudocyst formation may result in a pseudoaneurysm of the splenic artery. Similarly, aneurysms of the pancreatoduodenal, pancreatic, gastroduodenal, gastric, and gastroepiploic arteries are often pseudoaneurysms and may be associated with pancreatitis. Hepatic artery pseudoaneurysms may occur after blunt or penetrating liver trauma or after iatrogenic injury. Traumatic hepatic artery aneurysms are frequently intrahepatic in location and may be difficult to access surgically. The increasing use of percutaneous diagnostic and therapeutic biliary tract procedures has been implicated in the development of an increasing number of intrahepatic artery injuries and pseudoaneurysms. Similarly, iatrogenic gastroduodenal artery aneurysms have been seen in patients undergoing insertion of infusion pumps to administer chemotherapy for metastatic disease.

Several types of visceral aneurysms have been associated with hormonal changes and high flow states (Figure 12.2). Splenic artery aneurysms are more common in females and have been associated with multiparity.[10] Forty-five per cent of female patients with splenic artery aneurysms have had six or more pregnancies, 88% have had two or more pregnancies, and 98% have had at least one pregnancy.[2] In a report by Trastek *et al.*, women with splenic artery aneurysms who were pregnant at the time of diagnosis had an average of 4.5 previous pregnancies.[11] In addition to the increased blood flow to the spleen during pregnancy, it is possible that estrogen or the hormone relaxin (which is secreted during the last trimester of pregnancy) may have some effect on the elastic support of the arteries. The

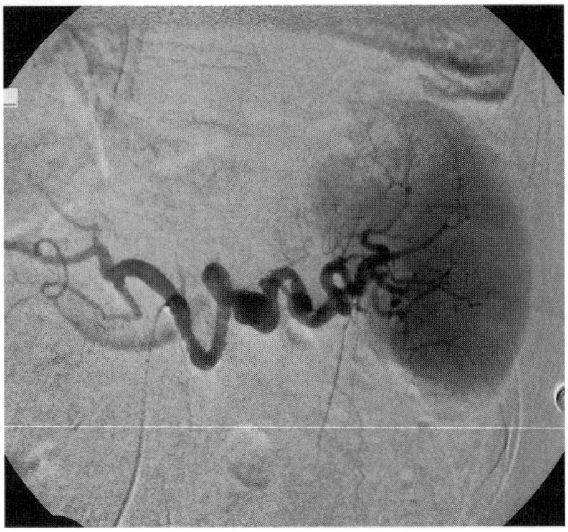

Figure 12.2 *Contrast arteriogram demonstrating a splenic artery aneurysm in a 52-year-old multiparous female patient.*

increased blood volume along with increased blood pressure and portal congestion may increase mechanical stress on the vessel.[12] The association of splenic artery aneurysms with portal hypertension and splenomegaly suggests that a hyperkinetic state in the spleen may result in the development of an aneurysm.[13, 14] These same hemodynamic factors may also account for the increasing recognition of splenic artery aneurysms in patients following orthotopic liver transplantation.[15, 16] The high blood flow state present in collateral vessels supplying blood to arterial beds in which the major arterial supply is stenotic or occluded may also predispose to the development of VAA in these vessels. In patients with SMA or celiac artery occlusive disease, the pancreatoduodenal and gastroduodenal arteries may become aneurysmal due to the increased blood flow in these important collateral vessels (Figure 12.3).[17]

Infection is an important etiological factor for the development of VAA. In the past, mycotic causes were the most common factor in the development of visceral aneurysms.[16] Mycotic aneurysms are often multiple and are found in the SMA, mesenteric branch vessels, and inferior mesenteric artery. Many of these mycotic aneurysms are the result of subacute bacterial endocarditis caused by non-hemolytic *Streptococcus.* In previous decades, infectious causes were thought to be involved in up to 60% of SMA aneurysms. Over the past decade, this proportion has decreased to approximately 30%.[18] Salmonella is another organism that is associated with aneurysm formation.[19] Prior to 1950, syphilis was an important cause of visceral aneurysms, especially in the celiac artery.[20]. Medial fibrodysplasia is thought to be an increasingly more common factor in the development of VAA. For example, aneurysms of the splenic artery occur six times more often in patients with medial fibrodysplasia than in a normal population.[21] Aneurysms of the renal and intracranial vessels may also be present in these patients (Figure 12.4). Medial degeneration is also an important cause of superior mesenteric and celiac aneurysms. Common pathological findings include disease in the media characterized by fragmentation of elastic fibers and loss of smooth muscle. Medial disease, with resulting vessel wall weakness, can be seen in fibromuscular dysplasia, Marfan's syndrome, and Ehlers–Danlos syndrome.

Patients with inflammatory vasculitis are also at risk for the development of VAA. Vasculitis develops by deposition of immune complexes in the vessel wall, leading to complement activation and infiltration by leukocytes. The media is subsequently weakened by the release

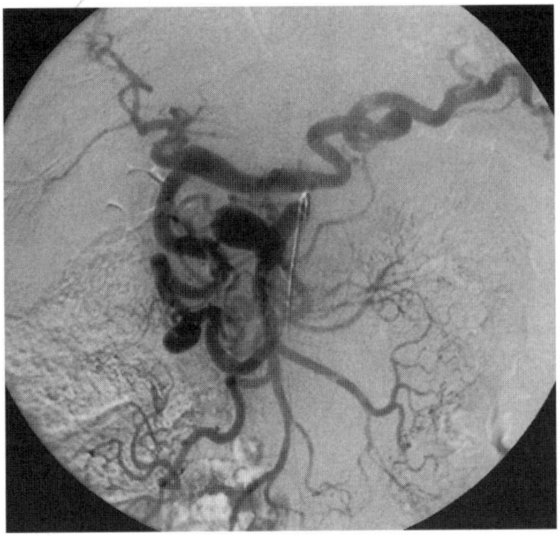

Figure 12.3 *Contrast arteriogram demonstrating an aneurysm of the pancreatoduodenal artery in a male patient with occlusion of the celiac artery and severe stenosis of the superior mesenteric artery. The aneurysm developed in an important collateral vessel.*

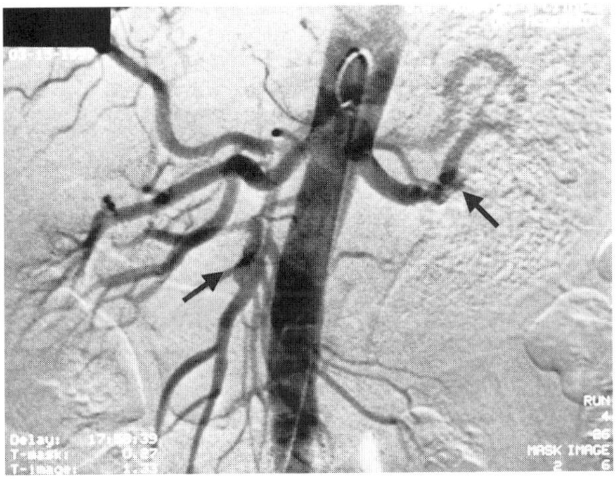

Figure 12.4 *Contrast arteriogram demonstrating aneurysms of both the superior mesenteric artery and the left renal artery in a young female patient with fibromuscular dysplasia.*

of proteases from the inflammatory cells. Polyarteritis nodosa is an inflammatory disease often found associated with aneurysm formation in the visceral arteries. These aneurysms are multiple in up to 90% of cases.[21] Systemic lupus erythematosus is another inflammatory condition that may be associated with cerebral or visceral aneurysms. Behçet's disease and Takayasu's disease are also associated with VAA formation. Rarely, patients with Kawasaki's disease may develop VAA.

As our population continues to increase in age, there has been an increase in the number of VAAs thought to be secondary to atherosclerotic disease. Common sites for atherosclerotic aneurysms include the splenic, common hepatic, celiac, and superior mesenteric arteries. Aortic dissection may propagate into the aortic branch vessels, leading to vessel wall weakness and the eventual development of visceral aneurysms. This can be seen in the common hepatic, celiac, and superior mesenteric arteries.

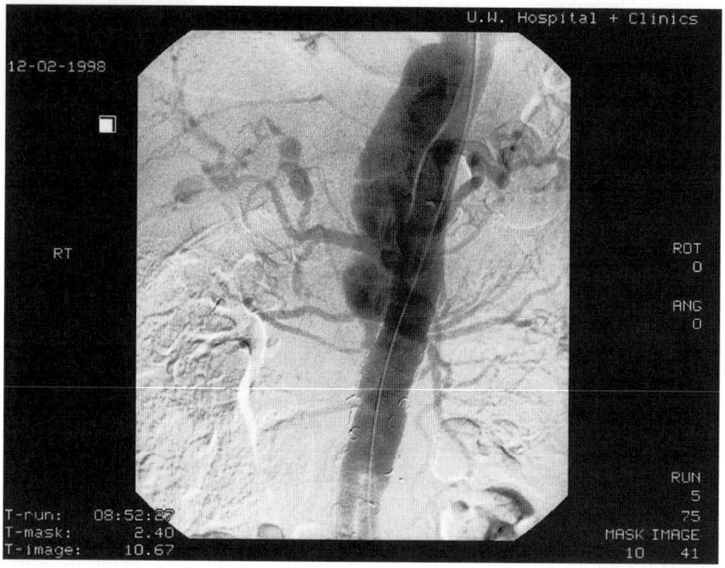

Figure 12.5 *Multiple hepatic artery aneurysms were discovered incidentally during evaluation of a thoracoabdominal aortic aneurysm.*

Fifteen to twenty per cent of patients with VAA have other associated visceral aneurysms. Multiple aneurysms may occur in the splenic arteries (18–30% are multiple) and hepatic arteries (10% are multiple), as well as in the mesenteric branch vessels. In one series, 31% of the patients with VAA were found to have other associated aneurysms involving the thoracic aorta, abdominal aorta, renal arteries, iliac arteries, and lower-extremity arteries (Figure 12.5).[22]

DIAGNOSIS

In the past, most VAAs were discovered at autopsy, with rupture often being the cause of death. Currently, the increased use of CT and angiography has led to the diagnosis of many VAAs in asymptomatic patients. Despite advances in diagnostic modalities and the wide availability of CT, a large proportion of VAAs still present with rupture.

Signs and symptoms of VAA vary depending on the artery that is involved. Most patients who are diagnosed with splenic artery aneurysms are asymptomatic. Symptomatic splenic aneurysms usually cause epigastric or left upper quadrant pain. Rupture of a splenic artery aneurysm may initially be confined to the lesser sac, known as the double rupture phenomenon. Eventually, the rupture releases from the lesser sac through the foramen of Winslow or freely into the peritoneal cavity, resulting in hypovolemic shock.[12] As with splenic artery aneurysms, the majority of hepatic artery aneurysms are asymptomatic at the time of diagnosis. About half of the symptomatic patients will present with abdominal pain and half with gastrointestinal hemorrhage or hemobilia.[23] Jaundice may occur in up to 10% of symptomatic patients; it is caused by extrinsic compression of the bile duct by the hepatic artery aneurysm. The classic triad of pain, hemobilia, and obstructive jaundice is seen in less than 30% of patients.

The majority (90%) of patients with SMA aneurysms are symptomatic at the time of diagnosis.[16] Patients usually complain of abdominal pain, which may mimic the postprandial symptoms of chronic mesenteric occlusive disease (Plate 15). Other symptoms include nausea, vomiting, gastrointestinal hemorrhage, hemobilia, and jaundice. Up to 50% of patients will be found to have a palpable mass on abdominal examination. Similar to SMA aneurysms, most patients (approximately 75%) with celiac artery aneurysms present with symptoms. Common symptoms include abdominal pain, gastrointestinal hemorrhage, jaundice, hemoptysis, and a palpable mass.

Aneurysms of the pancreaticoduodenal and gastroduodenal arteries are often associated with pancreatitis. Abdominal pain is present in 80% of patients and may be related to pancreatitis. Many patients (approximately 50%) will present with gastrointestinal hemorrhage, as the aneurysm erodes into an adjacent portion of the gastrointestinal tract.[23]

In addition to the clinical signs and symptoms, there are several diagnostic tests that can be used to evaluate VAAs. Many VAAs contain calcification that will appear on plain radiography of the abdomen. The aneurysm may appear as a curvilinear or signet-ring-shaped calcification on plain film. Many visceral aneurysms will produce a characteristic curvilinear eggshell pattern in the epigastrium right or left upper quadrant.[24]

Ultrasound or CT can size the aneurysm accurately, determine whether the aneurysm is thrombosed or active, and help to determine the visceral artery of origin. On ultrasound, the aneurysm may appear as an isolated hypoechogenic nodule or an anechoic cystic mass (Figure 12.6). CT has become a useful diagnostic modality. Not only are intact aneurysms demonstrated well, but ruptured aneurysms will also produce characteristic findings. On dynamic contrast CT, a VAA may appear as a brightly enhancing lesion with a variable amount of luminal thrombus (Figure 12.7). Three-dimensional spiral CT angiography has recently been used to diagnose VAA and define the arterial anatomy. Advances in CT angiography may eliminate the need for conventional contrast arteriography as a diagnostic tool in many cases.[25]

Magnetic resonance imaging (MRI) may also identify and characterize VAAs. The presence of a central flow void and thrombus are characteristic features. Magnetic resonance arteriography (MRA), using two-dimensional time of flight, or gadolinium-enhanced study, may provide some anatomic detail about the visceral circulation (Figure 12.8). As MRA technology continues to evolve, it may eventually replace many contrast arteriograms.

Currently, arteriography is the diagnostic modality of choice for VAAs and is recommended for all elective cases. Arteriography with selective visceral injection can determine the location and extent of the VAA, detect the presence of associated visceral aneurysms, visualize

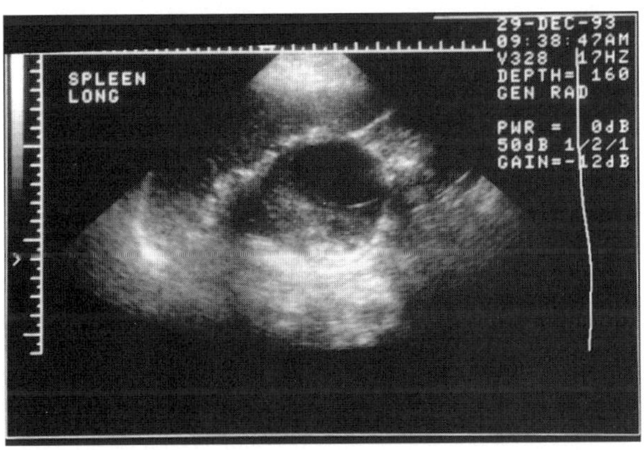

Figure 12.6 *Ultrasound of the left upper quadrant in a young female patient accurately demonstrates a large aneurysm of the splenic artery.*

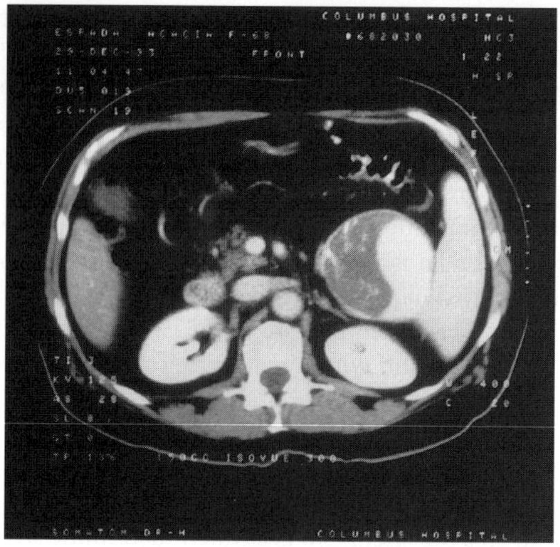

Figure 12.7 *Computed tomography scan of the abdomen with intravenous contrast demonstrates a large aneurysm of the splenic artery with intraluminal thrombus.*

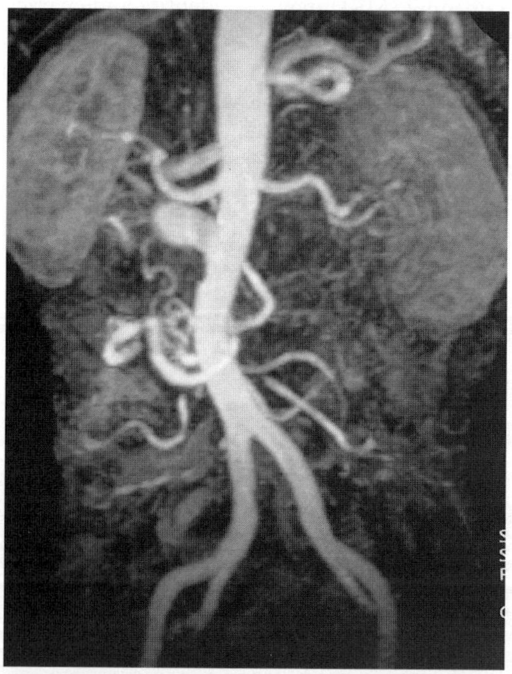

Figure 12.8 *Magnetic resonance arteriogram of the aorta and visceral vessels can help to visualize this pancreatoduodenal artery aneurysm. However, the origin of the aneurysm and the pattern of collateral flow could not be determined from this study.*

collaterals, and define the distal vasculature. Arteriography is often necessary to plan therapy, and it may also allow for therapeutic embolization.[26]

NATURAL HISTORY AND TREATMENT

Although the various types of VAAs have many features in common, the natural history and treatment options for each type of visceral aneurysm differs somewhat. The decision to treat an aneurysm depends greatly on what is known of the natural history of the disease process and the risks associated with intervention. Traditionally, surgery has been the major means by which visceral aneurysms were treated. Greater experience with the techniques of transcatheter embolization has made this an important treatment option for selected visceral aneurysms.

Splenic artery aneurysms

Splenic artery aneurysms have traditionally been thought to be the most common of the VAAs. They occur more commonly in women (66%) than men (34%). Most (72%) are true aneurysms. Approximately 30% of splenic aneurysms are multiple. These are usually small and saccular. About 44% are located in the mid-portion of the artery, and approximately 35% are located more distally towards the spleen.[23]

The reported risk of rupture of splenic artery aneurysms is thought to be low in cases not associated with pregnancy. Contemporary literature supports a rupture rate of only 2%. However, the mortality rate associated with rupture is high at approximately 36%.[23] Splenic artery aneurysms in women of childbearing age can be a serious and potentially life-threatening condition. Approximately 95% of splenic aneurysms diagnosed during pregnancy present with rupture, often during the last trimester. Rupture of a splenic artery aneurysm during pregnancy has been associated with a maternal mortality rate of 70% and a fetal death rate of 95%. Therefore, treatment of asymptomatic splenic artery aneurysms is recommended in pregnant women and women of childbearing age who might become pregnant.[10] Treatment is also indicated in patients with symptomatic or enlarging aneurysms. Although aneurysm size greater than 2.5 cm has been considered a relative indication for treatment, there is some evidence that small, asymptomatic splenic aneurysms in elderly patients may be observed safely.[26]

Surgical therapy for splenic aneurysms usually consists of splenectomy and removal of that portion of the splenic artery containing the aneurysm. Exposure is usually obtained through the lesser sac. Proximal splenic artery aneurysms may be treated with proximal and distal ligation with or without aneurysmectomy, thus preserving the spleen. Because of excellent collateral flow in most patients, it is usually not necessary to revascularize the spleen. The extensive blood supply from the stomach through the short gastric arteries usually prevents infarction in most cases. However, ligation of distally located splenic artery aneurysms without revascularization may result in splenic infarction. When extensive dissection of the pancreas or adjacent structures is required, aneurysm exclusion and ligation of contributing vessels is appropriate. Some pseudoaneurysms associated with pancreatitis may be treated safely by arterial ligation from within the aneurysm, especially if the lesion is located in the mid-portion of the splenic artery or it is embedded within the pancreas. Pseudoaneurysms of the distal splenic artery embedded in the pancreas may require distal pancreatectomy along with aneurysm resection.[21, 27]

Transcatheter embolization of splenic artery aneurysms has been used in the past in patients who are poor surgical risks. Currently, at least 12% of splenic artery aneurysms are treated in this way.[23] Treatment may consist of occlusion of the splenic artery itself or direct embolization of the aneurysm sac. The process involves selective catheterization of the splenic artery or the aneurysm sac, followed by the injection of Gelfoam particles or steel coils. Multiple coils placed in the neck of the aneurysm cause thrombosis of the sac and the more proximal splenic artery. In some cases, it is possible to fill the aneurysm with coils and maintain flow in the main splenic artery, thus preserving the spleen. Complications of this modality of treatment include embolization or ischemia to the spleen, resulting in areas of splenic infarction that are prone to abscess formation.[28]

With the recent advances in laparoscopic surgical techniques, and the improvement in optics and instrumentation, laparoscopic surgery is another form of treatment that has been applied to the treatment of splenic artery aneurysms. Patient position, trocar sites, and operative techniques vary somewhat between different published reports. Recently, Arca and co-workers described a 45-degree right semi-decubitus position and the use of a 10-mm 30-degree laparoscope.[29] The gastrocolic ligament is opened with the harmonic endoshears to provide access to the lesser sac for exposure of the splenic artery. Distal splenic artery aneurysms were removed along with the spleen. Their average intraoperative time was 150 min, and the average length of hospital stay was only 2.2 days. The laparoscopic approach will likely prove to be a good minimally invasive treatment option for splenic aneurysms not amenable to transcatheter embolization. Patients with aneurysm rupture and hemodynamic instability should not be treated laparoscopically.

Because of the high risk of splenic artery aneurysm rupture during pregnancy, especially in the third trimester, observation of these lesions in the pregnant patient is not recommended. Elective repair, usually ligation, during the second trimester is an option. Another option is to perform elective cesarean section with operative treatment of the splenic artery aneurysm at the same time.[10, 30, 31]

Hepatic artery aneurysms

The increasing use of percutaneous diagnostic and therapeutic biliary procedures and the routine use of CT to evaluate blunt liver trauma has led to the diagnosis of an increasing number of hepatic artery pseudoaneurysms. Approximately 50% of hepatic artery aneurysms are pseudoaneurysms, usually located intrahepatically.[23] The remainder are extrahepatic in location and are the result of many causes, including atherosclerotic disease. In up to 80% of patients with hepatic artery aneurysm, rupture is the reason for consultation. Patients present with pain or hemorrhage, often into the gastrointestinal tract. Rupture occurs with equal frequency into the peritoneal cavity or gastrointestinal tract. Because of the tendency for hepatic artery aneurysms to present with rupture and hemorrhage, most should be treated aggressively rather than be observed.[27]

Treatment for hepatic artery aneurysm depends upon its location. Aneurysms in the proximal hepatic artery, proximal to the gastroduodenal artery, can be treated with excision or exclusion without reconstruction (Plate 16). Usually, there is adequate collateral flow from the SMA through the gastroduodenal artery. Aneurysms involving the gastroduodenal artery, or cases in which collateral circulation is not sufficient, require arterial reconstruction. This may take the form of aortohepatic artery bypass, splenohepatic artery bypass, or interposition grafting using autogenous vein or prosthetic conduit. Large aneurysms and those associated with significant inflammation may be treated by opening the aneurysm and controlling feeding vessels from within the sac. Patients presenting with aneurysm rupture and hypotension are more vulnerable to hepatic ischemia following ligation, and may be more likely to require vascular reconstruction.[10, 21] For intrahepatic aneurysms, surgical exposure is much more

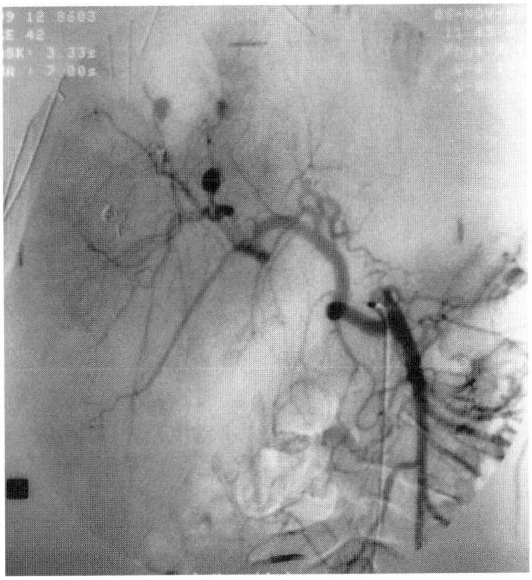

Figure 12.9 *Contrast arteriogram demonstrating multiple intrahepatic artery aneurysms in a young patient who previously suffered blunt liver trauma in an automobile accident.*

difficult. Surgical treatment may require hepatic resection of the area involving the aneurysm or ligation of the involved artery before it enters the liver parenchyma. Many of these intrahepatic pseudoaneurysms are small and multiple (Figure 12.9). Since the 1980s, transcatheter embolization of intrahepatic aneurysms has become a popular method of therapy and is thought to be the treatment of choice for many hepatic artery aneurysms.[5] Currently, at least 37% of hepatic artery aneurysms are treated by transcatheter embolization.[23] Transarterial access to the distal hepatic artery branches followed by coil embolization results in thrombosis of the aneurysm sac. It is preferable to thrombose the vessel feeding the aneurysm as distally as possible, as embolization of a more proximal vessel may result in incomplete thrombosis and recurrence. Often, a staged approach is required.[26] Percutaneous transhepatic embolization may be combined with transarterial ablation to achieve complete thrombosis (Figure 12.10).[31]

Superior mesenteric artery aneurysms

As the majority (90%) of patients with SMA aneurysms are symptomatic at presentation, intervention is indicated in most patients. Propagation of mural thrombus within the aneurysm or dissection resulting from acute expansion can cause occlusion of important collateral vessels, resulting in intestinal infarction. Rupture of the SMA aneurysm occurs in up to 50% of patients and is followed by acute, massive retroperitoneal, intra-abdominal, or gastrointestinal bleeding. Often, rupture of an SMA aneurysm results in shock and sudden death. The mortality rate following rupture of an SMA aneurysm is 30%. Because of the tendency for these lesions to produce life-threatening bowel ischemia, intervention is recommended for almost all SMA aneurysms.[21]

The usual treatment for SMA aneurysms involves surgical exposure through a midline or retroperitoneal approach. Transmesenteric exposure can be obtained by dissection at the base of the transverse colon mesentery. The SMA can also be exposed by a retroperitoneal approach using medial visceral rotation. Ligation of the aneurysm and all of its feeding vessels can be done if there is adequate collateral flow from the celiac axis or inferior mesenteric

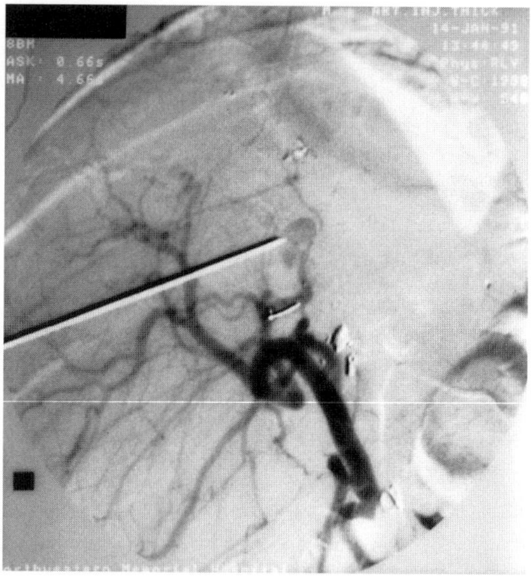

Figure 12.10 *This hepatic artery aneurysm was treated in a staged fashion with transcatheter embolization using a transarterial and percutaneous route. Note the needle approaching the aneurysm for direct puncture.*

artery. Most often, aneurysmectomy with vascular reconstruction is necessary to prevent bowel ischemia. For aneurysms of the proximal SMA, the proximal anastomosis is to the aorta, using operative techniques similar to those used for mesenteric occlusive disease. Reimplantation of the SMA into the aorta is another option. If the aneurysm is located more distally in the SMA, and the proximal SMA is normal, excision of the aneurysm followed by interposition grafting can be done (Plates 17 and 18). The authors' preference is to use autogenous saphenous vein for conduit. Prosthetic material or autogenous artery (hypogastric artery) can also be used.[22, 27]

Transcatheter embolization of SMA aneurysms is a treatment option for patients who are not good surgical candidates (i.e. hostile abdomen) or for aneurysms that are difficult to access surgically. Wires and coils have been used by several authors. Large aneurysms may be treated by filling the aneurysm itself, or by occluding only the neck of the lesion, thus maintaining patency of the main SMA. Submucosal aneurysms and small aneurysms that are difficult to access surgically may also be treated in this manner. In addition to coils, thrombin injection has been used to treat branch aneurysms of the SMA.[32] However, because of the risks of intestinal infarction, transcatheter embolization is not used often for the treatment of SMA aneurysms.

Celiac artery aneurysms

Celiac artery aneurysms are often associated with other aneurysms, such as abdominal aortic aneurysms (20% of patients) and other visceral artery aneurysms (40% of patients).[33] The natural history of celiac artery aneurysms appears to be one of expansion and rupture. Approximately 75% of patients are symptomatic at the time of diagnosis. The risk of rupture has been cited at about 13%. The mortality rate with rupture is very high – up to 100% in one series.[21] As with SMA aneurysms, thrombosis or rupture of celiac artery aneurysms can lead to serious end-organ ischemia, including intestinal infarction. Also similar to SMA aneurysms, treatment is recommended for almost all celiac artery aneurysms except in the very-high-risk patient.

Celiac artery aneurysms are usually treated operatively with approach through the lesser sac or using medial visceral rotation. Resection of the aneurysm with vascular reconstruction is used most often. Aneurysm resection is followed by aortoceliac bypass using autogenous or prosthetic conduit, usually originating from the supraceliac aorta.

Gastroduodenal, pancreatoduodenal, and pancreatic artery aneurysms

Gastroduodenal artery (GDA) aneurysms are quite rare, accounting for only 1.5% of all VAAs. Pancreatic and pancreatoduodenal artery (PDA) aneurysms together account for only 2% of all VAAs.[9] As mentioned previously, these aneurysms are often associated with pancreatitis. There have been a few reported cases of GDA and PDA aneurysms in patients with mesenteric occlusive disease, thought to be a consequence of the increased flow in the collateral vessels.[17,34] Up to 68% of GDA, PDA, and pancreatic artery aneurysms present with rupture. Rupture may occur into the retroperitoneum, the peritoneal cavity, an adjacent viscus, the portal vein, or the biliary tract. The mortality rate with rupture has been reported to be as high as 50%.[21]

Surgical treatment of GDA, PDA, and pancreatic artery aneurysms may involve concomitant treatment of the associated pancreatitis or pseudocyst. Because of the multiple communicating vessels, simple ligation of these aneurysms is often difficult. Opening of the aneurysm and suture ligation of all communicating vessels from with the sac ensures that all of the feeding vessels are identified and helps to prevent recurrence of the aneurysm after ligation. Aneurysms of the PDA or pancreatic arteries embedded in the middle or distal regions of the pancreas may be treated with distal pancreatectomy. Because of the serious associated medical conditions that these patients often have, the mortality rates associated with surgical treatment are high (up to 13%).[18]

The availability of wires and catheters for superselective catheterization of visceral vessels has permitted an increasing number of GDA and PDA aneurysms to be treated with transcatheter embolization. Because of the multiple feeding vessels, incomplete thrombosis is not uncommon and often a staged approach is required. As with hepatic artery aneurysms, the staged approach may require percutaneous puncture of an aneurysm in combination with transarterial embolization. It has been reported that approximately 15% of these aneurysms are treated with transcatheter embolization.[18] As experience increases, it is anticipated that this will be a more commonly used mode of therapy. In the rare patient who has developed a GDA or PDA aneurysm due to increased collateral flow associated with mesenteric occlusive disease, transcatheter embolization should be used with caution. Thrombosis of an important collateral vessel may lead to acute intestinal infarction. The authors prefer a surgical approach in these patients that involves mesenteric revascularization in addition to aneurysm ligation (Figure 12.11).

Gastric and gastroepiploic artery aneurysms

Gastric and gastroepiploic artery (GEA) aneurysms are rare, comprising only up to 4% of all VAAs. Ninety per cent of gastric and GEA aneurysms present emergently with rupture. Rupture is often accompanied by exsanguinating hemorrhage, with a 70% mortality rate. Most (70%) of these patients have gastrointestinal hemorrhage, while 30% present with intraperitoneal hemorrhage.[9,24]

Gastric and GEA aneurysms that occur in an extragastric location may be treated with ligation or excision of the aneurysm. Intramural lesions may be resected along with the involved portion of the stomach. As with many of the other VAAs, transcatheter embolization is

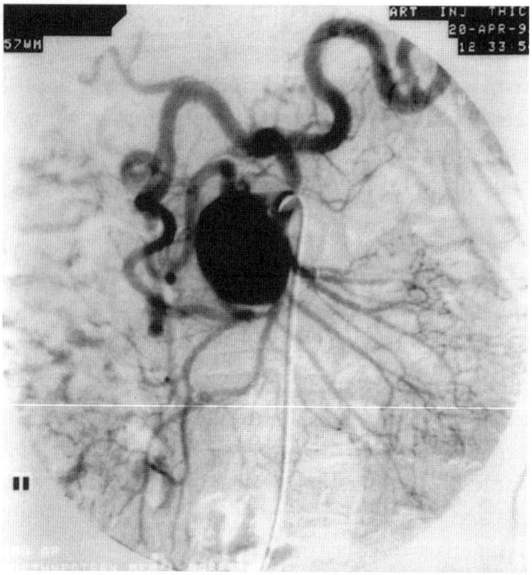

Figure 12.11 *A large aneurysm of the pancreatoduodenal artery was found in this patient presenting with symptoms of chronic mesenteric ischemia. The patient also had severe arterial occlusive disease of the celiac and superior mesenteric arteries with the development of the pancreatoduodenal artery aneurysm in an enlarged collateral vessel. He was treated with aortoceliac and aorta to superior mesenteric artery bypass using autogenous saphenous vein followed by ligation of the aneurysm.*

becoming a more popular method of treatment and may become the treatment of choice in the future.

Mesenteric branch artery aneurysms

Small aneurysms may occur in the jejunal, ileal, colic, and, most rarely, inferior mesenteric arteries. Jejunal aneurysms are most common, followed by aneurysms of the middle colic and ileal arteries. Seventy per cent of these aneurysms are symptomatic at the time of diagnosis, most (87%) with abdominal pain. Fifty-two per cent of patients present initially with hemorrhage and shock. Ruptured aneurysms often result in a large hematoma in the mesentery and can be complicated by venous stasis and bowel ischemia. The risk of rupture is thought to be low and is not predicted accurately by the size of the aneurysm. The mortality rate with rupture is reported to be approximately 20%.[9, 18]

Aneurysms of the mesenteric branch vessels can be treated with ligation or resection of the aneurysm. Distally located aneurysms may be excised along with a short segment of bowel. Because of the significant risk for intestinal infarction of bowel segments, transcatheter embolization has a limited role for the treatment of most of these aneurysms.

CONCLUSIONS

Aneurysms of the visceral arteries are rare but important vascular lesions. Their importance lies in their potential for rupture with life-threatening hemorrhage. Although VAA are being discovered more often incidentally through CT or ultrasound of the abdomen, a large number of these aneurysms present emergently with rupture. Common etiologies include atherosclerotic disease, trauma, medial degenerative disease, and arteritis. Infection, the most common cause of VAA in the past, is now seen less often. Associated VAAs and aneurysms in other locations are common.

Signs and symptoms of VAA vary depending on the location of the aneurysm. Thin patients may present with a palpable abdominal mass. Other patients may present with abdominal pain, nausea, vomiting, gastrointestinal bleeding, or hemobilia. Plain film of the abdomen and ultrasound may detect the aneurysm. CT is useful for assessing the size of the aneurysm and evaluating other intra-abdominal pathology. Contrast arteriography remains the diagnostic method of choice and is recommended for all elective cases. It is important to define the extent and location of the aneurysm, assess collateral flow, and look for multiple visceral aneurysms. MRA is an evolving technique that is likely to become more useful in the future.

Treatment options for VAA depend on the location of the aneurysm and the presentation of the patient. Patients presenting with rupture and hemodynamic instability from massive hemorrhage require emergent treatment, usually in the form of surgical ligation of the bleeding vessel. As more VAAs are being discovered in the asymptomatic state, elective treatment has allowed for more innovative and less invasive techniques. Excision of the VAA may be done with arterial reconstruction of the main vessel. Laparoscopic ligation is another option for elective treatment. As technology and experience evolve, many VAAs are being treated with transcatheter embolization. Difficult aneurysms may require a staged approach and sometimes combined percutaneous and transarterial embolization techniques.

An increased awareness of the various types of VAA, associated conditions, and diagnostic methods will lead to a larger number of VAAs being correctly diagnosed. Knowledge of the various treatment options will allow for more effective elective repair, thus avoiding the complications associated with aneurysm rupture.

REFERENCES

1 Bequssier M. Sur un aneurisme de l'artere splinque dont les parios se sont ossifiees. *Journal of Medicine Toulose* 1770; **32**: 157–157.
2 McGinnis HD, DeLuca SA. Splenic artery aneurysms. *Am Fam Physician* 1993; **47**: 1199–202.
3 Stevenson W. A case of abdominal aneurysm treated by laparotomy and the introduction of wire into sac; death. *Lancet* 1895; **1**: 22–3.
4 Lumsden A, Mattar S, Allen RC, Bacha E. Hepatic artery aneurysms: the management of 22 patients. *J Surg Res* 1996; **60**: 345–50.
5 Psathakis D, Muller G, Noah M, Diebold J, Bruch HP. Present management of hepatic artery aneurysms. Symptomatic left hepatic artery aneurysm; right hepatic artery aneurysm with erosion into the gallbladder and simultaneous colocholecystic fistula – a report of two unusual cases and the current state of etiology, diagnosis, histology and treatment. *Vasa* 1992; **21**: 210–15.
6 Lie M, Ertresvag K, Skjennald A. Rupture of a splenic artery aneurysm into the pancreatic duct. Case report. *Acta Chir Scand* 1990; **156**: 411–13.
7 DeBakey M, Cooley D. Successful resection of mylotic aneurysms of SMA. *Am Surg* 1953; **19**: 202–12.
8 Rokke O, Sondenaa K, Amundsen S, Bjerke-Larssen T, Jensen D. The diagnosis and management of splanchnic artery aneurysms. *Scand J Gastroenterol* 1996; **31**: 737–43.
9 Stanley JC, Wakefield TW, Graham LM, Whitehouse WM, Jr, Zelenock GB, Lindenauer SM. Clinical importance and management of splanchnic artery aneurysms. *J Vasc Surg* 1986; **3**: 836–40.
10 Holdsworth RJ, Gunn A. Ruptured splenic artery aneurysm in pregnancy. A review. *Br J Obstet Gynaecol* 1992; **99**: 595–7.

11 Trastek VF, Pairolero PC, Joyce JW, Hollier LH, Bernatz PE. Splenic artery aneurysms. *Surgery* 1982; **91**: 694–9.
12 Hallett JW, Jr. Splenic artery aneurysms. *Semin Vasc Surg* 1995; **8**: 321–6.
13 Ohta M, Hashizume M, Tanoue K, Kitano S, Sugimachi K, Yasumori K. Splenic hyperkinetic state and splenic artery aneurysm in portal hypertension. *Hepatogastroenterology* 1992; **39**: 529–32.
14 Lee PC, Rhee RY, Gordon RY, Fung JJ, Webster MW. Management of splenic artery aneurysms: the significance of portal and essential hypertension. *J Am Coll Surg* 1999; **189**: 483–90.
15 Kobori L, van der Kolk MJ, de Jong KP, Peeters PM, Klompmaker IJ, Kok T, *et al*. Splenic artery aneurysms in liver transplant patients. Liver Transplant Group. *J Hepatol* 1997; **27**: 890–3.
16 Robertson AJ, Rela M, Karani J, Heaton ND. Splenic artery aneurysm and orthotopic liver transplantation. *Transpl Int* 1999; **12**: 68–70.
17 Porcellini M, Iaccarino V, Bernardo B, Migliucci N, Capasso R. Visceral artery aneurysms secondary to collateral circulation. *Minerva Cardioangiol* 1996; **44**: 443–5.
18 Shanley CJ, Shah NL, Messina LM. Uncommon splanchnic artery aneurysms: pancreaticoduodenal, gastroduodenal, superior mesenteric, inferior mesenteric, and colic. *Ann Vasc Surg* 1996; **10**: 506–15.
19 Hammarsten O, Risberg B. Multiple aneurysms in a young patient – aetiological considerations. *Eur J Surg* 1996; **162**: 71–5.
20 Graham LM, Stanley JC, Whitehouse WM, Jr, Zelenock GB, Wakefield TW, Cronenwett JL, *et al*. Celiac artery aneurysms: historic (1745–1949) versus contemporary (1950–1984) differences in etiology and clinical importance. *J Vasc Surg* 1985; **2**: 757–64.
21 Messina LM, Shanley CJ. Visceral artery aneurysms. *Surg Clin North Am* 1997; **77**: 425–42.
22 Carr SC, Mahvi D, Hoch J, Acher C, Turnipseed W. Mortality of visceral aneurysm rupture. *J Vasc Surg*. In press.
23 Shanley CJ, Shah NL, Messina LM. Common splanchnic artery aneurysms: splenic, hepatic, and celiac. *Ann Vasc Surg* 1996; **10**: 315–22.
24 Jorgensen BA. Visceral artery aneurysms. A review. *Dan Med Bull* 1985; **32**: 237–42.
25 Howling SJ, Gordon H, McArthur T, Hatfield A, Lees WR. Hepatic artery aneurysms: evaluation using three-dimensional spiral CT angiography. *Clin Radiol* 1997; **52**: 227–30.
26 Carr SC, Pearce WH, Vogelzang RL, McCarthy WJ, Nemcek AA, Jr, *et al*. Current management of visceral artery aneurysms. *Surgery* 1996; **120**: 627–33.
27 Carr SC, Pearce WH. Management of visceral artery aneurysm. In: Yao JS, Pearce WH (eds). *Practical vascular surgery*. Stamford, CT: Appleton & Lange, 1999; 241–58.
28 Carr SC, Vogelzang R, Yao JS. Visceral artery aneurysms. In: Yao JS, Pearce WH (eds). *Arterial surgery: management of challenging problems*. Stamford, CT: Appleton & Lange, 1996; 417–31.
29 Arca MJ, Gagner M, Heniford BT, Sullivan TM, Beven EG. Splenic artery aneurysms: methods of laparoscopic repair. *J Vasc Surg* 1999; **30**: 184–8.
30 Herbeck M, Horbach T, Putzenlechner C, Klein P, Lang W. Ruptured splenic artery aneurysm during pregnancy: a rare case with both maternal and fetal survival. *Am J Obstet Gynecol* 1999; **181**: 763–4.
31 Rothbarth LJ, Redmond PL, Kumpe DA. Percutaneous transhepatic treatment of a large intrahepatic aneurysm. *Am J Roentgenol* 1989; **153**: 1077–8.
32 Kemmeter P, Bonnell B, VanderKolk W, Griggs T, VanErp J. Percutaneous thrombin injection of splanchnic artery aneurysms: two case reports. *J Vasc Interv Radiol* 2000; **11**: 469–472.
33 Graham LM, Rubin J. Visceral artery aneurysms. In: Strandness D, Breda A (eds). *Vascular diseases: surgical and interventional therapy*. New York: Churchill Livingstone, 1994; 811–21.
34 Imamura M, Shiiya N, Murashita T, Sasaki S, Matsui Y, Yasuda K. Pancreaticoduodenal artery aneurysm with celiac trunk occlusion – report of a case. *J Cardiovasc Surg* 1998; **39**: 565–7.

13

Mesenteric venous thrombosis

COREY J. JOST AND PETER GLOVICZKI

INTRODUCTION

Mesenteric venous thrombosis (MVT) is an uncommon cause of intestinal ischemia. In its advanced form, MVT results in bowel infarction and peritonitis, with high mortality. Thrombosis of the mesenteric venous system is difficult to diagnose because it occurs infrequently and has a varying clinical presentation. MVT presents with a spectrum of signs and symptoms ranging from mild or transient to severe and constant abdominal pain and septic shock.[1] In 1876, Fagge first reported thrombosis of the portomesenteric venous system resulting in intestinal infarction in a patient with multiple previous deep venous thromboses.[2] Elliot in 1895 was the first to successfully resect infarcted bowel secondary to MVT,[3] and Warren provided a detailed description and review of MVT in 1935.[4] Initial reports noted high mortality rates even when treated by bowel resection.

Although mortality is still high, progress has been made in the diagnosis and management of MVT. In contemporary series, MVT caused 5–15% of intestinal ischemia.[1,5] Earlier reports of acute mesenteric ischemia noted higher incidences of MVT, probably due to the inability to distinguish it from non-occlusive mesenteric ischemia.[6]

The management of MVT has evolved to allow us to diagnose this condition more accurately and offer treatment that decreases the need for bowel resection, with a decrease in morbidity, recurrence, and mortality.[7]

ETIOLOGY

Multiple events can result in MVT, all stimulating thrombogenesis through mechanisms identified by Virchow. Bussey defined MVT as primary or idiopathic and secondary if caused by a known pathological process.[8] Many patients previously classified as having primary MVT have been identified to have an underlying hypercoagulable state.[9] In the recent Mayo Clinic series of 72 patients, 15 (21%) patients had primary MVT with no identified hypercoagulability or secondary disorder.[1] In another series, a hypercoagulable state was present in 87% of patients with MVT.[10]

Reported underlying coagulation abnormalities include protein C deficiency,[11] protein S deficiency,[12] factor V Leiden mutation,[13] antiphospholipid antibodies (Lupus anticoagulant),[14] anticardiolipin antibody,[15] plasminogen activator deficiency,[16] hyperhomocysteinemia,[17] and anti-thrombin III deficiency.[18] Other spontaneous causes of MVT have been related to acquired pro-thrombotic states: pregnancy,[2] oral contraception,[19] nephrotic syndrome,[20] myeloproliferative disorder,[21] and polycythemia vera (Table 13.1).[22]

Secondary MVT can result from intra-abdominal pathology causing compression, stasis, or inflammation in the portomesenteric venous system (Table 13.2). Cirrhosis with subsequent portal hypertension can induce stasis, thus increasing the risk of MVT.[23] Another cause of MVT in cirrhotic patients has been treatment with sclerotherapy for esophageal or gastric varices. Concomitant vasopressin infusion has been reported to increase the incidence of complicating MVT.[24] Surgical intervention can also induce venous stasis and transient hypercoagulable states,[25] and even laparoscopic procedures can be complicated by MVT.[26] Certain surgical interventions are prone to MVT, specifically splenectomy.[27, 28] A long splenic vein remnant can create a cul de sac prone to thrombosis and nidus for thrombus propagation.

Inflammation has been hypothesized to cause mesenteric venous intimal injury. Reported acute inflammatory states causing MVT include acute pancreatitis,[29, 30] appendicitis,[31, 32] infected biliary prostheses,[33] and Mediterranean spotted fever.[34] Isolated reports have noted an association between MTV and human immunodeficiency virus (HIV) infection,[35] diabetes mellitus,[36] Klinefelter's syndrome,[37] and decompression sickness,[38] but the mechanism that caused the MVT was unknown.

Table 13.1 *Reported causes of 'spontaneous' mesenteric venous thrombosis*

Inherited hypercoagulable states
Factor V Leiden
Protein C deficiency
Protein S deficiency
Anti-hythrombin III deficiency
Plasminogen activator deficiency
Hyperhomocysteinemia
Prothrombin mutation
Aquired prothrombotic conditions
Anticardiolipin antibody
Antiphospholipid antibodies (Lupus anticoagulant)
Pregnancy
Oral contraceptives
Nephrotic syndrome
Myeloproliferative disorder
Polycythemia vera
Surgical stress
Maligancy
Miscellaneous
Klinefelter's syndrome
Decompression sickness
Diabetes mellitus
HIV infection

Table 13.2 *Etiologic factors in secondary mesenteric venous thrombosis*

Malignancy
Pancreatic cancer
Inflammation
Pancreatitis
Acute
Chronic
Biliary prosthesis – secondary infection
Mediterranean fever
Operative interventions
Splenectomy
Appendectomy
Pancreaticoduodenectomy
Laparoscopic colon resection
Venous stasis
Cirrhosis with portal hypertension
Spontaneous
Recent sclerotherapy
Increased by concomitant vasopressin

NATURAL HISTORY

The true incidence of this disease is not known, but a thrombus has been found in the mesenteric veins in 1.5% of individuals at autopsy.[39] The mean age of onset of acute MVT was 64 years (range 17–88 years) in a collected series of 112 patients published in the last decade.[1,10,40–42] The male-to-female ratio was 1.5 : 1 (69 : 47) with an increased risk of developing MVT in males.

The natural history of MVT is a spectrum of disease ranging from chronic occlusion with few symptoms to progressive acute thrombosis with bowel infarction. It is not known what percentage of patients will develop symptoms with mesenteric venous occlusion or progress to bowel infarction, or who will develop chronic or acute presentations. Symptoms and the degree of ischemia are related to the presence of collateral circulation and location of thrombosis.[43] A Mayo Clinic study from 1949 included pathological evaluation of infarcted bowel secondary to MVT. This study revealed that thrombosis of venous arcades and vasa recti were associated with infarction, and only 59% of the patients had large vein thrombus.[44] The pathophysiology of acute mesenteric venous occlusion was studied by Polk in 1964. In an experimental model of acute venous occlusion in dogs, he noted rapid fluid sequestration and capillary rupture resulting in submucosal and subserosal hemorrhage as well as serosanguineous ascites.[45] With progressive thrombosis and inadequate collateral drainage, infarction occurs in the ileum and jejunum in the majority of cases, with occasional involvement of the colon and duodenum (Figure 13.1).[1]

Acute and chronic MVT differ in presentation, prognosis, and management. Rhee *et al.* reviewed all cases of MVT at the Mayo Clinic and noted that 74% had acute MVT and 26% presented with chronic occlusions.[1] We will discuss these entities separately.

ACUTE MESENTERIC VENOUS THROMBOSIS

Clinical presentation

Acute MVT has the most variable presentation of all forms of mesenteric ischemia. Typically, symptoms are less acute than other forms of mesenteric ischemia. Insidious progression,

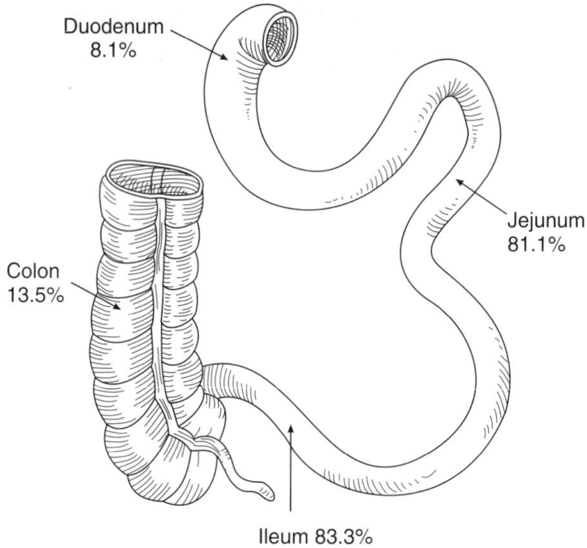

Figure 13.1 *Location of bowel infarction with mesenteric venous thrombosis from 29 patients. Reproduced with permission from Rhee RY, Gloviczki P, Mendonca CT, Petterson TM, Serry RD, Sarr MG,* et al. *Mesenteric venous thrombosis: still a lethal disease in the 1990s.* J Vasc Surg *1994;* ***20****: 688–97.*

Table 13.3 *Symptoms of mesenteric venous thrombosis (MVT)*

Symptom	Acute MVT	Chronic MVT	Total
Abdominal pain	83%	79%	82%
Anorexia	53%	37%	49%
Diarrhea	43%	37%	42%
Nausea and vomiting	42%	21%	36%
Upper GI bleeding	28%	37%	31%
Lower GI bleeding	23%	16%	21%
Constipation	13%	0%	7%

GI, Gastrointestinal. (Adapted with permission from Rhee RY, Gloviczki P, Mendonca CT, Petterson TM, Serry RD, Sarr MG, *et al.* Mesenteric venous thrombosis: still a lethal disease in the 1990s. *J Vasc Surg* 1994; **20**: 688–97.

however, is noted in most patients being diagnosed with MVT more then 48 h from onset of symptoms. Presenting symptoms include abdominal pain, anorexia, diarrhea, nausea, and vomiting in decreasing frequency (Table 13.3).[1] Signs of acute MVT are also somewhat variable, with abdominal distension being the most common (43%), followed by upper- or lower-gastrointestinal hemorrhage (28% and 23%, respectively).[1, 46] Another less commonly reported presentation is massive colonic distension.[47] Peritonitis is present in one-third of patients at initial presentation.[1]

Laboratory tests

Laboratory testing offers only supportive, non-diagnostic information. Leukocytosis, elevated serum lactate levels, and electrolyte abnormalities all suggest advanced bowel ischemia, but are

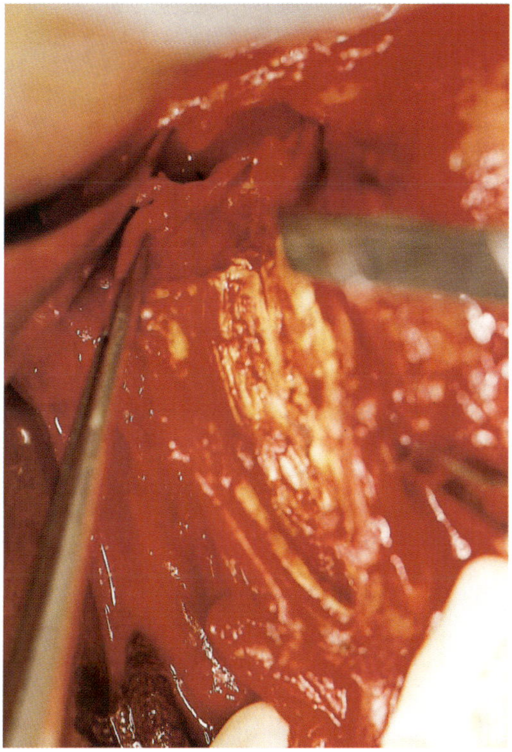

Plate 15 *Intraoperative photograph of a superior mesenteric artery aneurysm demonstrating intraluminal thrombus and debris. Distal embolism of this material over time can lead to symptoms of chronic or acute mesenteric ischemia.*

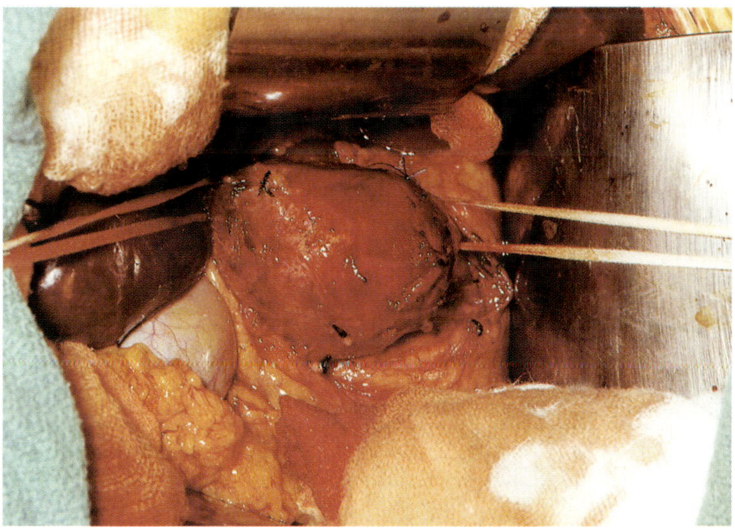

Plate 16 *Intraoperative photograph of a common hepatic artery aneurysm of the proximal aspect of the vessel. This lesion was treated with excision.*

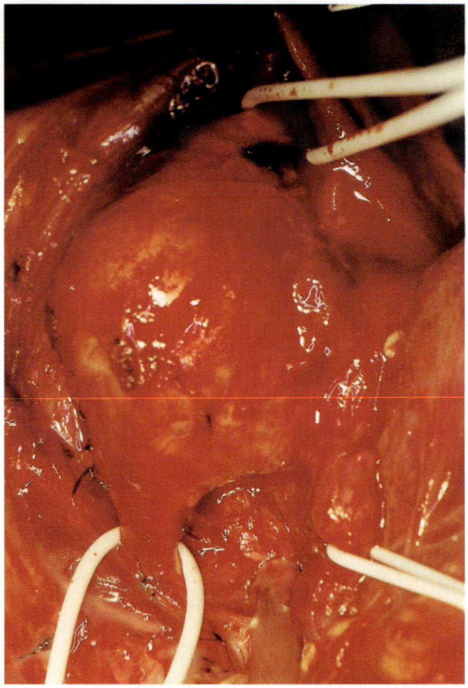

Plate 17 *Superior mesenteric artery aneurysm in a young female patient.*

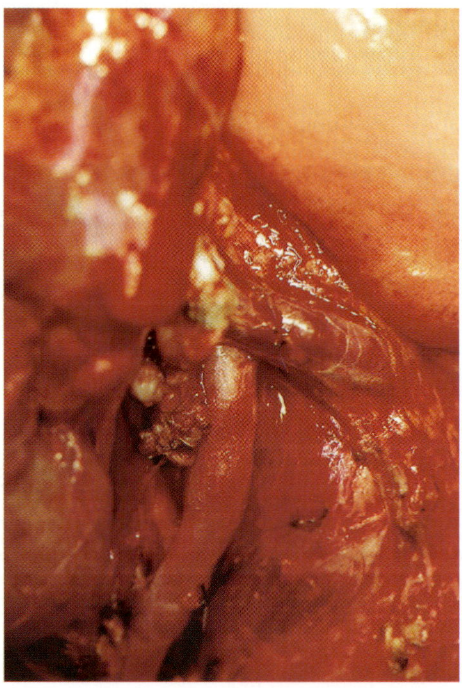

Plate 18 *The aneurysm shown in Plate 17 was excised and an interposition graft of autogenous saphenous vein was placed.*

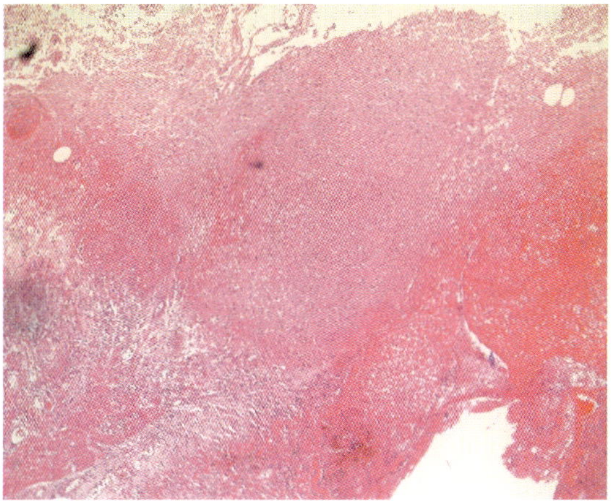

Plate 19 *Large-bowel mucosa showing transmural necrosis following arterial occlusion.*

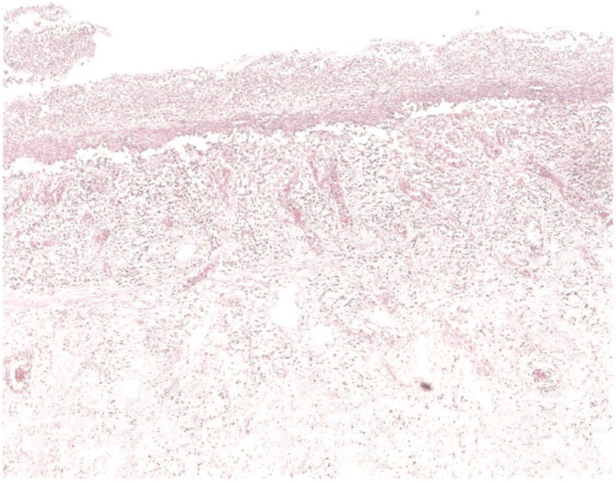

Plate 20 *Large-bowel mucosa showing mucosal necrosis following venous occlusion.*

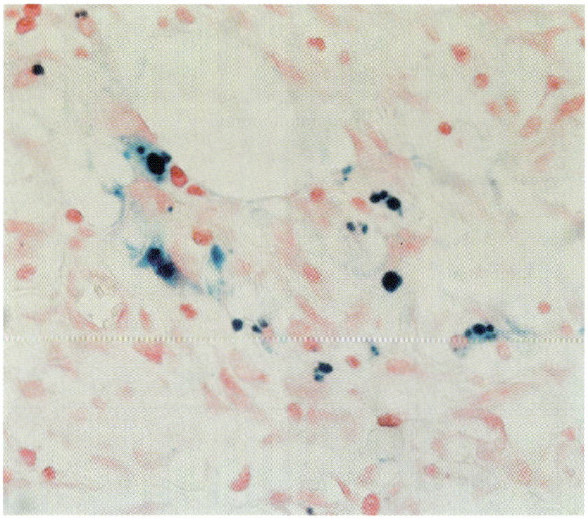

Plate 21 *The same case as shown in Plate 20, showing iron deposition in macrophages indicating underlying chronic ischemia.*

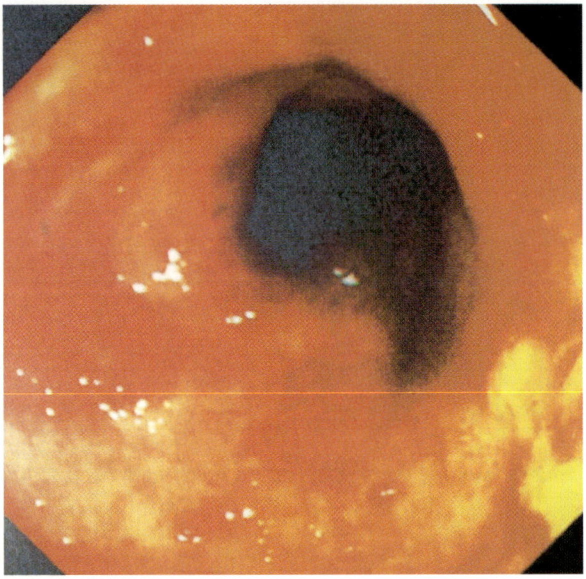

Plate 22 *Colonoscopic view in acute haemorrhagic ischemic colitis.*

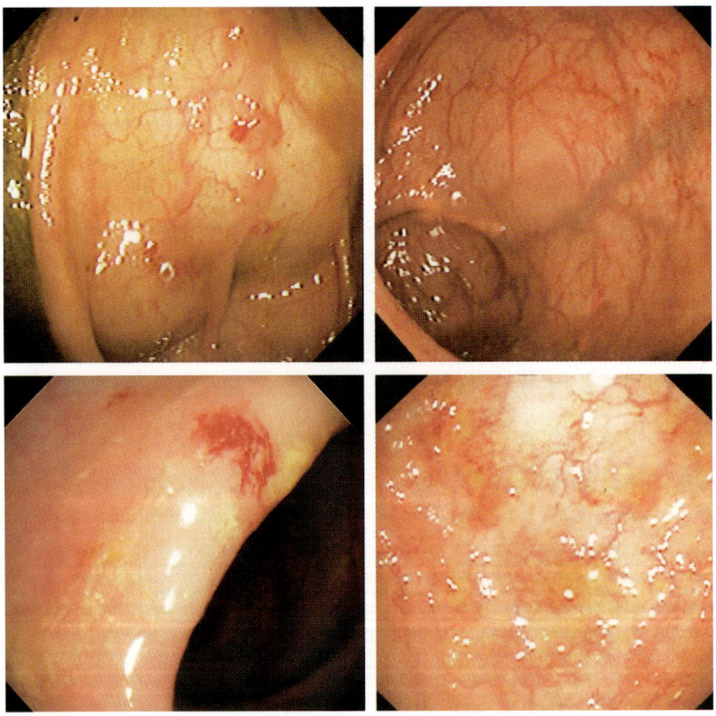

Plate 23 *The varied appearance of colonic angiodysplasia. Note that the draining vein is not always apparent.*
Courtesy of Paul A. Thodiyil, FRCS, Hammersmith Hospital, London.

non-specific. An elevated serum D-lactate level, however, appears to be highly suggestive for mesenteric ischemia.[48, 49]

Patients with MVT should undergo screening for hypercoagulable states, including factor V Leiden, Lupus anticoagulant, hyperhomocysteinemia, prothrombin mutation, plasminogen activator deficiency, antithrombin III deficiency, and proteins C and S deficiencies.

Diagnostic studies

Computerized tomography (CT) scan with intravenous contrast is the diagnostic test of choice for MVT (Figure 13.2). CT scan also aids in the diagnosis of secondary pathologies.[50] Rosen and co-workers described the CT findings of a dense venous wall and central lucency, which were present in all of their patients as confirmed by surgery or angiography.[51] In the Mayo Clinic series, when evidence of bowel ischemia and the presence of venous thrombus were considered together, CT scanning showed 100% sensitivity.[1] Magnetic resonance imaging (MRI) and magnetic resonance angiography (MRA) are also highly sensitive and specific, but these techniques are still limited by cost and availability.[52]

Duplex ultrasonography has been reported to have good sensitivity and specificity (83% and 93%, respectively) for portal vein thrombosis (Figure 13.3).[53] Limitations of ultrasonography included body habitus, bowel gas pattern, limited mesenteric vein imaging, and inter-sonographer variability. Miller and Berland noted the superiority of CT over Duplex scan for evaluation of the mesenteric and splenic veins, and the equivalence of the two modalities for evaluation of the portal vein.[54] Duplex scan has been noted to be of value as an initial, non-invasive diagnostic tool if positive.[55] However, this test should not delay surgical intervention in acutely ill patients or hinder timely CT scanning in stable patients.[56]

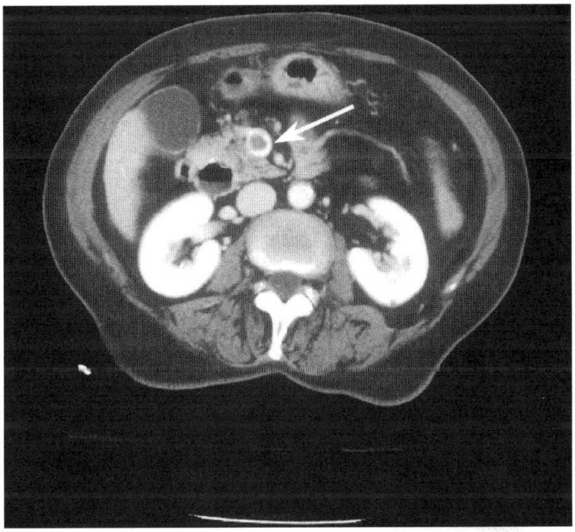

Figure 13.2 *Abdominal computerized tomography with intravenous contrast. Arrow denotes thrombus within the superior mesenteric vein, with associated edema.*

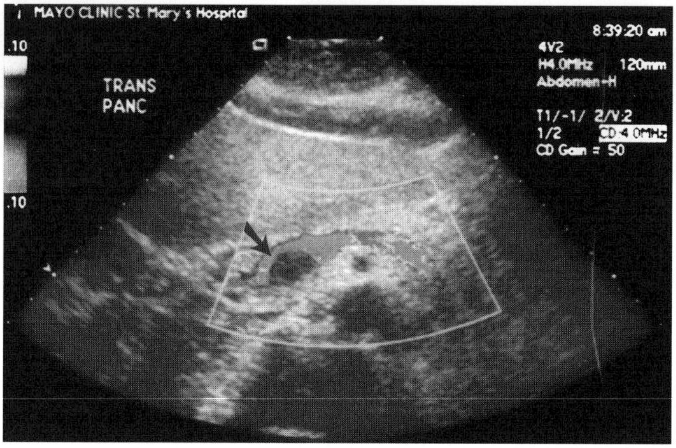

Figure 13.3 *Abdominal duplex ultrasonography noting thrombosis of the portomesenteric veins.*

Mesenteric arteriography with delayed venous imaging historically has been the diagnostic gold standard. However, it is time consuming and associated with its own (low) morbidity. Currently, angiography is not recommended as the initial investigation.

Reports have described diagnostic findings at small-bowel endoscopy. However, this does not identify the location of thrombosis.[57, 58] A potential future diagnostic modality is superconducting quantum interference device (SQUID) magnetometry. Experiments with this non-invasive device in a rabbit model of MVT showed promise for early rapid diagnosis of MVT prior to bowel necrosis.[59]

Treatment

The management goals for minimizing the morbidity and mortality of MVT are:

1 early diagnosis;
2 early anticoagulation;
3 early surgical intervention in patients with peritonitis.

The keys to early diagnosis are a high degree of clinical suspicion when evaluating patients with atypical abdominal pain and the liberal use of CT scans in this patient population. After making the diagnosis, anticoagulation and medical management is the first line of therapy in stable patients. If the patient shows signs of peritonitis, surgical exploration or laparoscopy is indicated to assess bowel viability directly.

MEDICAL MANAGEMENT

The initial management for all patients with MVT is prompt anticoagulation with intravenous heparin. Other supportive measures include aggressive fluid resuscitation and a broad-spectrum antibiotic, as the congested bowel might be vulnerable to bacterial translocation. Experience with low-molecular-weight heparin anticoagulation for MVT has been limited, and its role will be defined pending further study. After therapeutic intravenous anticoagulation is achieved, oral anticoagulation with warfarin sodium should be instituted when the patient is improving and not likely to require surgical intervention. Long-term/lifelong oral anticoagulation with warfarin sodium for MVT is advocated, especially for patients with hypercoagulable states.[60]

Table 13.4 *The management and outcome of 112 patients with mesenteric venous thrombosis*

Reference	Patients (*n*)	Operative	Bowel resection	Non-operative	Anti-coagulation	Mortality no. (%)
1	53	34	30	19	33	14 (27)
10	16	5	5	11	11	3 (19)
40	7	4	2	3	7	1 (14)
41	15	5	5	10	9	2 (13)
42	21	19	11	2	17	8 (38)
Total	112	67(60%)	53(47%)	45(40%)	77(69%)	28 (25%)

For patients with progressive symptoms without signs of peritonitis or bowel necrosis, thrombolytic therapy may be considered. Several reports have noted the success of thrombolytic therapy through arterial and transhepatic routes.[61,62] Thrombolytic therapy in patients with submucosal, subserosal, and free gastrointestinal hemorrhage is contraindicated. Further studies of this form of treatment are needed to define correct indications.

SURGICAL TREATMENT

Immediate surgical intervention is indicated when bowel viability is in question. Specific signs of non-viable bowel include signs of peritonitis, progressive sepsis, and increasing leukocytosis. Normal laboratory values or the presence of bowel sounds should not deter surgical exploration in the face of these indications. Imaging data support but do not replace clinical indications for exploration. About 60% of patients with acute MVT require surgical exploration (Table 13.4). Broad-spectrum antibiotics are given preoperatively to decrease infectious complications.

Surgical exploration allows for direct inspection of the bowel to assess viability. The goal is to prevent or limit bowel necrosis and septic complications. Exploration should assess intact pulsatile arterial inflow. Venous thrombosis typically involves the distal mesenteric vessels if associated with necrosis, and in one study only 12% of these patients have occlusion of the proximal superior mesenteric vein.[1] Other findings of MVT are a thickened cyanotic mesentery and bowel wall. Thrombosed veins in the cut edge of the mesentery are noted frequently. Some patients with MVT can present with extensive infarction of the entire jejunum and ileum; this either precludes resection, or at best requires lifelong intravenous nutrition. Laparoscopy has been utilized in critically ill patients with mesenteric ischemia for diagnostic purposes, but it is limited for resection and complete bowel and vascular assessment.[63]

Intraoperative heparin anticoagulation is indicated if not already initiated as this reduces the recurrence rate. Postoperative anticoagulation decreases recurrence of MVT from 26 to 14%.[43] Hemorrhagic complications with anticoagulation are increased at the benefit of increased bowel salvage.

We recommend minimal bowel resection to preserve bowel length and the liberal use of second-look laparotomy. Primary anastomosis can be performed in most of these patients. Bowel viability can be assessed with intravenous fluorescein and Wood's lamp inspection to aid in determination of resection margins.[43] The average length of bowel resected in the Mayo series was 121 cm.[1] The benefit of second-look laparotomy has been noted by Levy *et al.*[42] and Khodadadi *et al.*[64] Others have utilized second-look laparoscopy.[65]

Successful thrombectomy of the superior mesenteric vein, which is involved in 12–59%,[1,44] has been described in only a few patients.[66] Thrombosis is frequently extensive and involves the small distal veins as well as the intrahepatic portal outflow. Adjuncts of local thrombolytic therapy[67] and portovenous shunt decompression have been used with

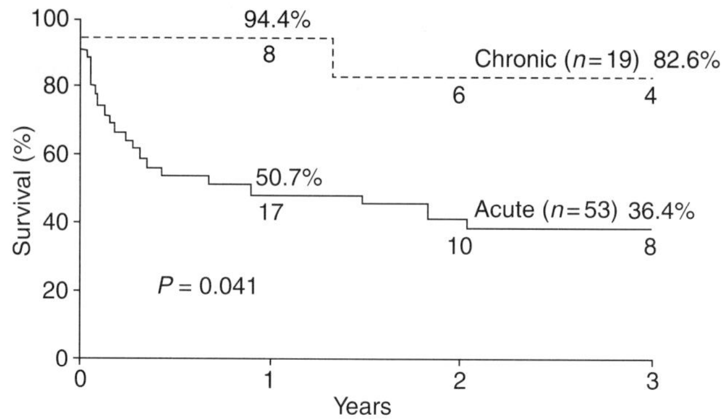

Figure 13.4 *Life-table analysis comparing patient survival of patients with acute and chronic mesenteric venous thrombosis.*

success.[40] Thrombectomy has poor results in the face of adherent thrombus that is older than 3–5 days.

Outcome

Early aggressive medical and surgical management of MVT reduces morbidity and mortality. However, the 30-day mortality even in contemporary series continues to be high, at 13–38% (Table 13.4). Deaths are frequently due to secondary concurrent severe medical conditions, which confounds the condition's mortality rate.[10, 41] The morbidity of MVT is high, with an average hospitalization of 22 days and a 55% complication rate, which included short bowel syndrome (23%), wound infections (21%), sepsis (17%), pneumonia (13%), pulmonary embolus (9%), renal failure (9%), and gastrointestinal bleeding (6%).[1] Patients presenting with necrotic bowel have worse prognoses and higher complication rates. The only factor that significantly improves survival and decreases recurrence is anticoagulation.[1, 43, 68]

Recurrence was noted in 27% of the 53 patients treated at the Mayo Clinic, and was higher in patients treated with anticoagulation alone as compared with anticoagulation and surgery (71% *v.* 32%).[1] Late survival in patients with acute MVT is good if they survive the first 3 months, and is predicated by other disease processes and length of remaining bowel (Figure 13.4).[1]

CHRONIC MESENTERIC VENOUS THROMBOSIS

Chronic MVT is frequently an incidental finding on imaging studies. Of the 19 patients reported by Rhee *et al.*, four had no symptoms, 15 had vague abdominal pain, and the only abnormal laboratory study was mild leukocytosis in 44%.[1] Other less commonly reported presentations include venous malformation of the colon,[69] protein-losing enteropathy,[70] and partial bowel obstruction secondary to stricture formation.[71]

Diagnosis is obtained with CT scanning with intravenous contrast, which can frequently detect even non-obstructive mesenteric venous thrombosis.[72] Many of these patients have underlying abdominal pathologies, such as chronic pancreatitis, pancreatic cancer, or portal hypertension. If not contraindicated by other medical conditions, anticoagulation is recommended to prevent further secondary thrombosis. Chronic MVT has a better long-term prognosis than acute MVT, as noted by the Mayo study (Figure 13.4).[1]

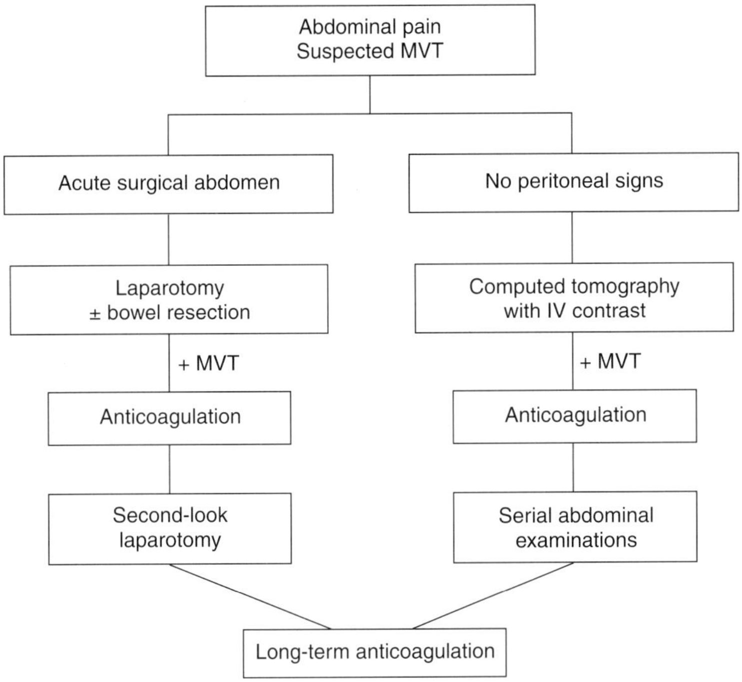

Figure 13.5 *Management algorithm for acute mesenteric venous thrombosis (MVT).*

SUMMARY

MVT is an infrequent clinical condition that requires a high index of clinical suspicion for timely diagnosis. When acute MVT is diagnosed, aggressive medical and surgical management should be initiated to prevent further bowel compromise. A logical algorithmic approach can be used in the management of these patients (Figure 13.5).

The diagnosis of MVT is achieved best by using CT with intravenous contrast. When the diagnosis of MVT is confirmed, all patients should be anticoagulated immediately with intravenous heparin or subcutaneous low-molecular-weight heparin. If the patient presents with signs of peritonitis, prompt surgical exploration is indicated to assess bowel viability. In the presence of bowel necrosis, all non-viable bowel should be resected. Fluorescein-assisted evaluation of the bowel for viability is helpful. If there are any questions of progression of ischemia or thrombosis, second-look laparotomy at 24 h should be performed. All patients should be maintained on heparin anticoagulation until they can be converted safely to oral warfarin sodium anticoagulation.

For patients that do not present with signs of peritonitis, we recommend a period of 48–72 h of close observation. These patients should be anticoagulated in a manner similar to the surgical patients. All patients without a contraindication to anticoagulation should undergo lifelong anticoagulation with warfarin sodium. Advances in diagnosis and management of patients with MVT have resulted in only modest improvements in survival. This is due largely to the indolent presentation of this disease and significant co-morbid conditions. Patients with chronic MVT fare better than those with acute MVT. This condition continues to present a challenge, and only by prompt diagnosis and skilled clinical judgment will its high morbidity and mortality be decreased.

ACKNOWLEDGMENT

The authors would like to thank Mrs Marcia Simonson for her editorial assistance.

KEY REFERENCES

Rhee RY, Gloviczki P, Mendonca CT, Petterson TM, Serry RD, Sarr MG, *et al*. Mesenteric venous thrombosis: still a lethal disease in the 1990s. *J Vasc Surg* 1994; **20**: 688–97.

This is the most recent large case series. It reports on both chronic and acute MVT.

Harward TR, Green D, Bergan JJ, Rizzo RJ, Yao JS. Mesenteric venous thrombosis. *J Vasc Surg* 1989; **9**: 328–33.

The first large clinical series that noted the strong association of MVT with hypercoagulable states.

Miller VE, Berland LL. Pulsed Doppler duplex sonography and CT of portal vein thrombosis. *Am J Roentgenol* 1985; **145**: 73–6.

Clear description of diagnostic modalities and their limitations.

Polk HC, Jr. Experimental mesenteric venous occlusion. *Am J Surg* 1964; **88**: 693–8.

Experimental description of the clinical and physiological manifestations of acute mesenteric venous occlusion.

REFERENCES

1 Rhee RY, Gloviczki P, Mendonca CT, Petterson TM, Serry RD, Sarr MG, *et al*. Mesenteric venous thrombosis: still a lethal disease in the 1990s. *J Vasc Surg* 1994; **20**: 688–97.

2 Fagge CH. Mesenteric venous thrombosis. *Trans Path Soc London* 1876; **27**: 124.

3 Elliot JW. The operative relief of gangrene of intestine due to occlusion of the mesenteric vessels. *Ann Surg* 1895; **21**: 9–23.

4 Warren S, Eberhard TP. Mesenteric venous thrombosis. *Surg Gynecol Obstet* 1935; **61**:102–21.

5 Howard TJ, Plaskon LA, Wiebke EA, Wilcox MG, Madura JA. Nonocclusive mesenteric ischemia remains a diagnostic dilemma. *Am J Surg* 1996; **171**: 405–8.

6 Boley SJ, Kaleya RN, Brandt LJ. Mesenteric venous thrombosis. *Surg Clin North Am* 1992; **72**: 183–201.

7 Rhee RY, Gloviczki P. Mesenteric venous thrombosis. *Surg Clin North Am* 1997; **77**: 327–38.

8 Bussey CD. Primary mesenteric venous thrombosis. *Arch Surg* 1955; **71**: 688–93.

9 Kitchens CS. Evolution of our understanding of the pathophysiology of primary mesenteric venous thrombosis. *Am J Surg* 1992; **163**: 346–8.

10 Harward TR, Green D, Bergan JJ, Rizzo RJ, Yao JS. Mesenteric venous thrombosis. *J Vasc Surg* 1989; **9**: 328–33.

11 Green D, Ganger DR, Blei AT. Protein C deficiency in splanchnic venous thrombosis. *Am J Med* 1987; **82**: 1171–4.

12 Broekmans AW, van Rooyen W, Westerveld BD, Briet E, Bertina RM. Mesenteric vein thrombosis as presenting manifestation of hereditary protein S deficiency. *Gastroenterology* 1987; **92**: 240–2.

13 Leebeek FW, Lameris JS, van Buuren HR, Gomez E, Madretsma S, Sonneveld P. Budd–Chiari syndrome, portal vein and mesenteric vein thrombosis in a patient homozygous for factor V Leiden mutation treated by TIPS and thrombolysis. *Br J Haematol* 1998; **102**: 929–31.

14 Lee HJ, Park JW, Chang JC. Mesenteric and portal venous obstruction associated with primary antiphospholipid antibody syndrome. *J Gastroenterol Hepatol* 1997; **12**: 822–6.

15 Blanc P, Barki J, Fabre JM, Larrey D, Domergue J, Michel H, Lavabre-Bertrand T. Superior mesenteric vein thrombosis associated with anticardiolipin antibody without autoimmune disease. *Am J Hematol* 1995; **48**: 137.

16 Boyko OB, Pizzo SV. Mesenteric vein thrombosis and vascular plasminogen activator. *Arch Path Lab Med* 1983;**107**: 541–2.

17 Marie I, Levesque H, Lecam-Duchez V, Borg JY, Ducrotte P, Philippe C. Mesenteric venous thrombosis revealing both factor II G20212A mutation and hyperhomocysteinemia related to pernicious anemia. *Gastroenterology* 2000; **118**: 237–8.

18 Gruenberg JC, Smallridge RC, Rosenberg RD. Inherited antithrombin-III deficiency causing mesenteric venous infarction: a new clinical entity. *Ann Surg* 1975; **181**: 791–4.

19 Graubard ZG, Friedman M. Mesenteric venous thrombosis associated with pregnancy and oral contraception. A case report. *S Afr Med J* 1987; **71**: 453.

20 De Luca M, Dugo M, Arduini R, Liessi G. Acute venous thrombosis of spleno-mesenteric portal axis: an unusual localization of thromboembolism in the nephrotic syndrome. *Am J Nephrol* 1991; **11**: 260–3.

21 Tossou H, Iglicki F, Casadevall N, Delamarre J, Dupas JL, Capron JP. Superior mesenteric vein thrombosis as a manifestation of a latent myeloproliferative disorder. *J Clin Gastroenterol* 1991; **13**: 597–8.

22 Whelan MA, Kaim P. Mesenteric complications in a patient with polycythemia vera. *Am J Gastroenterol* 1982; **77**: 526–8.

23 Venturini I, Cioni G, Turrini F, Gandolfo M, Modonesi G, Cosenza R, *et al*. Mesenteric vein thrombosis: a rare cause of abdominal pain in cirrhotic patients – two case reports. *Hepatogastroenterology* 1998; **45**: 44–7.

24 Goldberg H, Fabry TL. Mesenteric thrombosis following sclerotherapy during vasopressin infusion: mechanism and therapeutic implications. *J Clin Gastroenterol* 1989; **11**: 56–7.

25 Sanabria JR, Hiruki T, Szalay DA, Tandan V, Gallinger S. Superior mesenteric vein thrombosis after the Whipple procedure: an aggressive, combined treatment approach. *Can J Surg* 1997; **40**: 467–70.

26 Millikan KW, Szczerba SM, Dominguez JM, McKenna R, Rorig JC. Superior mesenteric and portal vein thrombosis following laparoscopic-assisted right hemicolectomy. Report of a case. *Dis Colon Rectum* 1996; **39**: 1171–5.

27 Kowal-Vern A, Radhakrishnan J, Goldman J, Hutchins W, Blank J. Mesenteric and portal vein thrombosis after splenectomy for autoimmune hemolytic anemia. *J Clin Gastroenterol* 1988; **10**: 108–10.

28 Rattner DW, Ellman L, Warshaw AL. Portal vein thrombosis after elective splenectomy. An underappreciated, potentially lethal syndrome. *Arch Surg* 1993; **128**: 565–9.

29 Krummen DM, Cannova J, Schreiber H. Conservative management strategy for pancreatitis-associated mesenteric venous thrombosis. *Am Surg* 1996; **62**: 432–4.

30 Cornu-Labat G, Kasirajan K, Simon R, Smith DJ, Herman ML, Rubin JR. Acute mesenteric vein thrombosis and pancreatitis. A rare association. *Int J Pancreatol* 1997; **21**: 249–51.

31 Farin P, Paajanen H, Miettinen P. Intraoperative US diagnosis of pylephlebitis (portal vein thrombosis) as a complication of appendicitis: a case report. *Abdom Imaging* 1997; **22**: 401–3.

32 Heresbach D, Pagenault M, Gueret P, Crenn P, Heresbach-Le BN, Malledant Y, *et al*. Leiden factor V mutation in four patients with small bowel infarctions. *Gastroenterology* 1997; **113**: 322–5.

33 Capone A, Cassone M, Tarasi D, Brandimarte C, Tarasi A, Galie M, Venditti M. Fatal suppurative mesenteric thrombophlebitis caused by an *Enterococcus faecalis* highly resistant to aminoglycosides: case report. *J Chemother* 1998; **10**: 128–31.

34 Landau Z, Feld S, Kunichezky S, Grinspan M, Gorbacz M. Thrombosis of the mesenteric vein as a complication of Mediterranean spotted fever. *Clin Infect Dis* 1992; **15**: 1070–1.

35 Narayanan TS, Narawane NM, Phadke AY, Abraham P. Multiple abdominal venous thrombosis in HIV-seropositive patient. *Indian J Gastroenterol* 1998; **17**:105–6.

36 Myszor MF, Calam J, Hodgson HJ. Portal and mesenteric vein thrombosis in two diabetic patients. *Diabet Med* 1987; **4**: 485–7.

37 Murray FE. Mesenteric vein thrombosis associated with Klinefelter's syndrome – a case report. *Angiology* 1988; **39** (1 Pt 1):45–8.

38 Gertler SL, Stein J, Simon T, Miyai K. Mesenteric venous thrombosis as sole complication of decompression sickness. *Dig Dis Sci* 1984; **29**: 91–5.

39 Grendell JH, Ockner RK. Mesenteric venous thrombosis. *Gastroenterology* 1982; **82**: 358–72.

40 Gertsch P, Matthews J, Lerut J, Luder P, Blumgart LH. Acute thrombosis of the splanchnic veins. *Arch Surg* 1993; **128**: 341–5.

41 Grieshop RJ, Dalsing MC, Cikrit DF, Lalka SG, Sawchuk AP. Acute mesenteric venous thrombosis. Revisited in a time of diagnostic clarity. *Am Surg* 1991; **57**: 573–7.

42 Levy PJ, Krausz MM, Manny J. The role of second-look procedure in improving survival time for patients with mesenteric venous thrombosis. *Surg Gynecol Obstet* 1990; **170**: 287–91.

43 Abdu RA, Zakhour BJ, Dallis DJ. Mesenteric venous thrombosis – 1911 to 1984. *Surgery* 1987; **101**: 383–8.

44 Johnson CC, Baggenstoss AH. Mesenteric vascular occlusion I. Study of 99 cases of occlusion of veins. *Mayo Clin Proc* 1949; **24**: 628–36.

45 Polk HC, Jr. Experimental mesenteric venous occlusion. *Am J Surg* 1964; **88**: 693–8.

46 Miyaki CT, Park YS, Gopalswamy N. Upper gastrointestinal bleeding in acute mesenteric thrombosis. *J Clin Gastroenterol* 1988; **10**: 84–7.

47 Roman RJ, Loeb PM. Massive colonic dilatation as initial presentation of mesenteric vein thrombosis. *Dig Dis Sci* 1987; **32**: 323–6.

48 Murray MJ, Gonze MD, Nowak LR, Cobb CF. Serum D-lactate levels as an aid to diagnosing acute intestinal ischemia. *Am J Surg* 1994; **167**: 575–8.

49 Gunel E, Caglayan O, Caglayan F. Serum D-lactate levels as a predictor of intestinal ischemia-reperfusion injury. *Pediatr Surg Int* 1998; **14**: 59–61.

50 Vogelzang RL, Gore RM, Anschuetz SL, Blei AT. Thrombosis of the splanchnic veins: CT diagnosis. *Am J Roentgen* 1988; **150**: 93–6.

51 Rosen A, Korobkin M, Silverman PM, Dunnick NR, Kelvin FM. Mesenteric vein thrombosis: CT identification. *Am J Roentgenol* 1984; **143**: 83–6.

52 Gehl HB, Bohndorf K, Klose KC, Gunther RW. Two-dimensional MR angiography in the evaluation of abdominal veins with gradient refocused sequences. *J Comput Assist Tomogr* 1990; **14**: 619–24.

53 Alpern MB, Rubin JM, Williams DM, Capek P. Porta hepatis: duplex Doppler US with angiographic correlation. *Radiology* 1987; **162** (1 Pt 1): 53–6.

54 Miller VE, Berland LL. Pulsed Doppler duplex sonography and CT of portal vein thrombosis. *Am J Roentgenol* 1985; **145**: 73–6.

55 Verbanck JJ, Rutgeerts LJ, Haerens MH, Tytgat JH, Segaert MF, Tytgat HJ, Afschrift MB. Partial splenoportal and superior mesenteric venous thrombosis. Early sonographic diagnosis and successful conservative management. *Gastroenterology* 1984; **86** (5 Pt 1): 949–52.

56 Matos C, Van Gansbeke D, Zalcman M, Ansay J, Delcour C, Engelholm L, Struyven J. Mesenteric vein thrombosis: early CT and US diagnosis and conservative management. *Gastrointest Radiol* 1986; **11**: 322–5.

57 Leiser A, Wysjenbeek A, Kadish U. Mesenteric venous infarction presenting as an upper GI bleeding and diagnosed by upper GI endoscopy. *Endoscopy* 1985; **17**: 119–20.

58 Wade TP, Jewell WR, Andrus CH. Mesenteric venous thrombosis. Modern management and endoscopic diagnosis. *Surg Endosc* 1992; **6**: 283–4.

59 Allos SH, Staton DJ, Bradshaw LA, Halter S, Wikswo JPJ, Richards WO. Superconducting quantum interference device magnetometer for diagnosis of ischemia caused by mesenteric venous thrombosis. *World J Surg* 1997; **21**: 173–7.

60 Bemelman WA, Butzelaar RM, Khargi K, Keeman JN. Mesenteric venous thrombosis caused by deficiency of physiologic anti-coagulants: report of a case. *Neth J Surg* 1990; **42**: 16–19.

61 Rivitz SM, Geller SC, Hahn C, Waltman AC. Treatment of acute mesenteric venous thrombosis with transjugular intramesenteric urokinase infusion. *J Vasc Interv Radiol* 1995; **6**: 219–23.
62 Poplausky MR, Kaufman JA, Geller SC, Waltman AC. Mesenteric venous thrombosis treated with urokinase via the superior mesenteric artery. *Gastroenterology* 1996; **110**: 1633–5.
63 Orlando R, Crowell KL. Laparoscopy in the critically ill. *Surg Endosc* 1997; **11**: 1072–4.
64 Khodadadi J, Rozencwajg J, Nacasch N, Schmidt B, Feuchtwanger MM. Mesenteric vein thrombosis. The importance of a second-look operation. *Arch Surg* 1980; **115**: 315–17.
65 Spittler C, Chari V, Husni E, Patzakis J, Li P, Zelis J, Chung R. Second-look laparoscopy for visceral ischemia facilitated by preinstalled ports. *Am Surg* 1997; **63**: 732–4.
66 Inahara T. Acute superior mesenteric venous thrombosis: treatment by thrombectomy. *Ann Surg* 1971; **174**: 956–61.
67 Klempnauer J, Grothues F, Bektas H, Pichlmayr R. Results of portal thrombectomy and splanchnic thrombolysis for the surgical management of acute mesentericoportal thrombosis. *Br J Surg* 1997; **84**: 129–32.
68 Dada FB, Balan AD. Recurrent primary mesenteric venous thrombosis. *South Med J* 1987; **80**:1329–30.
69 Clements WD, Cunnick G, Hull D. Superior mesenteric vein occlusion presenting with a massive vascular malformation of the colon. *Br J Clin Pract* 1997; **51**: 189–91.
70 Honbo K, Kawaji H, Tsuchimochi M, Hasui K, Tanaka K, Taira A. Protein-losing enteropathy due to thrombophlebitis of the mesenteric vein: report of a case. *Surg Today* 1998; **28**: 932–6.
71 Eugene C, Valla D, Wesenfelder L, Fingerhut A, Bergue A, Merrer J, *et al.* Small intestinal stricture complicating superior mesenteric vein thrombosis. A study of three cases. *Gut* 1995; **37**: 292–5.
72 Boverie JH, Counet D, Meunier P, Dondelinger RF. Small bowel enema and diagnosis of chronic nonischemic disturbance of superior mesenteric venous blood flow. *Abdom Imaging* 1993; **18**: 265–70.

14

Management of portal hypertension

JOHN R. GALLOWAY AND ATEF SALAM

INTRODUCTION

The clinical entity of portal hypertension remains one of the most vexing problems in medicine today and is a continued source of major morbidity and mortality. In order to achieve acceptable patient outcome, a thorough knowledge and understanding are desired of the etiology, the important anatomical and pathophysiological considerations, and the natural history of portal hypertension, as well as the various treatment modalities available. In this chapter, we examine these areas with a particular emphasis on the management of variceal bleeding. It is also important to realize that a paradox exists in the management of patients with portal hypertension: it is understood that portal hypertension, while often a result of hepatic disease, also serves as the driving force to maintain prograde portal blood flow into the liver and thereby prevent further deterioration in liver function. This is especially true in cirrhosis. Unfortunately, the two most common complications of portal hypertension – variceal bleeding and ascites – are often best treated by relieving this elevation in portal pressure. This chapter will hopefully shed light on how to best balance the need to maintain portal perfusion and prevent hepatic decompensation while at the same time achieving control of variceal bleeding and ascites.

ETIOLOGY

The normal portal venous pressure is less than 10 mm Hg and is necessary to ensure portal blood flow through the hepatic sinusoids at a rate that exceeds 1 l/min. Portal hypertension occurs when the portal venous pressure exceeds 10 mm Hg. Elevated pressures in the portal venous system occur as a result of obstruction to prograde portal blood flow in the presinusoidal, sinusoidal, or post-sinusoidal locations. Regardless of the location of obstruction, pressures often rise to 30 mm Hg.[1] Presinusoidal portal hypertension occurs at either the extrahepatic or intrahepatic location. Extrahepatic portal vein occlusion is caused by thrombosis of the extrahepatic portal vein or one of its major branches. Obstruction of the intrahepatic portal vein before the sinusoids occurs most commonly as a result of schistosomiasis or congenital hepatic fibrosis. Hepatocellular function is well preserved in presinusoidal portal hypertension and is an important consideration in management and prognosis. Sinusoidal obstruction is usually the result of cirrhosis, with alcohol and viral hepatitis being the most likely etiologies. Less common forms of cirrhosis include alpha-1-anti-trypsin deficiency, hemochromatosis, Wilson's disease, primary biliary cirrhosis, and sclerosing cholangitis. In contrast to extrahepatic portal hypertension, the hepatocellular function in sinusoidal portal hypertension is almost always compromised. It is, therefore, important to define accurately the cause of the underlying liver disease as it impacts on the long-term management and prognosis.

Post-sinusoidal portal hypertension is due to hepatic venous outflow obstruction. This more rare form of portal hypertension occurs either as a result of major vessel occlusion, known as Budd–Chiari syndrome, or as hepatic venule obstruction, which is seen in veno-occlusive disease associated with bone marrow transplant patients. Table 14.1 lists the various causes of portal hypertension according to the site of portal vein block.[2]

Table 14.1 *Etiology of portal hypertension*

Presinusoidal obstruction
Portal vein thrombosis
Congenital
Sepsis
Trauma
Malignant
Hypercoagulable states
Periportal inflammation
Intrahepatic
Congenital hepatic fibrosis
Schistosomiasis
Idiopathic non-cirrhotic portal hypertension
Sarcoidosis
Myeloproliferative disorders
Sinusoidal obstruction
Cirrhosis
Nodular regenerative hyperplasia
Fatty liver
Wilson's disease
Post-sinusoidal obstruction
Budd–Chiari syndrome
Hypercoagulable states
Congenital venous webs
Malignant obstruction
Parasitic infestation
Veno-occlusive disease
Increased flow
Hepatic arterial-portal venous fistula
Splenic or mesenteric arterial-venous fistula
Massive splenomegaly

Anatomical and pathophysiological considerations

The portal vein arises behind the neck of the pancreas from the confluence of the superior mesenteric vein (SMV), which drains the splanchnic venous system and the splenic vein. It runs through the posterior aspect of the hepatoduodenal ligament for a distance of approximately 6–8 cm before bifurcating into the right and left branches at the liver hilum. The SMV and splenic veins are also relatively constant in their respective routes through the base of the mesentery and behind the pancreas. The other portal venous tributaries are numerous and subject to significant anatomic variations. The inferior mesenteric vein, which drains the left hemicolon and upper rectum, enters into either the splenic vein or the SMV. The left gastric vein or coronary venous system drains the lesser curve of the stomach and the lower esophagus and usually enters either the splenic or portal vein close to the superior angle of their confluence behind the neck of the pancreas. The main tributaries to the SMV are the right colic, middle colic, and ileocolic veins, which drain the more proximal aspects of the colon and the distal small bowel, along with several jejunal branches draining the middle and proximal small bowel. The umbilical vein, when present, consistently enters the left portal vein branch.[3]

In portal hypertension, the increased pressure causes the development of several collateral pathways. Gastroesophageal varices develop in the gastric fundus and distal esophagus as a portal azygous connection. These varices are fed primarily by the left gastric or coronary venous system and the short gastric veins; they are the collaterals that most frequently cause variceal bleeding. The umbilical vein collaterals can become quite large, giving rise to the well-known sign of caput medusae. The inferior mesenteric vein collaterals form around the rectum and anal canal as the high-pressure superior and middle hemorrhoidal veins decompress into low-pressure systemic inferior hemorrhoidal veins. Very large hemorrhoids and distal colonic varices are formed in response to this phenomenon.[4] Finally, collateral pathways develop after previous surgery where vascular adhesions serve as a connection between the high-pressure venous tributaries of the portal system to the low-pressure bed of the systemic circulation. These connections can form across the diaphragm, to the retroperitoneum, and to the anterior abdominal wall, particularly at the mucocutaneous junctions of an ileostomy or colostomy.[5, 6]

As these portosystemic collateral pathways mature, a significant portion of portal blood bypasses the sinusoidal circulation. Nevertheless, portal pressure remains remarkably elevated despite the decompression resulting from portosystemic collaterals and intrahepatic shunting. The compensatory increase in hepatic arterial perfusion, which offsets the loss of portal blood flow and oxygen to the liver, does not alone explain the maintenance of portal hypertension. Previously, two major hypotheses were advanced to explain this phenomenon.[7, 8] The backward theory postulated that portal hypertension was due to increased hepatic vascular resistance, so that in the presence of normal portal blood flow, pressure necessarily increased. Hypertrophy of myelofibroblasts within the sinusoidal bed was the possible mechanism by which this could be achieved. However, further studies showed that this mechanism alone could not account for maintenance of portal hypertension. The forward theory supposed that an increased splanchnic blood flow was the major feature of portal hypertension. Studies suggested further that the increased and hyperdynamic portal blood flow together with decreased splanchnic precapillary resistance maintained portal hypertension even in the face of extensive portosystemic shunting. Over the past decade, the pathophysiological steps in the development and maintenance of portal hypertension have been clarified further and appear to incorporate components of both the backward and forward flow theories of portal hypertension into a unifying hypothesis.[9, 10] These steps are as follows:

1 The primary event is an increased vascular resistance to portal blood flow with a resultant back pressure on the splanchnic venous bed.

2 Portosystemic collaterals develop to bypass this blockage.
3 A compensatory increase in splanchnic arterial inflow along with splanchnic vasodilation contributes to a further rise in portal pressure.
4 The multifactorial expansion in plasma volume increases circulating volume within the splanchnic bed and aggravates the portal hypertension.
5 Peripheral vasodilation occurs and the associated hyperdynamic systemic circulation increases blood flow to the splanchnic circulation by as much as 50%.
6 The intensity of this hyperdynamic state seems to be directly proportional to the extent of portosystemic shunting. Various local vasoactive substances produced in the splanchnic circulation escape via collaterals to the systemic circulation where their vasoactive effects cause a hyperdynamic circulation. The most important of these vasoactive substances include bile acids, serotonin, glucagon, and prostaglandin.[11]

NATURAL HISTORY OF PORTAL HYPERTENSION

The natural history of portal hypertension depends primarily on the underlying functional hepatic reserve. Conditions causing presinusoidal block, where hepatocellular function remains nearly normal, carry the best prognosis. The major lethal complication is hemorrhage from esophageal varices rather than from progressive liver failure. The mortality for each bleeding episode is 5%, and long-term survival can be excellent if variceal hemorrhage is controlled.[12] In post-sinusoidal block, as seen in Budd–Chiari syndrome and veno-occlusive disease, severe hepatocellular congestion and necrosis can result in death from liver failure rather than from hemorrhage.[13] In patients with sinusoidal block caused by cirrhosis, variceal bleeding contributes significantly to mortality. The mortality rate from the initial variceal bleed approaches 50%. If the patient survives the initial variceal hemorrhage, then the risk of re-bleeding remains very high; therefore, understanding of the risks for the initial bleed as well as re-bleeding determines subsequent evaluation and treatment. Unfortunately, patients with cirrhosis may still succumb to a slow, inexorable decline in hepatic function even when the risk of variceal hemorrhage has been minimized.[14]

At the time of diagnosis, 30% of patients with compensated liver disease will have varices, while 60% of those with poor hepatic function will have varices. Therefore, poor liver function, as well as continued alcohol abuse, is associated with a more frequent variceal bleeding. Limited data suggest that varices will develop in approximately 8% of patients with cirrhosis each year. Further evidence demonstrates that small varices will develop into large varices. Large varices do have a higher risk of an initial bleed (30% per year) than do small varices (15% per year).[15, 16] An additional risk factor for the first variceal hemorrhage is the presence of red-wale markings at the time of endoscopy (also known as red-dot sign). These cherry-red spots are well-recognized predictors of impending rupture. A final risk factor for the initial variceal bleed is a portal pressure greater than 12 mm Hg.[17, 18]

After the initial episode of bleeding, the re-bleeding rate is 75% over 1 year, with re-bleeding occurring in 30–50% of patients during the first 6 weeks. Large varices also predict further bleeding episodes. Subsequent mortality in the first year approaches 30%.[19] Ultimately, one-third of all cirrhotic patients experience variceal bleeding, and it is clear that the natural history of portal hypertension can be modified significantly by appropriate recognition and management of variceal bleeding.[20] Control of variceal hemorrhage is now more important than ever before, since in the past 25 years the availability of liver transplant has offered a way to restore normal liver function.

CLINICAL EVALUATION OF THE PATIENT WITH PORTAL HYPERTENSION

The clinical evaluation of the patient with portal hypertension should focus on the assessment of varices and hepatocellular function. Endoscopy is done when cirrhosis is discovered to determine the patient's future risk for variceal bleeding. At the time of acute variceal bleeding, endoscopy confirms the diagnosis and variceal ablation is performed. After bleeding ceases, repeat endoscopy is performed in 24–48 h to reassess the varices in the patient who now has a stable intravascular volume and a stomach and esophagus empty of blood. This endoscopic assessment should include variceal size, extent, and associated risk factors. The presence of gastric varices and portal hypertensive gastropathy is also noted.[21, 22] Finally, upper endoscopy also excludes other gastroesophageal or duodenal pathology as a source of upper-gastrointestinal bleeding.[23]

It is also necessary to determine the etiology of the portal hypertension and to assess hepatocellular function. The clinical parameters of jaundice, ascites, encephalopathy, and nutritional status, as well as the serum levels of bilirubin and albumin, and the prothrombin time determine the Child–Pugh classification (Table 14.2).[24, 25] Patients with Child's A liver disease have well-preserved liver function and will tolerate most treatment modalities. In contrast, patients who are a Child's C have advanced disease and probably should be managed only by liver transplantation.[26] Additional laboratory studies used to identify the exact underlying liver disease include:

- specific serum autoantibodies as markers of primary biliary cirrhosis and autoimmune chronic active hepatitis;
- serological markers of hepatitis B and C;
- serum ceruloplasmin and iron metabolism studies to rule out Wilson's disease and hemochromatosis, respectively;
- serum alpha-1-antitrypsin levels.

In some cases, liver biopsy may be important to confirm cirrhosis, evaluate the activity of disease, and document the degree of fibrosis.[27, 28]

The final step in evaluation is radiographic imaging. Multiple modalities are available, but Doppler ultrasonography and visceral angiography play the dominant role.[29] Doppler ultrasonography provides detailed information non-invasively. The main portal vein, the splenic vein, and the SMV can be visualized for size, patency, and direction of flow. The hepatic and splenic parenchymal can also be scanned for hemangiomas, cysts, tumors, and cirrhosis. All patients being considered for transjugular intrahepatic portosystemic shunts (TIPS) should undergo Doppler ultrasound.[30, 31] In patients being considered for surgical shunting, visceral angiography is usually required. This procedure is accomplished by venous phase imaging

Table 14.2 *Child–Pugh classification*

Parameter	1 Point	2 Points	3 Points
Bilirubin (mg/dl)	<2	2–3	>3
Albumin (g/dl)	>3–5	2.8–3.5	<2.8
Prothrombin time	1–3	4–6	>6
Ascites	None	Slight	Moderate
Encephalopathy	None	1–2	3–4

Grades: A, 5–6 points; B, 7–9 points; C, 10–15 points.

following SMA and splenic artery injections of radiographic contrast. It remains the gold standard for direct visualization of the portomesenteric venous system and its major collaterals. Finally, measurement of the hepatic venous pressure gradient may be useful in establishing the presence of portal hypertension and, in some instances, the response to therapy. Drawbacks to angiography include its invasive nature in a patient with coagulopathy, as well as large dye loads in patients who may have inadequate intravascular volume following a variceal bleed or hepatorenal syndrome.[32] Finally, both computerized axial tomography (CAT) and magnetic resonance imaging (MRI) can be utilized for the detection of cirrhosis, hepatosplenomegaly, and focal hepatic lesions. CAT assessment of liver volume is also used in some centers as an indirect assessment of hepatic reserve and for determining liver size in potential liver transplant candidates. Magnetic resonance angiography (MRA) is a rapidly advancing technique that is continuing to improve in its ability to yield accurate vascular images of the portal circulation. As such, it may eventually supercede conventional angiography.[33]

TREATMENT METHODS FOR VARICEAL BLEEDING

Prophylactic treatment of variceal bleeding

The mortality rate of an initial variceal bleeding episode approaches 30–50%. This has generated a significant interest in prevention of the initial hemorrhage. Patients with compromised liver function, large varices with red-dot sign, and elevated portal pressures are candidates for prophylactic therapy.[34] Shunt surgery, devascularization procedures, TIPS, endoscopic ablation, and pharmacotherapy have all been examined in a prophylactic role against variceal bleeding. Unfortunately, in the early 1960s prophylactic portosystemic shunt trials showed a significant increase in encephalopathy and a reduction in survival, even though the first variceal bleed was prevented.[35–37] Only a more recent study from Japan showed significant improval in the survival rate of patients undergoing gastric devascularization, while preventing initial variceal hemorrhage.[38] However, when all surgical series are balanced together, there is insufficient support for prophylactic surgery in the 1990s. The role of the TIPS procedure has not been examined well in the prevention of first bleeding episodes. Placement of TIPS in high-risk patients awaiting liver transplantation to prevent variceal bleeding and control intractable ascites may be reasonable, but it is currently unsupported with firm data.[39] Nineteen sclerotherapy trials to prevent a first variceal hemorrhage show mixed results. A recent meta-analysis of these trials shows an extreme heterogeneity, which makes meaningful analysis difficult.[40] There may be a benefit of sclerotherapy in patients who are at extremely high risk for bleeding. Nevertheless, reliable selection criteria to determine which patients would benefit most from prophylactic sclerotherapy are currently not available.

The only modality recommended to prevent the initial variceal bleeding episode is pharmacological therapy. Non-selective beta-adrenergic receptor antagonists, such as propranolol or nadolol, have been studied in nine trials. The risk of re-bleeding was reduced significantly in four of the studies; in three studies, a trend towards significant reduction occurred. A meta-analysis of all the trials showed a significant decrease in the risk of bleeding in patients treated with these agents. Although it did not reach statistical significance, mortality also appeared to be decreased.[41–43] An alternative pharmacological treatment is the administration of long-acting nitrates, such as isosorbide-5-mononitrate. This has been shown to produce a sustained decrease in portal pressures resulting in a similar efficacy in decreasing bleeding rates and improving survival.[44, 45] Accordingly, many hepatologists now use both of the drugs in combination to prevent variceal hemorrhage.[46]

Emergency treatment of acute variceal bleeding

The various treatment options for control of acute variceal hemorrhage include pharmacotherapy, endoscopic ablation, balloon tamponade, TIPS, and, rarely, operative decompression or devascularization procedures. The immediate goal of emergency treatment of the patient with bleeding gastroesophageal varices is stabilization so that they can be brought to elective evaluation. This elective evaluation, as outlined previously, determines which available option – medical, surgical, or radiographic – offers the best possible survival advantage by controlling variceal bleeding while maintaining underlying hepatic function.

GENERAL PRINCIPLES OF RESUSCITATION

The patient with massive variceal bleeding is at significant risk of dying from hypovolemic shock. Therefore, the principles of managing hypovolemic shock govern the resuscitation of the patient with variceal bleeding.

The patient is best managed in an intensive care environment. The initial insertion of multiple large-bore intravenous lines for volume resuscitation is required. A central venous line is used in unstable patients. Most patients with cirrhosis have a hyperdynamic state obviating the need for routine pulmonary artery catheterization unless the patient has known cardiac disease. A urinary catheter should be placed to monitor urine output and ensure adequate volume resuscitation. Careful observation of the patient's airway and oxygen administration to maintain arterial oxygen saturation above 94% is also appropriate. The risk of respiratory failure and death from aspiration during massive variceal bleeding is significant. Early intubation of the patient is appropriate, particularly when subsequent endoscopic ablation therapy is anticipated. The judicious administration of crystalloid solutions should maintain blood pressure and urine output. The temptation to use vasopressors should be resisted. The appropriate use of fresh frozen plasma, platelets, and cryoprecipitate to correct coagulation defects is very important. The role of 5% albumin is less clear, but its use can be justified in the presence of significant ascites and severe hypoproteinemia. Packed red blood cells are transfused to obtain adequate oxygen delivery by maintaining a hemoglobin of 9–10 g/dl. Finally, over-resuscitation of the patient should be avoided as it leads to a hypervolemic state that will not only worsen ascites but can quite possibly lead to increased portal pressures with persistent or recurrent variceal bleeding.

Pharmacotherapy

The drugs for treatment of acute variceal bleeding are vasopressin and a synthetic analogue of somatostatin, octreotide. Both drugs are potent splanchnic vasoconstrictors administered intravenously to reduce portal pressure and stop the acute bleeding episode. Vasopressin is successful in approximately 70% of patients. The recurrence of bleeding is high when the infusions are discontinued, and survival is not necessarily improved. Vasopressin should be administered in the intensive care unit with appropriate hemodynamic monitoring. It is best given through a central venous line because of the infrequent but significant complication of soft-tissue necrosis when administered peripherally. Nitroglycerin is administered concurrently to reduce the systemic vasoconstrictive effects of vasopressin, and may further reduce portal pressure. Vasopressin is delivered initially as 20 units in 200 cm^3 dextrose 5% in water over 20 min, and is followed by a continuous infusion in a dose range of 0.1–0.4 units/min. Nitroglycerin infusion begins at 40 μg/min and is titrated to maintain a mean arterial blood pressure of 60–70 mm Hg. Once the cessation of bleeding has been documented by endoscopy, the vasopressin and nitroglycerin drips are terminated. It is not necessary to wean the patient from the vasopressin. The side effects associated with vasopressin include fluid retention, volume overload, hyponatremia, and vasoconstriction.[47–49] Analogues of vasopressin, such as terlipressin, have been used widely in Europe because of a longer half-life and less diffuse systemic vasoconstriction. Terlipressin is currently not available in the USA.[50, 51] Octreotide has largely replaced the use of vasopressin and nitroglycerin. This is also a potent splanchnic vasoconstrictor when administered in a supraphysiological dose. It is equally effective as the vasopressin/nitroglycerin combination in control of variceal bleeding control without the hemodynamic side effects. Octreotide is administered continuously in a dose range of 50–100 μg/h after an initial intravenous bolus of 50 μg.[52–54]

Endoscopic therapy

Early upper endoscopy is essential to document the site of variceal bleeding and potentially stop the hemorrhage. The therapeutic role of endoscopic variceal ablation is well established. Endoscopic sclerotherapy consists of the injection of a sclerosing agent (sodium morruate or ethanolamine) into either the lumen of the varix (intravariceal) or the submucosal tissue surrounding the varix (paravariceal) in order to clot or tamponade the varix, respectively (Figure 14.1). In acute variceal bleeding, endoscopic sclerotherapy stops bleeding in 90% of the cases.[55, 56] The technical aspects of variceal sclerotherapy have been studied extensively. No one technique or agent has been found to be significantly better than the others in controlling bleeding.[57] The known risks of endoscopic sclerotherapy include esophageal perforation, esophageal ulceration with bleeding, stricture formation, and, rarely, sclerosant embolization to the pulmonary circulation resulting in respiratory distress.[58] In 1986, the technique of endoscopic variceal band ligation was introduced. This procedure uses a triggering device at the end of the endoscope to deploy small rubber bands around the base of the varix causing interruption of variceal blood flow (Figure 14.2).[59] The recent development of a multiple band ligating device, which allows banding without reinserting the endoscope repeatedly, has led to a preference over variceal sclerotherapy. Multiple controlled trials comparing sclerotherapy with banding suggest that variceal band ligation controls bleeding equally but with lower complication rates.[60–63] As such, it has become the endoscopic therapeutic procedure of choice in most portal hypertensive centers.[64]

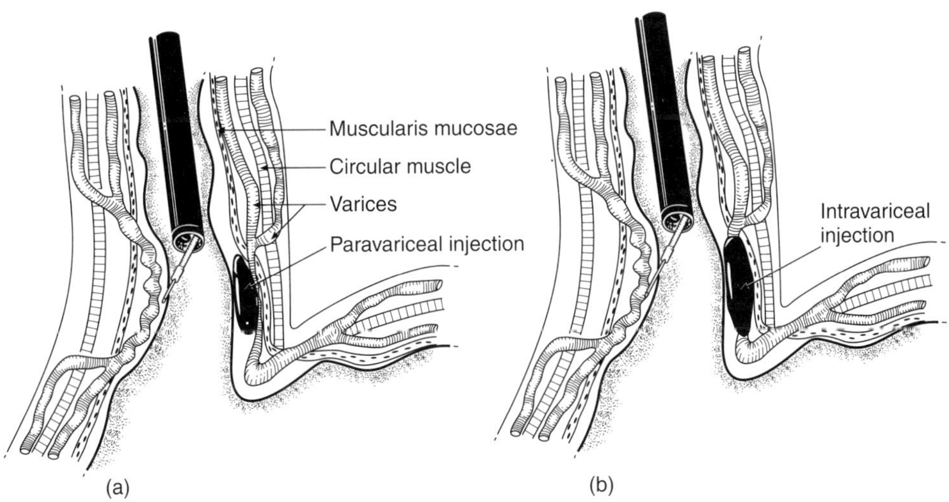

Figure 14.1 *Endoscopic variceal sclerosis. (a) Paravariceal injections tamponade the varix. (b) Intravariceal injections thrombose the varix.*

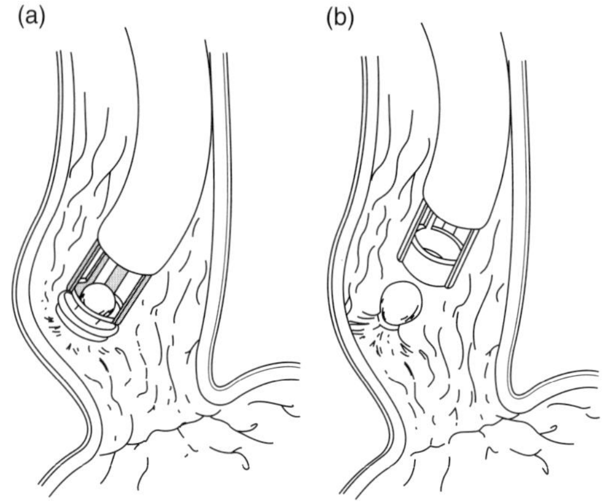

Figure 14.2 *Endoscopic variceal band ligation. (a) Suctioning of varix into ligating device. (b) Ligation of variceal base with O-ring.*

Balloon tamponade

There are rare occasions when neither pharmacological therapy nor endoscopic ablation will stop acute variceal bleeding. In these cases, bleeding tends to be torrential and resuscitation continues to fail. When this occurs, balloon tamponade can be life saving by stabilizing the patient until definitive decompression can be performed. There are two balloon tamponade devices. The older, three-lumen Sengstaken–Blakemore tube and the newer, four-lumen Minnesota tube are currently available (Figure 14.3). Both tubes have a gastric balloon, an esophageal balloon, and a distal gastric port for aspiration of gastric contents. The fourth lumen of the Minnesota tube is a proximal aspiration port above the esophageal balloon for salivary secretions from the cervical esophagus. While the concept of tamponade is simple, actual placement of these tubes can be difficult. By the time the decision is made to institute tamponade, the patient is usually already intubated and ventilated due to continued bleeding and hemodynamic instability. If not, the patient must undergo urgent intubation before insertion of either tube in order to prevent the risk of aspiration. Placement of a well-lubricated tube through the oral pharynx rather than the nasopharynx is recommended, as shown in the package insert for both tubes. It is important to place the gastric balloon completely within the stomach. After inserting the gastric balloon more than 45 cm into the stomach from the incisors, approximately 50 cm^3 of air is injected into the gastric balloon. An upper-abdominal radiograph confirms that the balloon is well within the stomach. 250–350 cm^3 of air is then placed into the balloon. The gastric balloon is gently pulled superiorly against the gastroesophageal junction to tamponade the varices. Care must be taken not to pull the balloon too tightly against the gastroesophageal junction as it is unncessary for stopping the bleeding and only increases the risk of mucosal ischemia and ulceration by the balloon. Inflation of the gastric balloon alone interrupts the inflow of blood from the variceal channels below the gastroesophageal junction to the varices in the distal esophagus where bleeding usually occurs. Bleeding ceases in nearly all cases. If bleeding continues, then inflation of the esophageal balloon becomes necessary. In most patients, 100 cm^3 of air instilled into the esophageal balloon will obtain an appropriate amount of tamponade. The pressure within the esophageal balloon should be monitored every 2 h with a blood-pressure manometer. Air is released or added from the esophageal balloon to keep pressure in the range of 40–45 mm Hg. Balloon tamponade is potentially dangerous. Complications include aspiration, ulceration, and

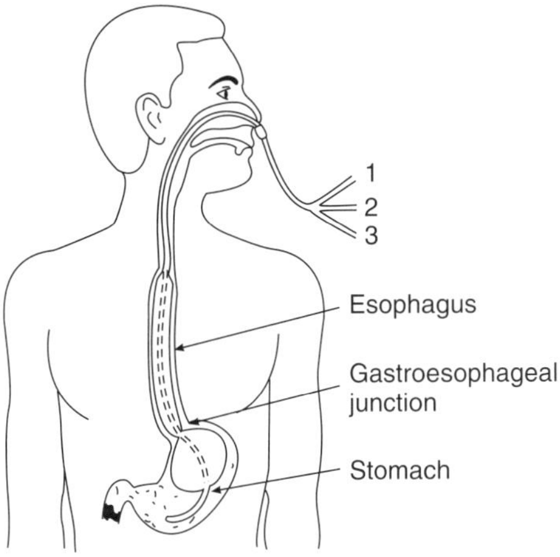

Figure 14.3 *Sengstaken–Blakemore tube for tamponade of acutely bleeding gastroesophageal lumen. 1, gastric balloon lumen; 2, gastric aspiration lumen; 3, esophageal balloon lumen.*

esophageal rupture. Nevertheless, it is an effective means of controlling massive variceal bleeding when all else has failed. Balloon tamponade is a temporizing measure, not a definitive therapy. The balloon should not be left inflated for more than 48 h. Institution of balloon tamponade warrants decompression of the varices via TIPS or surgical shunting once the patient has been stabilized and any coagulopathy and metabolic derangements corrected.[65, 66]

Emergency decompression

When less invasive procedures have failed to stop the variceal bleeding, emergency decompression of the varices becomes necessary. Although surgical shunts have been utilized in the emergency setting in the past, TIPS has now replaced surgical decompression as the procedure of choice for control of acute variceal bleeding unresponsive to pharmacological or endoscopic methods.

TIPS essentially functions as a side-to-side portosystemic shunt as it decompresses not only the varices but the entire portal venous circulation. It does this by diverting portal blood flow away from the hepatic sinusoids through a hepatic parenchymal tunnel created between the high-pressure portal venous system and the low-pressure hepatic veins (Figure 14.4).[67–69] When a TIPS is anticipated, Doppler ultrasonography is necessary to ensure portal vein patency. The TIPS is performed in the angiography suite equipped with good digital imaging. Cannulation of the right internal jugular vein provides the most direct route to the hepatic veins. A diagnostic angiographic catheter is then advanced under fluoroscopic guidance into the hepatic veins. The right hepatic vein is usually selected to create the shunt. A 16- or 21-gauge needle is then used to access the portal venous system by puncturing the hepatic parenchymal through the right hepatic vein into the right portal vein. A guidewire is advanced into the portal vein followed by passage of an angiographic catheter. Venography is performed to define better the portomesenteric venous anatomy. The transhepatic parenchymal tract is then dilated with an angioplasty balloon followed by deployment of a balloon-expandable metallic wall stent. After positioning of the wall stent, the tract is dilated sequentially with larger balloons until the portosystemic gradient has dropped to less than 12 mm Hg. Usually, a 10-mm wall stent is necessary to obtain this drop in portal pressure. Additional stents placed in parallel fashion are often necessary to sufficiently lower portal pressure. A final portal venogram is obtained to assess the stent position and patency, and to identify any persistent

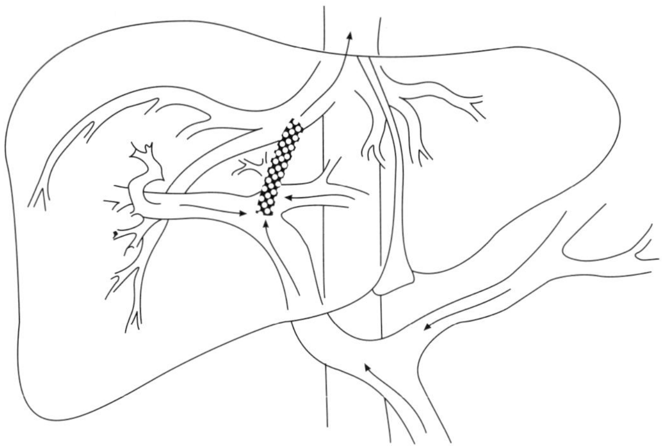

Figure 14.4 *Transjugular intrahepatic portosystemic shunts (TIPS). Arrows indicate direction of porto-mesenteric blood flow through the shunt.*

filling of gastroesophageal varices. Embolization of persistent variceal inflow collaterals is sometimes added to the procedure, although clinica efficacy of this additional step is unproven.

The non-invasive nature of TIPS compared with surgical shunts has made it very popular. Currently, it can be performed with a technical success rate of approximately 90% and a procedural complication rate of 10%. Technical failures occur primarily due to hard, atrophic livers, occluded portal or hepatic veins, and massive ascites that should be drained before placement of the TIPS. Procedural and technical complications are listed in Table 14.3.[70] The long-term stenosis rate is 50% at 12 months, and a 20% re-bleeding rate is seen during the first year. A 30% encephalopathy rate occurs after TIPS since it is a total shunt. Most patients require medical therapy. A rigorous program of follow-up with Doppler ultrasonography every 4–6 months is required after TIPS.[71] Catheterization of the TIPS with angioplasty of any stenosed shunts is necessary to maintain an adequate patency rate.[72] Contraindications to TIPS include right-sided heart failure, severe hepatic failure, polycystic liver disease, and large hepatic neoplasms.[73–75]

The success of TIPS has markedly reduced the need for surgical shunts in the emergency setting. Surgical shunts are now restricted to those patients in whom TIPS is technically unfeasible, or when a trained interventional radiologist is not available. Historically, the most common surgical shunt for emergency control of variceal bleeding has been the portacaval H-graft shunt, which can be performed rapidly with excellent decompression of the portal venous system.[76] However, because of its interference with the portal dissection in subsequent liver transplantation, the mesocaval C-graft shunt is currently recommended and is equally effective.[77, 78] The distal splenorenal shunt (DSRS) is technically demanding and does not lend itself to the emergency setting. A final algorithm utilizing all these modalities for control of acute variceal bleeding is presented in Figure 14.5.

Definitive treatment of recurrent variceal bleeding

From the1940s to the early 1990s, surgical shunt procedures played a dominant role in the management of both acute and recurrent variceal bleeding. With the technical advances in endoscopy, endoscopic variceal ablation as a means for long-term control of recurrent variceal bleeding was considered. Unfortunately, with a re-bleeding rate of 40–50%, its long-term

Table 14.3 *Complications of transjugular intrahepatic portosystemic shunts (TIPS)*

Procedural
Subcapsular hematoma
Hemorrhage from capsular rupture
Misplacement of wall stent
Neck hematoma
Hemobilia
Gallbladder puncture
Acute stent thrombosis
Acute portal vein thrombosis
Systemic
Cor pulmonale
Adult respiratory distress syndrome
Sepsis
Acute hepatic failure
Myocardial infarction
Bacterial endocarditis
Contrast-induced renal failure
Long-term
TIPS stenosis
TIPS thrombosis
Recurrent variceal bleeding
Hepatic encephalopathy

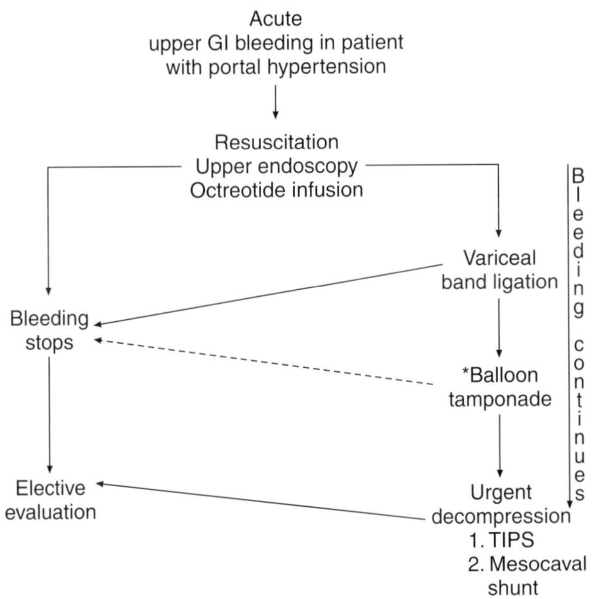

Figure 14.5 *Algorithm for the emergency treatment of bleeding from gastroesophageal varices. *Balloon tamponade usually mandates urgent decompression. GI, gastrointestinal; TIPS, transjugular intrahepatic portosystemic shunt.*

benefits have not been realized fully. In the mean time, technical advances in orthotopic liver transplantation along with improved immunosuppression led to its acceptance in the treatment of patients with end-stage liver disease. Liver transplantation offered the advantage of controlling variceal bleeding by removing portal hypertension altogether as well as restoring liver function to normal. Unfortunately, a limited donor pool has made transplantation unacceptable as a primary means of controlling variceal bleeding. Since the 1990s, the use of TIPS has become ubiquitous. Its original role in acute variceal bleeding and as a bridge to transplantation in patients with poor liver function has expanded and led to its overuse, even though it has a relatively high re-bleeding rate and requires constant monitoring.

As a result of these developments, surgical shunts now appear to play a more secondary role.[79] They are performed rarely in the community, and even in major portal hypertensive centers the number of shunts has decreased from well over 100 a year in the early 1980s to less than 25 per year currently. Furthermore, these shunts are being performed in a more highly selected group of patients than ever before. Nevertheless, a working knowledge of the various shunt procedures is necessary since they still play an important, though less dominant, role in the long-term control of variceal bleeding.

The goal of all shunts is to reduce variceal pressure and, thereby, stop acute variceal bleeding and prevent recurrent variceal bleeding. All shunts, both total and selective, accomplish this goal more than 90% of the time, even though long-term shunt patency rates vary. The difference between total, partial, and selective surgical shunts is the degree in which total perfusion is maintained to the cirrhotic liver, and the effect that altered perfusion may ultimately have on liver function. Total shunts are large-diameter shunts that decompress the entire portomesenteric venous system by diverting all blood flow out of and away from the liver. Total shunting occurs because the patent shunt offers a pathway of lower resistance to portal flow than the higher resistance of the diseased liver. Partial shunts are similar to the TIPS in that they are 8–10 mm in diameter, and they propose to reduce variceal pressures while maintaining some degree of portal flow. Selective shunts lower the variceal pressures by decompressing only the gastric and splenic components of the portomesenteric venous circulation, while maintaining portal perfusion into the cirrhotic liver. In most cases, the surgeon considering surgical decompression should perform the type of shunt that he or she knows how to do best. Being concerned about the debated pros and cons of each shunt is appropriate if the surgeon can perform all shunts equally well. However, the advantages of one particular shunt over another are potentially lost if the surgeon is considering a shunt in which they have little experience.

TOTAL SHUNTS

A total portosystemic shunt can be accomplished in several ways. The classic end-to-side portacaval shunt divides the portal vein just below its bifurcation, and turns the mesenteric end of the portal vein down to the infrahepatic inferior vena cava (IVC) where an end-to-side anastomosis is created.[80] This end-to-side shunt completely diverts portal blood flow away from the liver. Unlike other total shunts, it does not decompress the obstructed sinusoids because the portal vein is unable to act as an outflow tract. As such, the end-to-side portacaval shunt controls variceal bleeding extremely well but does not alleviate ascites. It has an excellent long-term patency rate, exceeding 90%. This shunt has been abandoned due to its anatomical interference with subsequent liver transplantation.[81]

All other types of total portosystemic shunts are variations of the side-to-side portacaval shunt. Each of these shunts involve joining the portal vein or one of its major tributaries to the IVC or one of its major tributaries. They include the side-to-side portacaval shunt, the portacaval H-graft shunt, the mesocaval C-graft shunt, the mesorenal shunt, and the central splenorenal shunt (Figure 14.6). In general, the closer the shunt is to the liver, the greater the patency rate because the veins involving the anastomosis are larger and have a greater blood flow. While prosthetic materials may increase the risk of thrombosis, their use may obviate the need for extensive mobilization of veins used to form the shunt. Mobilization of these veins

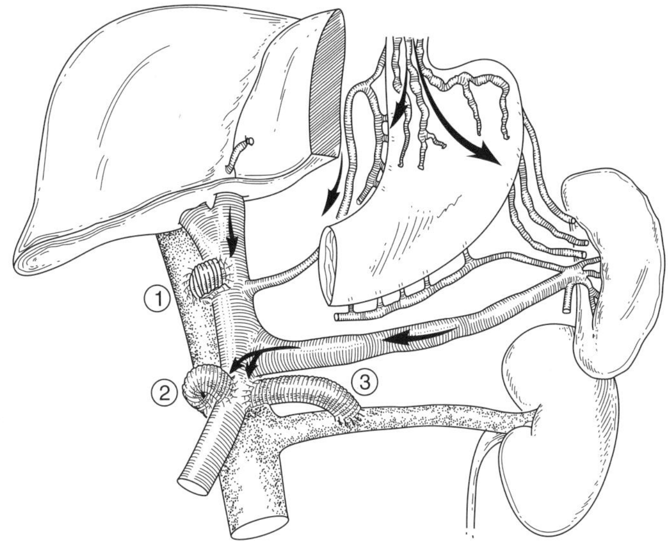

Figure 14.6 *Various interposition total portosystemic shunts. 1, portacaval H-graft shunt; 2, mesocaval C-graft shunt; 3, mesorenal shunt.*

can be technically demanding and hazardous in the portal hypertensive field. Factors that determine the choice of shunt include the technical feasibility based on preoperative radiographic evaluation of the relevant vessels, the surgeon's experience and preference, and the potential need for subsequent liver transplantation.

The side-to-side portacaval shunt is the most difficult total shunt to perform, because of the distance between the IVC and the portal vein. Extensive dissection and mobilization of both veins is required to create the anastomosis. An enlarged caudate lobe may contribute to its technical difficulty. Thrombosis is also a problem when too much tension on the side-to-side anastomosis occurs.[82] These technical difficulties led to the interpositioning of a vascular prosthesis between the portal vein and IVC, the so-called portacaval H-graft shunt. Placement of a short piece of Gore-TexTM (polytetrafluoroethylene, PTFE) graft 18 mm or more in size obtains a reasonably low thrombosis rate and accomplishes the same decompressive goals as a direct side-to-side portacaval shunt.[83] This shunt has become the procedure of choice in many portal hypertensive centers. It is technically easier and can be performed rapidly. It is not feasible in portal vein thrombosis. Both the side-to-side portacaval shunt and the portacaval H-graft shunt alleviate ascites extremely well. The extensive portal dissection may make future liver transplantation more difficult, therefore other total shunt alternatives should be considered if the patient is felt to be a potential liver transplant candidate.[84, 85] These technical concerns have made the interposition mesocaval C-shunt the total shunt of choice for control of variceal bleeding. It is perhaps the easiest, most rapid, and safest of all total decompressive shunt procedures. As a functional side-to-side total shunt, it consistently stops variceal bleeding and controls ascites. As the shunt is performed below the transverse mesocolon, away from the porta hepatis, it does not interfere with future liver transplantation. A previously unacceptable shunt thrombosis rate of 20% with earlier graft materials has been improved with autologous vein grafts or ringed PTFE. The gentle C-configuration of the shunt also decreases shunt thrombosis. These technical advances have yielded a very acceptable long-term patency rate of 90%.[86, 87] Additional variations of the side-to-side portacaval shunt include the central splenorenal shunt and the mesorenal shunt.[88] These shunts are technically more difficult and have a higher thrombosis rate – approaching 30–40% – thus they are used rarely.

The technical aspects of the mesocaval shunt are considered here briefly. The mesocaval shunt can be performed through either a bilateral subcostal or long midline incision. The transverse mesocolon is elevated and the middle colic vein is traced to its junction with the SMV. Palpation of the SMA may help locate the SMV. However, anatomic relationships are inconsistent, and a wide transverse incision at the base of the mesentery best identifies the SMV. The SMV is freed from the surrounding connective tissue and lymphatics before it passes beneath the pancreas. Small venous tributaries of the SMV are divided, but the major branches are left intact. Division of the middle colic vein facilitates a more proximal exposure to the SMV. At least 6 cm of the vein is exposed. The IVC is identified through the right colonic mesentery adjacent to the duodenum. The anterior surface of the IVC must be cleared for several centimeters below the renal veins. Mobilization of the right colon may facilitate adequate exposure of the inferior vena cava, but this may be a bloody exercise in portal hypertension. The C-loop of the duodenum is partially mobilized and retracted superiorly for several centimeters to further exposure of the IVC. A 16–20-mm ringed PTFE graft is trimmed to appropriate length. The IVC is occluded partially with a side-biting vascular clamp, and a small button or ellipse approximating the size of the graft is cut from the IVC. The IVC anastomosis is performed with a continuous 4-0 vascular suture being careful to evert the anastomosis. Proximal and distal control of the exposed SMV is then obtained. The relatively large venous tributaries entering the posterior lateral and posterior medial aspects of the SMV are also controlled. Small vascular clamps are placed proximally and distally on the SMV, and the graft is then anastomosed to the right anterolateral side of the SMV using a continuous 4-0 suture. Care is taken to avoid twisting or kinking of the graft. The final graft configuration takes on a gentle C-shape as it arises from the anterior surface of the IVC and circles around the third portion of the duodenum to enter into the right anterolateral side of the SMV. This C-configuration of a shunt allows for a relatively easy and large anastomosis to be performed between the graft and the main portion of the SMV, thus contributing to an improved patency rate.[89, 90]

Total portosystemic shunts have, to a large extent, been replaced by the TIPS procedure. Nevertheless, historically, they have contributed significantly to the surgical control of variceal bleeding, with a combined success rate that has exceeded 90% over the long term in most patient populations.[91] Their additional value in controlling ascites formation is well described. The primary disadvantage of these shunts has been the significant encephalopathy rate approaching 30–40% due to diversion of portal blood flow from the hepatic parenchyma. Patient outcome has been determined ultimately by the severity of the underlying liver disease and its response to this loss of portal perfusion. These shunts have changed the mechanism of death from variceal bleeding to progressive hepatic failure, and long-term survival rates have failed to improve.[92–95] Despite these unfortunate facts, the use of total shunts can be recommended in the following:

- urgent control of acute variceal bleeding if TIPS is unavailable or technically impossible;
- as a long-term bridge to hepatic transplantation in patients with recurrent, medically intractable variceal bleeding, and compromised liver function, if TIPS is not available;
- as a method of surgical decompression to relieve intractable ascites in patients where a TIPS cannot be performed;
- in patients who bleed from rectal, colonic, or stomal varices when a TIPS cannot be performed;
- for acute decompression of the congested liver in Budd–Chiari syndrome.

PARTIAL SHUNTS

Partial portacaval shunts developed as a compromise resulting from the technical demands of selective shunts and the metabolic problems from portal blood flow deprivation seen in total shunts. Partial portacaval shunts attempt to maintain portal perfusion while decompressing the varices and avoid the technical difficulties seen in selective shunts. The operation, as described by Sarfeh and Reopens, is placement of a small 8–10-mm interposition portacaval shunt of ringed PTFE graft combined with ligation of the coronary vein and other collateral vessels.[96] Postoperative portal perfusion is best preserved when a smaller shunt is utilized. Eighty per cent of patients with an 8-mm H-graft shunt maintain portal perfusion. When a 10-mm graft is placed, portal perfusion is maintained in only 50%. Control of variceal bleeding is approximately 90%, but unfortunately a 20% shunt stenosis or occlusion is related directly to the small size of the shunt. Overall patency rate can be improved in stenosed or occluded shunts when they are managed by radiographic interventional techniques. Since portal perfusion is maintained to some extent, encephalopathy rate is 10–15%.[97–99] It appears that the partial shunts are very similar in their hemodynamics and physiology to the radiographically placed TIPS. Their use may be questioned since a major operation is required.[100] Indeed, like other shunts, TIPS has relegated partial shunts to a secondary role.

SELECTIVE SHUNTS

Selective shunts are designed to lower the pressure within the gastroesophageal varices while maintaining portal hypertension and prograde portal blood flow into the liver. The most common of these operations is the distal splenorenal shunt, originally reported by Warren and colleagues in 1967 (Figure 14.7).[101] Variations of this shunt exist, including the distal splenorenal shunt with splenopancreatic disconnection, the distal splenocaval shunt, and the coronary caval shunt. The concern over the increased incidence of hepatic encephalopathy, progressive liver failure, and poor survival in patients undergoing total shunts led to the concept of selective variceal decompression. It was postulated that by maintaining portal perfusion to the diseased liver, then the risk of encephalopathy and liver failure would decrease while control of variceal bleeding would be obtained. The ultimate goal was to improve the long-term survival rate after selective shunting in patients with cirrhosis who would otherwise die of progressive liver failure due to loss of portal perfusion from a total shunt.[102–104]

The DSRS has been used in clinical practice for over 30 years and is the most common selective shunt performed. It consists of an anastomosis between the distal end of the splenic vein to the side of the left renal vein (LRV). This results in selective decompression of the gastroesophageal varices by the trans-splenic flow of blood from the high-pressure varices through the short gastric veins, across the splenic parenchyma, into the splenic vein, and out of the low-pressure LRV. The additional ligation of the coronary vein, the gastroepiploic vein, and the gastrocolic veins in the splenic flexure serves to disconnect the right side of the portal system from its gastrosplenic components, leading to maintenance of portal hypertension and prograde flow into the liver.[105] Anatomically, the DSRS is contraindicated when the splenic vein is thrombosed. Therefore, suitable anatomy to perform the DSRS must be defined by preoperative angiography, which includes a selective splenic arteriogram with venous phases to visualize the patency, size, and course of the splenic vein. Selective cannulation of the LRV is necessary to determine the juxtapositioning between the splenic vein and the LRV as well as anatomical variations, such as a retroaortic or circumaortic LRV, which may contribute to a difficult anastomosis. Finally, a selective SMA arteriogram with venous phase defines the presence of portal vein or SMV thrombosis and identifies the size, number, and location of coronary vein collaterals that must be ligated to ensure selectivity of the DSRS.

The DSRS is performed through a left bilateral subcostal incision. The lesser sac is entered through the gastrocolic ligament with the greater curvature of the stomach being devascularized extensively from the pylorus to the short gastric veins. Several short gastric veins must be left intact as they are a critical component to trans-splenic decompression of the gastro-

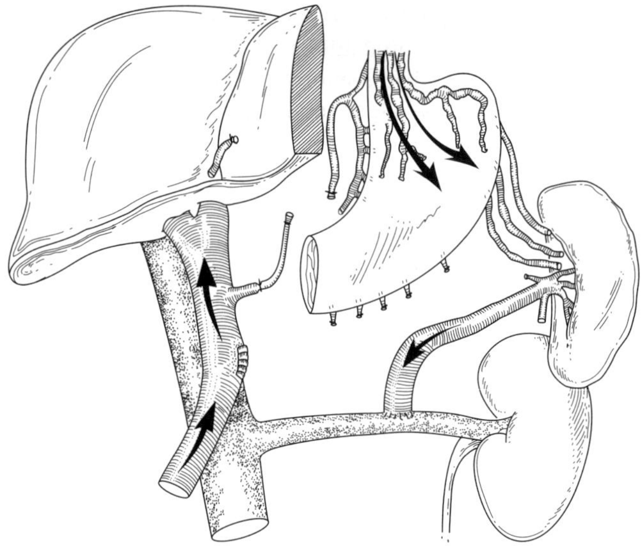

Figure 14.7 *Distal splenorenal shunt.*

esophageal varices. Division of the splenocolic ligament allows greater access to the tail of the pancreas and the proximal splenic vein, which is valuable if extensive dissection of the splenic vein is required. The retroperitoneal tissues along the inferior border of the pancreas are incised. The pancreas is elevated anteriorly and rotated superiorly to expose the splenic vein, which lies on the posterior aspect of the pancreas. The junction of the splenic vein with the SMV is then defined and the splenic vein encircled with a vessel loop. The anterior and posterior surfaces of the SMV and portal vein axis are cleared, which identifies any coronary veins that may enter either the portal vein or splenic near this junction. When identified, it should be ligated carefully. Dissection of the splenic vein from the pancreatic groove is the most difficult aspect of the shunt. It proceeds from the junction of the splenic vein with the SMV towards the tail of the pancreas. Dissection should occur directly on the vein with clearance of all perivenular tissues after completely freeing the posterior aspect of the splenic vein from the retroperitoneum. Exact identification and careful ligation of the multiple pancreatic venous tributaries is performed. The length of vein removed from the pancreatic groove is determined by the amount required to ensure a tension-free anastomosis, which is estimated by the splenic vein's juxtapositioning with the LRV. The search for the LRV begins by dividing the retroperitoneal tissues inferior to the pancreas. The surrounding lymphatics should be ligated meticulously in order to avoid postoperative chylous ascites. The LRV is identified as it courses over the aorta and is elevated out of the retroperitoneum with careful dissection. The adrenal vein is ligated and divided, but other tributaries of the LRV are left intact. The distal splenic vein is divided near its junction with the SMV, and the stump of the splenic vein is oversewn with a continuous 5-0 vascular suture. The LRV is occluded partially with a side-biting vascular clamp, and a longitudinal venotomy is made. The splenic vein is then anastomosed to the anterior superior surface of the LRV using a continuous 5-0 vascular suture for the posterior row and multiple interrupted 5-0 vascular sutures anteriorly to allow the shunt to balloon open when blood flow is restored. It is usually necessary to trim a portion of the splenic vein to prevent redundancy of the vein, which can lead to kinking or twisting, and early shunt thrombosis. Following completion of the DSRS, the coronary-azygos disconnection is performed by further identification and ligation of the coronary vein and several of its tributaries above the pancreas as they enter into the lesser curve of the stomach.[106, 107]

In highly selected patients with good liver function, i.e. Child's A and B cirrhotics, the overall mortality rate in nine published series of DSRS is 6%. The variceal re-bleeding rate is 5.9%. and the shunt thrombosis rate is 4.4%. The overall incidence of hepatic encephalopathy at 4 years is 15%, which is about 50% less than total shunts. The overall long-term survival rate, after distal splenorenal shunt, unfortunately, has not improved when all patients are considered.[108] However, there is a tendency towards improved survival in patients with good hepatic function and patients that have non-alcoholic cirrhosis. Abstinence from alcohol in patients with compensated alcoholic cirrhosis may also yield a survival advantage after DSRS.[109, 110] As with the total shunts, the survival for patients undergoing selective shunting depends on the underlying hepatic function, with 1-year survival rates ranging from 75 to 90%, and 5-year survival rates ranging from 50 to 70%.[111, 112] The key to obtaining good results with the DSRS is to select patients with adequate liver function in whom there is a reasonable life expectancy from their liver disease. Postoperative complications after the DSRS include intra-abdominal bleeding requiring re-exploration, infections, and early postoperative liver failure, which should be low if the patient has been chosen well. Postoperative complications unique to the DSRS include worsening of cirrhotic ascites, chylous ascites, and non-occlusive portal vein thrombosis, which usually resolves over several weeks. Splenic vein thrombosis can occur, resulting in loss of the shunt, re-exploration, and splenectomy with devascularization.[113] Gastric paresis due to a temporary loss of the antral pump is seen occasionally. Surprisingly, acute pancreatitis after DSRS is rare, despite the significant manipulation of the gland during surgery.

Variations of the distal splenorenal shunt have also been devised to address several physiological and technical problems. The DSRS with splenopancreatic disconnection was advanced in the early 1980s by Warren and colleagues to address the facts that 60% of patients with alcoholic cirrhosis undergoing selective shunting lost portal perfusion within 1 year after DSRS, and that long-term survival was no better than total portosystemic shunting.[114] It was observed that this loss of portal perfusion after DSRS occurred primarily via a system of transpancreatic collaterals, known as the pancreatic siphon, which developed between the high-pressure portal system and the low-pressure splenorenal anastomosis.[115] It was hypothesized that loss of hepatotropic factors through this pancreatic siphon would result in decreased liver size and function and poor survival. Disconnection of the splenic vein completely from the pancreas would disrupt the pancreatic siphon and preserve portal perfusion and selectivity of the DSRS.[116] Inokuchi and Sugimachi described a similar collateralization after selective shunting, which they termed portal malcirculation, and corroborated its physiological significance to the loss of portal perfusion.[117] Studies showed that the DSRS with splenopancreatic disconnection did maintain the selectivity of the shunt and was associated with improved maintenance of hepatic function and good control of variceal bleeding. The 4 year survival rate of 70% in both the alcoholic and the non-alcoholic populations suggested that the DSRS with splenopancreatic disconnection had achieved its goal.[118] Unfortunately, this advancement in the DSRS remains rarely used as it presents an increased technical challenge, particularly in those patients who need it the most, i.e. the alcoholic population.

The selective splenocaval shunt was developed in response to the presence of congenital abnormalities of the left renal vein, such as a completely retroaortic vein or a circumaortic renal vein with a very small anterior component.[119] In these cases, left renal vein outflow can be compromised, leading to ineffective decompression of the gastroesophageal varices. This problem can be overcome and the principals of selective shunting maintained by constructing a direct splenocaval shunt when the left renal vein is not suitable for the DSRS. Finally, selective decompression of gastroesophageal varices can be achieved by creation of a coronary caval shunt, also termed the left gastric venocaval shunt. This operation, which was developed in Japan, is performed primarily in the post-hepatitis cirrhotic population, where the coronary vein tends to be a single, large tributary feeding into the gastroesophageal junction. Because of its significant size and venous wall thickness, it is very suitable for dissection and anastomosis into the inferior vena cava. Operative mortality rate is 3%, and the variceal re-bleeding rate is approximately 8%.[120] Ultimate survival rates, once again, are dependent on underlying liver function and are not significantly different from distal splenorenal shunt.

Devascularization procedures

These procedures consist of splenectomy, gastric and esophageal devascularization, and occasionally esophageal transection. Their purpose is to decrease inflow of portal blood into the gastroesophageal varices. A willingness to perform a very extensive devascularization of the stomach and the distal 8 cm of the esophagus will achieve reasonable bleeding control. Nevertheless, re-bleeding rates vary from 10% to 40% and are probably dependent on patient anatomy, technical factors, and the experience of the surgeon. Clearly, the risk of re-bleeding is probably higher than seen in decompression procedures. In favor of devascularization is the maintenance of portal hypertension and portal perfusion. Therefore, encephalopathy rates mimic those seen with selective shunts. Devascularization procedures enjoy wide popularity outside the USA and Europe, where shunts are preferred. The indication for these procedures is for patients who have extensive portal and mesenteric venous thrombosis and in whom no usable veins for formation of a shunt exist.[121, 122]

RECURRENT VARICEAL BLEEDING: DEFINITIVE MANAGEMENT

Definitive treatment of recurrent variceal bleeding should be considered a multidisciplinary approach involving both non-operative and operative alternatives. Non-operative alternatives include pharmacotherapy, endoscopic ablation, balloon tamponade, and TIPS. Operative approaches include total and selective shunting procedures, devascularization procedures, and ultimately liver transplantation. Liver transplantation provides the most ideal means for portal decompression because it also restores normal hepatic function to the patient. Unfortunately, a limited donor pool precludes this treatment modality from being used as a primary treatment for variceal bleeding alone, particularly if the patient has well-compensated liver disease. The question arises as to the role each of these modalities actually plays in the treatment of variceal bleeding. Randomized trials in the literature would lead one to believe that these are competing therapies, but clinical experience has shown that the various treatments have distinct and complementary roles. The selection of a specific therapy is based on numerous factors, including type of liver disease present, underlying hepatic function, location of the bleeding varices, candidacy for liver transplant, where the patient lives, availability of tertiary health care, and the associated co-morbidity factors.

Ultimately, the definitive therapy chosen should balance the risk of re-bleeding should that therapy fail versus the risk of liver failure brought on by the institution of that therapy and/or by the natural history of the underlying liver disease. Many management algorithms have been devised that have taken all of these factors into consideration. A typical algorithm is offered in Figure 14.8. This emphasizes the need for a multidisciplinary team approach, which, if followed, can lead to reasonable control of variceal bleeding. This algorithm has been generated from certain guiding principles:

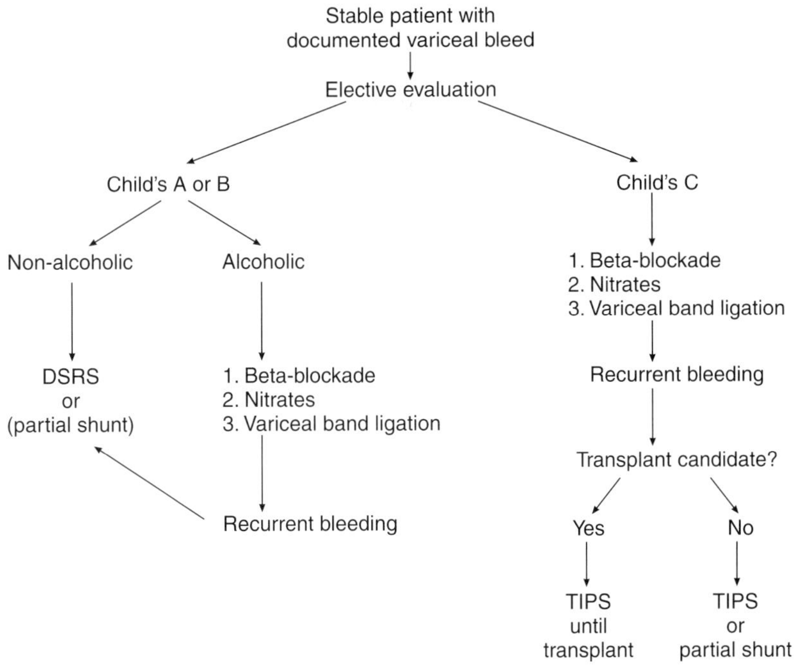

Figure 14.8 *Algorithm for definitive management of recurrent variceal bleeding. DSRS, distal splenorenal shunt; TIPS, transjugular intrahepatic portosystemic shunt.*

1 full evaluation of the patient after stabilization from the acute variceal bleeding episode;
2 institution as first-line therapy of both beta blockade and endoscopic ablation, which should lower the re-bleeding from 80% to approximately 40%;[123–130]
3 radiographic decompression via TIPS in those patients who fail first-line therapy if their hepatic function is poor;[131–136]
4 surgical decompression, preferably with selective shunting for first-line therapy failures, if hepatic function is well compensated;[137–139]
5 liver transplantation for appropriate candidates with end-stage liver disease, using the TIPS as a bridge to transplantation, if first-line therapy fails to control variceal bleeding;[140–146]
6 utilization of TIPS as a salvage procedure in patients with recurrent variceal bleeding and poor function who are not liver transplantation candidates.

When considering what type of shunt surgery should be performed, the above algorithm shows a preference for selective shunting when possible. However, the data supporting the concept of selective shunts over total shunts are mixed at best.[147–155] Nevertheless, it is thought that maintaining portal perfusion is beneficial in preventing encephalopathy and progressive liver failure. It is also apparent that patients with well-compensated liver function can do well in the short term, regardless of the type of shunt performed, with variceal bleeding being controlled at least 90% of the time. Experienced portal hypertensive surgeons therefore tend to perform the shunt in which they have the most experience. With the advent of TIPS, however, shunt surgery has been pushed into a more secondary role. Perhaps one of the most important questions remaining is whether TIPS, in its avoidance of major surgical stress, offers any short- or long-term benefits over selective shunting in good-function cirrhotic patients. Hopefully, an ongoing multicenter, randomized, controlled trial in the USA will yield a definitive answer.

ASCITES

The presence of ascites in the patient with portal hypertension is usually a sign of hepatic decompensation, and is a significantly negative prognostic factor worse than variceal bleeding. The patient with tense ascites is not only subject to abdominal discomfort, anorexia, and respiratory compromise, but may also exhibit other major complications, such as spontaneous bacterial peritonitis and hepatorenal syndrome. Ascites almost always places a patient in the Child–Pugh C category. The incidence of variceal bleeding is also increased when ascites is present. The problem of ascites is enhanced further by overzealous attempts at medical management resulting in severe electrolyte disturbances, renal failure, and hepatic encephalopathy. Therefore, the presence of poorly controlled ascites should initiate a liver transplant evaluation in appropriate candidates.

Pathogenesis of ascites

Ascites develops in cirrhosis because of disorders in renal function and the formation and reabsorption of hepatic and splanchnic lymphatic fluid.[156] In the kidney, a fall in glomerular filtration rate occurs secondary to a decrease in renal blood flow due to active renal vasoconstriction. These changes are mediated by the sympathetic nervous system, which appears to be sensitive to the rise in sinusoidal pressure seen in cirrhosis. In addition, there are increased levels of vasoconstrictors in the blood, including catecholamines, angiotensin, thromboxane A2, and endothelin. As ascites develops, a decrease in effective plasma volume leads to further increases in the level of these vasoconstrictors as well as elevated plasma renin and aldosterone levels. Ultimately, tremendous sodium and water retention occurs in response to these humoral changes.[157–159] The increased sinusoidal pressure of portal hypertension also results in exudation of lymphatic fluid from the surface of the liver into the peritoneal cavity.[160] The microcirculation of the liver is different from other capillary beds in that there are large pores in the endothelial lining of the sinusoids and no anatomic basement membrane. Therefore, as the venous pressure increases, large amounts of lymphatic fluid weep from the surface of the liver. Both protein and water are lost through this very porous capillary system. This is especially true in Budd–Chiari syndrome, in which the sinusoidal pressure is the highest and the loss of protein the greatest. In cirrhosis, capillarization of the sinusoids prevents some protein loss from the sinusoids, but it does not prevent the formation of low-protein ascites.[161] The gradual increase in peritoneal lymph by the above mechanisms eventually overwhelms the ability of the peritoneum to absorb all the fluids, resulting in ascites.[162]

Evaluation and management of ascites

The presence of ascites can be difficult to determine on physical examination. The finding of shifting dullness is neither sensitive nor specific. While tense ascites is readily obvious, many patients require ultrasonography or computerized tomography (CT) scan to determine the presence and amount of ascites. Paracentesis is also an excellent and safe method for characterizing the fluid. Diagnostic paracentesis should measure the cell count, cytology, total protein and albumin content, lactate dehydrogenase, and, if chylous ascites is suspected, triglycerides.[163] If the white blood cell count exceeds 250 cells/mm^3, then the patient most likely has subacute bacterial peritonitis, even if there are minimal symptoms and negative subsequent cultures.[164] Further evaluation of the patient with ascites should also include an assessment of renal function, including a 24-h urine collection to measure creatinine clearance and urinary sodium and potassium.

The management of ascites is primarily medical.[165] Patients with ascites are in positive sodium balance, so reduction in sodium intake will slow the rate of ascites formation. Diets containing 2 g sodium are generally tolerated. Diets containing less than 2 g sodium are unpalatable, leading to a decrease in oral intake and exacerbation of any pre-existing protein energy malnutrition. In most cases, diuretics are required to achieve negative sodium and fluid balance. Diuretics cause a natriuresis with loss of fluid from the plasma. Ascites is then mobilized by the body to maintain effective plasma volume. The goal should be gentle diuresis of approximately 2–2.5 kg/week in order to avoid major electrolyte and renal disturbance. Control of ascites will be achieved in 90% of patients using a combination of aldactone to counteract the secondary hyperaldosteronism present in cirrhosis and the judicious use of a loop diuretic, such as frusemide. When effective diuresis cannot be achieved, then removal of a large volume of ascites via therapeutic paracentesis has been shown to be an effective way to provide symptomatic relief to patients with massive ascites. If less than 5 l of ascites is removed, then large-volume paracentesis is well tolerated without the infusion of 5% albumin.[166, 167] Once a patient requires weekly paracenteses to keep the patient comfortable, the morbidity rate increases and the acceptability rate decreases. At this point, the patient has true medically intractable ascites and should be referred for liver transplantation evaluation. The use of TIPS to alleviate the portal hypertension and decrease the rate of ascites formation should also be considered at this time. Sixty-five per cent of patients undergoing TIPS will completely respond and no longer require paracentesis. An additional 15% will require less frequent paracentesis after TIPS.[168–171] The use of peritoneovenous shunts, such as the LeVeen valve or Denver shunt, for the management of intractable ascites has largely disappeared and is primarily of historical interest. Their use should be limited to patients with truly intractable ascites in whom the TIPS procedure cannot be performed and who are not considered transplant candidates.[172–175]

SPECIAL TOPICS IN PORTAL HYPERTENSION

Budd–Chiari syndrome

Budd–Chiari syndrome is hepatic venous outflow obstruction leading to progressive liver damage and severe portal hypertension. This post-sinusoidal venous obstruction can occur from the efferent venules of the hepatic lobule to the suprahepatic IVC. The resultant back pressure on hepatic parenchyma results initially in centrilobular sinusoidal dilation and congestion, which may progress to hepatocyte necrosis. These changes are reversible if the venous outflow obstruction is relieved. However, progressive hepatocyte death, hepatic fibrosis, and liver failure will occur if the obstruction continues.[176] The etiology of Budd–Chiari syndrome includes local mechanical factors, such as congenital membranous webs of the hepatic veins, as well as obstruction of the venous outflow tracts by malignant neoplasms, such as hepatoma, renal cell carcinoma, or right atrial tumors. Schistosomiasis, hydatid disease, and amebic abscesses are manifestations of parasitic infections, which may cause hepatic venous outflow obstruction. Polycythemia rubra vera and myeloproliferative disorders result in hypercoagulable states that can lead to Budd–Chiari syndrome. Hypercoagulable states resulting from deficiencies in antithrombin III, protein C, and protein S have also been implicated.[177] A special form of Budd–Chiari syndrome is the veno-occlusive disease occurring at the microvascular level in patients receiving antineoplastic therapy or undergoing bone marrow transplantation.[178]

The clinical presentation of Budd–Chiari syndrome varies significantly.[179] Malaise, abdominal pain, abdominal distention with ascites, and hepatomegaly are common findings early on in the disease. Jaundice, increased liver enzyme levels, and coagulopathy occur as progressive hepatic failure ensues. The severity of presentation correlates well with the rapidity of onset as well as the number and extent of hepatic vein involvement. A gradual onset of mild symptoms is seen when obstruction is limited to a single hepatic vein. In this form, the process may be self-limited, and hepatic function is maintained with little therapeutic intervention. In contrast, sudden occlusion of all major hepatic veins can lead to rather acute onset of severe symptoms, as associated hepatocellular necrosis rapidly develops and progresses to fulminant hepatic failure. If the patient survives the acute necrosis phase, then the hepatic parenchyma is replaced irreversibly by fibrosis, and chronic liver failure results. Jaundice, coagulopathy, intractable ascites, variceal bleeding, and hepatic encephalopathy then dominate the clinical picture.

The diagnosis of Budd–Chiari syndrome, particularly in its insidious form, requires a high index of suspicion.[180] Doppler ultrasonography suggests the diagnosis by showing the triad of absent hepatic vein flow, sluggish portal vein flow, and an increased hepatic artery resistance index. CT scans and MRI show a grossly swollen liver with compensatory hypertrophy of the caudate lobe. CT and MRI also rule out other clinical diagnoses, define defects in the inferior vena cava or hepatic veins, and visualize predisposing factors such as tumors or liver abscesses. Angiography remains the gold standard test for diagnosis. A hepatic vein study that establishes a spider-web pattern of small hepatic veins or demonstrates a membranous venous web is diagnostic. If surgical decompression is anticipated, angiography also helps to define the patency of the portomesenteric venous axis and determine the degree of IVC compression by the enlarged caudate lobe. Bilobar liver biopsies confirm the diagnosis of Budd–Chiari syndrome. Historical quantification of the degree of parenchymal damage also determines therapeutic management. Pathological features, which may vary between lobes, include sinusoidal dilation and congestion, hepatocyte necrosis, and hepatic fibrosis. Finally, a complete hematological assessment is mandatory, and aggressive treatment of any underlying hypercoagulable state should parallel surgical therapy.[181]

Management of Budd–Chiari syndrome is dictated by the above evaluation, and is summarized in the algorithm in Figure 14.9. Once the diagnosis has been confirmed radiographically, bilobar liver biopsies are necessary to determine whether surgical decompression of the obstructed liver is required. If no centrilobular necrosis or fibrosis is evident, then medical management of underlying symptoms such as ascites is appropriate. Close observation and periodic re-evaluation are essential. If centrilobular necrosis is present, then surgical decompression of the congested liver becomes mandatory. Venous webs can be excised by transcardiac membraneotomy.[182] In most cases, however, conversion of the portal vein to an outflow tract is necessary, which is accomplished by a total portosystemic shunt. Radiological assessment of the patency of the IVC is necessary to determine what type of shunt to perform. If IVC obstruction is present, then a suprahepatic shunt, such as a mesoatrial shunt, is required to decompress the liver. If the IVC is unobstructed, then an infrahepatic shunt, such as a portacaval H-graft shunt or a mesocaval C-graft shunt, is appropriate.[183–185] Once hepatic decompression has been achieved, the histopathology of the liver and hepatic function should return to near normal over a period of weeks. Hepatic encephalopathy due to total shunting may occur early postoperatively, but this usually resolves as the liver heals. The use of warfarin and anti-platelet drugs lessen the risk of thrombosis, especially if a hypercoagulable state is present. Limited experience with TIPS in Budd–Chiari syndrome does exist.[186, 187] Transsuprahepatic venocaval access into the portal system through the hepatic parenchyma can be achieved. Placement of the TIPS leads to direct decompression of the sinusoids into the suprahepatic IVC. Unfortunately, the small size of the stent and the relatively long length of the tunnel increase the risk of occlusion, particularly in the presence of a hypercoagulable state. Finally, if significant hepatic fibrosis is present on biopsy, then the liver damage is

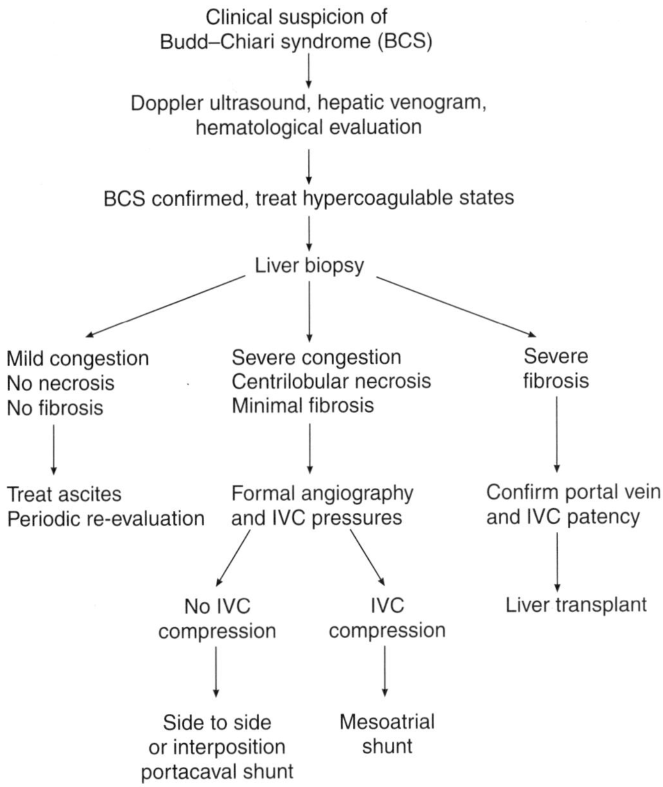

Figure 14.9 *Algorithm for management of Budd–Chiari syndrome. IVC, inferior vena cava.*

irreversible and surgical decompression is of no value. Liver transplant should then be considered. Survival rates following transplantation for Budd–Chiari syndrome exceed 70% at 3 years, with a satisfactory quality of life despite the inherent problems of rejection, immunosuppression, and infection.[188, 189]

Extrahepatic portal vein thrombosis

Extrahepatic portal vein thrombosis (PVT) is a relatively uncommon cause of portal hypertension and variceal bleeding. Management is distinctly different from that of patients with cirrhosis, since these patients usually have a normal liver. Therefore, survival is not limited by the progression of the underlying liver disease. The major goal of therapy is control of variceal bleeding with markedly less concern for liver dysfunction. The causes of PVT are divided into three categories:

- direct causes, including neonatal peritonitis, abdominal trauma, iatrogenic surgical trauma, and cysts or tumors of the porta hepatitis;
- rare congenital abnormalities of the portal vein, such as portal vein stenosis, atresia, and agenesis;
- factors that cause PVT indirectly, such as neonatal systemic sepsis, dehydration, multiple exchange transfusions, and hypercoagulable states.

Finally, approximately 50% of all cases of PVT are idiopathic.[190]

The pathophysiology of portal hypertension caused by PVT is different from that seen in cirrhosis. The sinusoidal block in cirrhosis leads to elevated sinusoidal pressures and the formation of hepatofugal collaterals. In contrast, the presinusoidal block of PVT is associated with a normal hepatic sinusoidal pressure and a high pressure in the obstructed splanchnic bed. This results in the formation of multiple hepatopedal collaterals from the high-pressure splanchnic bed to the low-pressure sinusoidal bed of the liver. This has been known as 'cavernous transformation' of the portal vein, which is a misnomer and should not be used. A significant amount of total hepatic blood flow is provided through these collaterals, therefore PVT is not synonymous with the absence of portal perfusion.[191]

PVT can present soon after birth, or it may not manifest until adulthood. In infancy, variceal bleeding, ascites, and growth failure are the predominant symptoms. As the child matures, collateralization to the low-pressure hepatic sinusoids allows the ascites and growth failure to resolve. However, there is still significant splanchnic hypertension, which may result in variceal bleeding and hypersplenism. Clinical evaluation of patients suspected of having PVT and normal liver function includes endoscopy to define the presence of gastroesophageal varices and Doppler ultrasonography to detect hepatopedal collateralization. Angiography is diagnostic, as it demonstrates a normal hepatic wedge pressure and clarifies the location and extent of the PVT and resultant collateral development. Liver biopsy is occasionally necessary to document that the patient has a normal liver, as PVT also occurs in patients with cirrhosis. Once the diagnosis of PVT is made, a complete hematological evaluation is mandatory to rule out a hypercoagulable state.[192] Definition and appropriate management of such disorders may prevent subsequent problems of shunt thrombosis, venous intestinal infarction, or other spontaneous thrombosis.

Experience in the treatment of variceal bleeding due to PVT indicates there are three treatment options available:

1 expectant medical therapy;
2 endoscopic variceal ablation;
3 operative intervention, including total shunts, selective shunts, and devascularization procedures.

This group of patients tolerates bleeding episodes very well because of their normal liver; therefore, it is tempting to manage each acute bleeding episode without providing definitive therapy. Expectant conservative medical management may be appropriate in infants with variceal bleeding. Their bleeding seems to be self-limited, and in 50% of cases they outgrow their variceal bleeding without definitive therapy. Surgical intervention, especially shunts, is also technically demanding due to small stature. This conservative approach is not justified in the adult population, where mortality from variceal bleeding and PVT ultimately approaches 20%. It must be remembered that the same normal liver that allows patients to tolerate conservative therapy also makes them excellent candidates for more aggressive treatment. Acute variceal bleeding can be managed initially by endoscopic ablation with sclerotherapy or variceal banding.[193] However, if varices are not obliterated or if there is recurrent bleeding, then there should be a low threshold to proceed with definitive decompression. The use of endoscopic therapy in preventing recurrent variceal bleeding should be confined to those patients in whom a definitive decompression procedure cannot be performed.

Surgical shunts are quite successful in this patient population. If portal vein occlusion is completely extrahepatic, and the patient's left portal vein branch is patent, then an interposition graft between the SMV and the left portal vein is feasible.[194] Otherwise, these patients are ideal candidates for the distal splenorenal shunt if they have a patent splenic vein. Variceal bleeding is controlled and patients maintain good prograde portal perfusion through hepatopedal collaterals. A long-term patency rate exceeding 90% can be obtained, especially

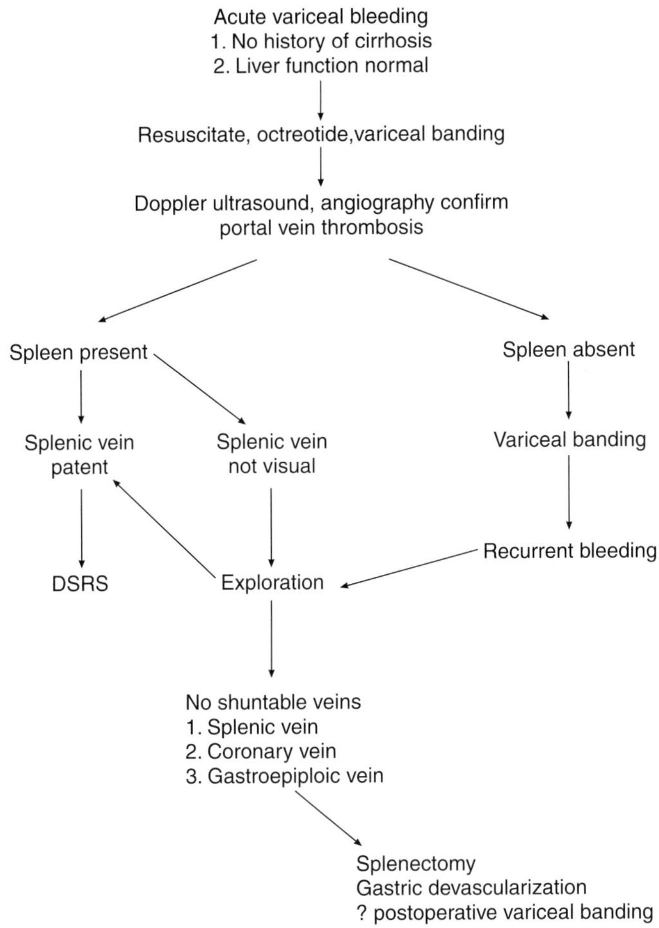

Figure 14.10 *Algorithm for management of variceal bleeding from portal vein thrombosis. DSRS, distal splenorenal shunt.*

when the occasional shunt stenosis seen in this patient population is treated with balloon angioplasty.[195] Total portosystemic shunts have also been used in this group of patients. The central splenorenal shunt and the mesocaval shunt have been the most commonly used. Bleeding control is acceptable, but whether post-shunt encephalopathy develops is unclear.[196] Splenectomy with extensive gastric devascularization has been used with reasonable success. However, the re-bleed rate is higher than in decompression therapy. Therefore, it is not recommended as a primary procedure, unless shuntable veins are not available or in patients with failed shunts.[197] Finally, the patient with PVT who presents with splenomegaly and thrombocytopenia but not variceal bleeding should not undergo prophylactic splenectomy for their low platelet count. The thrombocytopenia is more alarming to the physician than to the patient. Although reduced in number, these platelets function normally and do not contribute significantly to the patient's risk of bleeding. The tragedy of splenectomy in this setting is that these patients lose their best option for definitive operation (DSRS) if they subsequently bleed from their varices. In patients who have undergone prior splenectomy, control of bleeding tends to be more difficult and is more likely to require multiple or continued sclerotherapy.[198] The algorithm shown in Figure 14.10 outlines the management of variceal bleeding in patients with PVT.

FURTHER READING

Cameron JL. Portasystemic shunts. In: Cameron JL (ed.) *Atlas of surgery*, Vol. 1. Toronto: BC Decker, 1990; 252–325.

Henderson JM, Barnes DS, Geisinger MA. Portal hypertension. *Curr Probl Surg* 1998; **35**: 379–452.

Henderson JM, Millikan WJ (eds). W. Dean Warren MD memorial issue on portal hypertension. *Am J Surg* 1990; **160**: 1–138.

Rikkers LF. The changing spectrum of treatment of variceal bleeding. *Ann Surg* 1988; **228**: 536–46.

Rikkers LF, Jin G. Surgical management of acute variceal hemorrhage. *World J Surg* 1994; **18**: 193–9.

Vargas HE, Gerber D, Abu-Elmagd K. Management of portal hypertension-related bleeding. *Surg Clin North Am* **79**: 1–22.

REFERENCES

1 Reynolds T, Ito S, Iwatsuki S. Measurement of portal pressure and its clinical application. *Am J Med* 1970; **49**: 649–57.

2 Boyer TD, Henderson JM. Portal hypertension and bleeding esophageal varices. In: Zakin D, Boyer TD (eds). *Hepatology: a textbook of liver disease*, 3rd edn. Philadelphia: W. B. Saunders, 1996; 720.

3 Douglass BC, Baggenstoss A, Holinshed W. The anatomy of the portal vein and its tributaries. *Surg Gynecol Obstet* 1950; **91**: 562–70.

4 Rousselot L, Moreno A, Panke W. Studies on portal hypertension IV: the clinical and physiopathologic significance of self-established (non-surgical) portal venous shunts. *Ann Surg* 1959; **150**: 384–96.

5 Cappell MS, Price JB. Characterization of the syndrome of small and large intestinal variceal bleeding. *Dig Dis Sci* 1987; **32**: 422–7.

6 Graeber GM, Ratner MH, Ackerman NB. Massive hemorrhage from ileostomy and colostomy stomas due to muco-cutaneous varices in patients with cirrhosis. *Surgery* 1976; **79**: 107–10.

7 Benoit JN, Womack WA, Hernandez L, Granger DN. 'Forward' and 'backward' flow mechanisms of portal hypertension. *Gastroenterology* 1985; **89**: 1092–6.

8 Benoit JN, Granger N. Splanchnic hemodynamics in chronic portal hypertension. *Semin Liver Dis* 1986; **6**: 287–98.

9 Mahl TC, Groszmann RJ. Pathophysiology of portal hypertension and variceal bleeding. *Surg Clin North Am* 190; **70**: 251–66.

10 Groszmann RJ. Hyperdynamic circulation of liver disease 40 years later: pathophysiology and clinical consequences. *Hepatology* 1994; **20**: 59–63.

11 Sitzman JV, Li S-S, Adkinson F. Evidence for role of prostacyclin as a systemic hormone in portal hypertension. *Surgery* 1991; **109**: 149–53.

12 Webb LJ, Sherlock S. The aetiology, presentation and natural history of extrahepatic portal venous obstruction. *QJM* 1979; **48**: 627–39.

13 Klein AS, Cameron JL. Diagnosis and management of the Budd–Chiari syndrome. *Am J Surg* 1990; **160**: 128–33.

14 Galambos JT. *Cirrhosis: major problems in internal medicine*. Philadelphia: W. B. Saunders, 1979.

15 Christensen E, Faverholdt L, Schlichting P, Juhl P, Poulsen H, Tygstrup N. Aspects of natural history of gastrointestinal bleeding in cirrhosis and the effect of prednisolone. *Gastroenterology* 1981; **81**: 944–52.

16 Lebrec D, Fleury P, Rueff B, Nahum H, Benhamou JP. Portal hypertension size of esophageal varices, and risk of gastrointestinal bleeding in alcoholic cirrhosis. *Gastroenterology* 1980; **79**: 1139–44.

17 Garcia-Tsao G, Groszmann RJ, Fisher RL, Conn HO, Atterbury CE, Glickman M. Portal pressure, presence of gastroesophageal varices and variceal bleeding. *Hepatology* 1985; **5**: 419–24.

18 North Italian Endoscopic Club for the Study and Treatment of Esophageal Varices. Prediction of the first variceal hemorrhage in patients with cirrhosis of the liver and esophageal varices. *N Engl J Med* 1988; **319**: 983–9.

19 Graham DY, Smith JL. The course of patients after variceal hemorrhage. *Gastroenterology* 1981; **80**: 800–9.

20 Degradia A. The natural history of esophageal varices in patients with alcoholic liver disease. *Am J Gastroenterol* 1972; **72**: 520–40.

21 Sarin SK, Lahoti D, Saxena SP, Murthy S, Makwana UK. Prevalence, classification and natural history of gastric varices: a long-term follow-up study in 568 portal hypertension patients. *Hepatology* 1992; **16**: 1343–9.

22 D'Amico G, Montalbano L, Traina M, Pisa R, Menozzi M, Spano C, Pagliaro L. Natural history of congestive gastropathy in cirrhosis. *Gastroenterology* 1990; **99**: 1558–64.

23 Cales P. Gastroesophageal endoscopic features in cirrhosis. *Gastroenterology* 1990; **98**: 56–62.

24 Child CG. *The liver and portal hypertension*. Philadelphia: W. B. Saunders, 1964; 50.

25 Pugh RNH, Murry-Lyon M, Dawson JL, Pietroni MC, Williams R. Transection of the oesophagus for bleeding esophageal varices. *Br J Surg* 1973; **60**: 646–9.

26 De Franchis R, Pascal JP, Ancona E, Burroughs AK, Henderson M, Fleig W, *et al*. Definitions, methodology, and therapeutic strategies in portal hypertension. *J Hepatol* 1992; **15**: 256–61.

27 Sherlock S. Chronic hepatitis. In: Sherlock S (ed.). *Diseases of the liver and biliary system*. London: Blackwell, 1989; 348.

28 Conn HO, Atterbury CE. Cirrhosis. In: Schiff L, Schiff ER (eds). *Diseases of the liver*, 6th edn. Philadelphia: Lippincott, 1987; 725.

29 Oliver TW, Sones PJ. *Hepatic radiology*. New York: Macmillan, 1984; 243–75.

30 Bolondi L, Gandolfi L, Arienti V, Caletti GC, Corcioni E, Gasbarrini G, Labo G. Ultrasonography in the diagnosis of portal hypertension: diminished response of portal vessels to respiration. *Radiology* 1982; **142**: 167–72.

31 Burns P, Taylor K, Blei AT. Doppler flowmetry and portal hypertension. *Gastroenterology* 1987; **92**: 824–6.

32 Oliver TW, Sones PH. Hepatic angiography: portal hypertension. In: Bernardino ME, Sones PH (eds). *Hepatic radiology*. New York: Macmillan, 1984; 243–75.

33 Grace ND. Management of portal hypertension. *Gastroenterologist* 1993; **1**: 39–58.

34 Burroughs AK. Primary prevention of bleeding from esophageal varices. *N Engl J Med* 1999; **340**: 1033–5.

35 Jackson FC, Perrin EB, Smith AG, Dagradi AE, Nadal HM. A clinical investigation of the portacaval shunt. II: survival analysis of the prophylactic operation. *Am J Surg* 1968; **115**: 22–42.

36 Resnick RH, Chalmers TC, Ishihara AM, Garceau AJ, Callow AD, Schimmel EM, O'Hara ET. A controlled trial of the prophylactic portacaval shunt: a final report. *Ann Intern Med* 1969; **70**: 675–88.

37 Conn HO, Lindenmuth WW, May CJ, Ramsby GR. Prophylactic portacaval anastomosis: a tale of two studies. *Medicine* 1972; **51**: 27–40.

38 Inokuchi K, Cooperative Study Group of Portal Hypertension in Japan. Improved survival after prophylactic portal nondecompressive surgery for esophageal varices: a randomized controlled trial. *Hepatology* 1990; **12**: 1–6.

39 Kerlan RK, LaBerge JM, Gordon RL, Ring EJ. Transjugular intrahepatic portosystemic shunts: current status. *Am J Roentgenol* 1995; **164**: 1059–66.

40 D'Amico G, Pagliano L, Bosch J. The treatment of portal hypertension: a meta-analytic review. *Hepatology* 1995; **22**: 332–54.

41 Poynard T, Cales P, Pasta L, Ideo G, Pascal JP, Pagliaro L, Lebrec D. Beta-adrenergic-antagonist drugs in the prevention of gastrointestinal bleeding in patients with cirrhosis and esophageal varices: an analysis of data and prognostic factors in 589 patients from four randomized clinical trials. *N Engl J Med* 1991; **324**: 1532–8.

42 Grace ND. Prevention of initial variceal hemorrhage. In: Grossman RJ, Grace ND (eds). *Complications of portal hypertension: esophagogastric varices and ascites*. Philadelphia: W. B. Saunders, 1992; 149–61.
43 Grace ND. Prophylaxis of first variceal bleeding: drug treatment. In: Holstege A, Hahn EG, Scholmerich J. *Portal hypertension (Folk Symposium 79)*. Dordrecht: Kluwer Academic Publishers, 995: 269–77.
44 Navasa M, Chesta J, Bosch J, Rodes J. Reduction of portal pressure by isosorbide-5-mononitrate in patients with cirrhosis. *Gastroenterology* 1989; **96**: 1110–18.
45 Ikigami M, Toyonaga A, Tanikawa K. Reduction of portal pressure by chronic administration of isosorbide dinitrate in patients with cirrhosis: effects on systemic and splanchnic hemodynamics and liver function. *Am J Gastroenterol* 1992; **87**: 1160–4.
46 Angelico M, Carli L, Piat C, Gentile S, Rinaldi V, Bologna E, Capocaccia L. Isosorbide-5-mononitrate versus propranolol in the prevention of first bleeding in cirrhosis. *Gastroenterology* 1993; **104**: 1460–5.
47 Groszmann RJ, Kravetz D, Bosch J, Glickman M, Bruix J, Bredfeldt J, *et al*. Nitroglycerin improves the hemodynamic response to vasopressin in portal hypertension. *Hepatology* 1982; **2**: 757–62.
48 Tsai YT, Lay CS, Lai KH, Ng WW, Yeh YS, Wang JY, *et al*. Controlled trial of vasopressin plus nitroglycerin vs vasopressin alone in the treatment of bleeding esophageal varices. *Hepatology* 1986; **6**: 406–9.
49 Gimson AES, Westaby D, Hegarty J, Alastair W, Williams R. A randomized trial of vasopressin plus nitroglycerin in the control of acute variceal hemorrhage. *Hepatology* 1986; **6**: 410–13.
50 Freeman JG, Cobden I, Lishman AH, Record CO. Controlled trial of terlipressin ('glypressin') versus vasopressin in the early treatment of esophageal varices. *Lancet* 1982; **2**: 66–8.
51 Patch D, Burroughs AK. Advances in drug therapy for acute variceal hemorrhage. *Baillieres Clin Gastroenterol* 1998; **11**: 311.
52 Valenzuela JE, Schubert T, Fogel MR, Strong RM, Levine J, Mills PR, *et al*. A multi-center randomized, double-blind trial of somatostatin in the management of acute hemorrhage from esophageal varices. *Hepatology* 1989; **10**: 958–61.
53 Burrough AK, McCormick A, Hughes MD, Sprengers D, D'Heygere F, McIntyre N. Randomized, double-blind, placebo-controlled trial of somatostatin for variceal bleeding: emergency control and prevention of early variceal rebleeding. *Gastroenterology* 1990; **99**: 1388–95.
54 Hwang JS, Lin CH, Chang CF. A randomized controlled trial comparing octreotide and vasopressin in the control of acute esophageal variceal bleeding. *J Hepatol* 1992; **16**: 320–5.
55 Johnson GW, Roger HW. A review of 15 years experience in the use of sclerotherapy in the control of acute hemorrhage from esophageal varices. *Br J Surg* 1973; **60**: 797–800.
56 Paquet K-J, Oberhammer E. Sclerotherapy of bleeding oesophageal varices by means of endoscopy. *Endoscopy* 1978; **10**: 7–12.
57 Terblanche J, Krige JE. Emergency sclerotherapy. In: Bosch J, Groszmann RJ (eds). *Portal hypertension: pathophysiology and treatment*. Oxford: Blackwell Scientific, 1994; 140.
58 Schyman BM, Berkman JW, Tedesco FJ, Griffin JW, Jr, Assad RT. Complications of endoscopic injection sclerotherapy: a review. *Am J Gastroenterol* 1987; **82**: 823–9.
59 Van Stiegmann G, Goff JS, Sun J, Hruza D, Reveille RM. Endoscopic ligation of esophageal varices. *Am J Surg* 1990; **159**: 21–5.
60 Lane L, Cook D. Endoscopic ligation compared with sclerotherapy for treatment of esophageal variceal bleeding: a meta-analysis. *Ann Intern Med* 1995; **123**: 280.
61 Laine L, Stein C, Sharma V. Randomized comparison of ligation versus ligation plus sclerotherapy in patients with bleeding esophageal varices. *Gastroenterology* 1996; **110**: 529.
62 Slosbert EA, Keefe EK. Sclerotherapy versus banding in the treatment of variceal bleeding. *Clin Liver Dis* 1997; **1**: 77.
63 Saeed ZA, Stiegmann GV, Ramirez FC, Reveille RM, Goff JS, Hepps KS, Cole RA. Endoscopic variceal ligation is superior to combined ligation and sclerotherapy for esophageal varices: a multicenter prospective randomized trial. *Hepatology* 1997; **25**: 71–4.
64 Laine L. Management of actively bleeding esophageal varices. *Gastrointest Endosc* 1997; **46**: 83.
65 Panes J, Teres J, Bosch J, Rodes J. Efficacy of balloon tamponade in treatment of bleeding gastric and esophageal varices. Results in 151 consecutive episodes. *Dig Dis Sci* 1988; **33**: 454–9.

66 Avgerinos A, Klonis C, Rekoumis G, Gouma P, Papadimitriou N, Raptis S. A prospective randomized trial comparing somatostatin, balloon tamponade and the combination of both methods in the management of acute variceal hemorrhage. *J Hepatol* 1991; **13**: 78–83.

67 Colapinto RF, Stronell RD, Gildiner M, Ritchie AC, Langer B, Taylor BR, Blendis LM. Formation of intrahepatic portosystemic shunts using a balloon dilatation catheter: preliminary clinical experience. *Am J Roentgenol* 1983; **140**: 709–14.

68 Richter GM, Palmaz JC, Noeldge G, Garcia F, Sibitt RR. The intrahepatic portosystemic stent-shunt: a new nonsurgical percutaneous method. *Radiology* 1989; **114**: 1511–16.

69 Rossle M, Haag K, Ochs A, Sellinger M, Noldge G, Perarnau JM, *et al.* The transjugular intrahepatic portosystemic stent-shunt procedure for variceal bleeding. *N Engl J Med* 1994; **330**: 165–71.

70 Freedman AM, Sanyal AJ, Tisnado J, Cole PE, Shiffman ML, Luketic VA, *et al.* Complications for transjugular intrahepatic portosystemic shunt: a comprehensive review. *Radiographics* 1993; **13**: 1185–210.

71 Dodd GD, Zajko AB, Orons PD, Martin MS, Eichner LS, Santaguida LA. Detection of transjugular intrahepatic portosystemic shunt dysfunction: value of duplex Doppler sonography. *Am J Roentgenol* 1985; **164**: 1119–24.

72 Fillmore DJ, Miller FJ, Fox LF, Dasyure JA, Tzetze CC. Transjugular intrahepatic portosystemic shunt: mid-term clinical and angiographic follow-up. *J Vasc Interv Radiol* 1996; **7**: 255–61.

73 Coldwell DM, Ring EJ, Rees CR, Zemel G, Darcy MD, Haskal ZJ, *et al.* Multicenter investigation of the role of the transjugular intrahepatic portosystemic shunt in management of portal hypertension. *Radiology* 1995; **196**: 335–40.

74 LaBerge JM, Somberg KA, Lake JR, Gordon RL, Kerlan RK, Jr, Ascher NL, *et al.* Two-year outcome following transjugular intrahepatic portosystemic shunt for variceal bleeding: results in 90 patients. *Gastroenterology* 1995; **108**: 1143–51.

75 Barton RE, Roach J, Saxon RR, Zakin PC, Petersen BD, Keller KS. TIPS: short-and long-term results: a survey of 1750 patients. *Semin Intervent Radiol* 1995; **12**: 364–7.

76 Airlift J, Airlift MS, Airlift SL, Rhomboid M, Girard B. Three decades of experience with emergency portacaval shunt for acutely bleeding esophageal varices in 400 unselected patients with cirrhosis of the liver. *J Am Coll Surg* 1995; **179**: 257–72.

77 Maxillaria V, Too S, Czechs A, Mazzaferro V, Todo S, Tzakis A, *et al.* Liver transplantation in patients with previous portosystemic shunt. *Am J Surg* 1990; **160**: 111.

78 Grace ND. The side-to-side portacaval shunt revisited. *N Engl J Med* 1994; **330**: 208–9.

79 Vader Viet JA, de Viser E, Buskers FG. Changing pattern of portosystemic shunt surgery. *Eur J Surg* 1995; **161**: 877–80.

80 EcK NV. On the question of ligature of the portal vein. *Voen Med Zh* 1877; **130**: 1–22.

81 Brems JJ, Hiatt JR, Klein AS, Millis JM, Colonna JO, Quinones-Baldrich WJ, *et al.* Effect of prior portasystemic shunt on subsequent liver transplantation. *Ann Surg* 1989; **209**: 51–6.

82 Grace ND. The side-to-side portacaval shunt revisited. *N Engl J Med* 1994; **330**: 208–9.

83 Sarfeh IB, Reopens EB, Mason GR. A systematic appraisal of portacaval H-graft diameters. Clinical and hemodynamic perspectives. *Ann Surg* 1986; **204**: 356.

84 Mazzaferro V, Todo S, Czechs AG, Stieber AC, Makowka L, Starzl TE. Liver transplantation in patients with previous portasystemic shunt. *Am J Surg* 1990; **160**: 111–14.

85 Crass RA, Keeffe EB, Pinson CW. Management of variceal hemorrhage in the potential liver transplant candidate. *Am J Surg* 1989; **157**: 476–8.

86 Drapanas T. Interposition mesocaval shunt for treatment of portal hypertension. *Ann Surg* 1972; **176**: 435.

87 Soutter DI, Langer B, Taylor BR, Greig P. Emergency portosystemic shunting in cirrrhotics with bleeding varices: a comparison of portacaval and mesocaval shunts. *HPB Surg* 1989; **1**: 107.

88 Linton RR, Ellis DS, Geary JE. Critical comparative analysis or early and late results of splenorenal and direct portacaval shunts performed in 169 patients with portal cirrhosis. *Ann Surg* 1961; **154**: 446.

89 Cameron JL. Interposition mesocaval shunt. In: *Atlas of surgery*, Vol. I. Toronto: Becher, 1990; 288–97.

90 Henderson JM. Portal hypertension and shunt surgery. *Adv Surg* 1993; **26**: 233.
91 Knechtle SJ, Kalayoglu M, D'Alessandro AM, Pirsch JD, Armbrust MJ, Sproat IA, *et al.* Portal hypertension: surgical management in the 1990s. *Surgery* 1994; **116**: 687–93.
92 Jackson FC, Perrin EB, Felix WR, Smith AG. A clinical investigation of the portacaval shunt II. Survival analysis of the therapeutic operation. *Ann Surg* 1971; **174**: 672–701.
93 Resnick RH, Iber FL, Ishihara AM, Chalmers TC, Zimmerman H. A controlled study of the therapeutic portacaval shunt. *Gastroenterology* 1974; **67**: 843–57.
94 Rueff B, Prandi D, Degos F, Sicot J, Degos JD, Sicot C, *et al.* A controlled study of therapeutic portacaval shunt in alcoholic cirrhosis. *Lancet* 1976; **1**: 655–9.
95 Reynolds TB, Donovan AJ, Mikkelsen WP, Redeker AG, Turrill FL, Weiner JM. Results of a 12-year randomized trial of portacaval shunt in patients with alcoholic liver disease and bleeding varices. *Gastroenterology* 1981; **80**: 1005–11.
96 Sarfeh IB, Reopens EB. Partial versus total portacaval H-graft diameters. *Ann Surg* 1986; **204**: 56–62.
97 Sarfeh IB, Reopens EB. Partial versus total portacaval shunt in alcoholic cirrhosis. *Ann Surg* 1994; **219**: 353–61.
98 Collins JC, Reopens EB, Sarfeh IB. Narrow diameter portacaval shunts for management of variceal bleeding. *World J Surg* 1994; **18**: 211.
99 Adam R, Diamond T, Bismuth H. Partial portacaval shunt: renaissance of an old concept. *Surgery* 1992; **111**: 610.
100 Rosemurgy AS, Goode SE, Zwiebel BR, Black TJ, Brady PG. A prospective trial of transjugular intrahepatic portosystemic stent shunts versus small-diameter prosthetic H-graft portacaval shunts in the treatment of bleeding varices. *Ann Surg* 1996; **224**: 378–84.
101 Warren WD, Zeppa R, Foman JS. Selective transsplenic decompression of gastroesophageal varices by distal splenorenal shunt. *Ann Surg* 1967; **166**: 437–55.
102 Warren WD, Foman JS, Zeppa R. Further evaluation of selective decompression of varices by distal splenorenal shunt. *Ann Surg* 1969; **169**: 652.
103 Bismuth H, Franco D, Hepp J. Portal-systemic shunt in hepatic cirrhosis: does the type of shunt decisively influence the clinical result? *Ann Surg* 1974; **179**: 209.
104 Rikkers LF. Definitive therapy for variceal bleeding: a personal view. *Am J Surg* 1990; **160**: 80–5.
105 Henderson JM. Role of distal splenorenal shunt for long-term management of variceal bleeding. *World J Surg* 1994; **18**: 205.
106 Warren WD, Millikan WJ. Selective transplenic decompression procedure: changes in technique after 300 cases. *Contemp Surg* 1981; **18**: 11.
107 Henderson JM, Warren WD. Portal hypertension. *Curr Probl Surg* 1988; **25**: 177–83.
108 Iannitti DA, Henderson JM. The role of surgery in the treatment of portal hypertension. *Clin Liver Dis* 1997; **1**: 99.
109 Henderson JM, Millikan WJ, Jr, Wright-Bacon L, Kutner MH, Warren WD. Hemodynamic differences between alcoholic and nonalcoholic cirrhotics following distal splenorenal shunt: effect on survival? *Ann Surg* 1983; **198**: 325–34.
110 Zeppa R, Hensley GT, Levi JU, Bergstresser PR, Hutson DG, Sr, Livingstone AS, *et al.* The comparative survival of alcoholics versus non-alcoholics after distal splenorenal shunt. *Ann Surg* 1978; **187**: 510–14.
111 Warren WD, Millikan WJ, Jr, Henderson JM, Wright L, Kutner M, Smith RB 3rd, *et al.* Ten year's portal hypertensive surgery at Emory. *Ann Surg* 1982: **195**: 530–42.
112 Henderson JM, Gilmore GT, Hooks MA, Galloway JR, Dodson TF, Hood MM, *et al.* Selective shunt in the management of variceal bleeding in the era of liver transplantation. *Ann Surg* 1992; **216**: 248–55.
113 Richards WO, Pearson TC, Henderson JM, Millikan WJ, Jr, Warren WD. Evaluation and treatment of early hemorrhage of the alimentary tract after selective shunt procedures. *Surg Gynecol Obstet* 1987; **164**: 530–6.
114 Belghiti J, Grenier P, Nouel O, Nahum H, Fekete F. Long-term loss of Warren's shunt selectivity: angiographic demonstration. *Arch Surg* 1981; **116**: 1121.

115 Henderson JM, Gong-Liang J, Galloway J, Millikan WJ, Jr, Sones PJ, Warren WD. Portaprival collaterals following distal splenorenal shunt: Incidence magnitude, and associated portal perfusion changes. *J Hepatol* 1985; **1**: 649.

116 Warren WD, Millikan WJ, Jr, Henderson JM, Abu-Elmagd KM, Galloway JR, Shires GT 3rd, *et al.* Splenopancreatic disconnection: improved selectivity of distal splenorenal shunt. *Ann Surg* 1986; **204**: 346–54.

117 Inokuchi K, Sugimachi K. The selective shunt for variceal bleeding: a personal perspective. *Am J Surg* 1990; **160**: 48.

118 Henderson JM, Warren WD, Millikan WJ, Galloway JR, Kawasaki S, Kutner MH. Distal splenorenal shunt with splenopancreatic disconnection: A four-year assessment. *Ann Surg* 1989; **210**: 332–9.

119 Atta HM, Henderson JM, Galloway JR, Millikan WJ. Selective splenorenal shunt. *Arch Surg* 1992; **126**: 582–5.

120 Inokuchi K, Kobayashi M, Kusaba A, Ogawa Y, Saku M, Shiizaki T. New selective decompression of esophageal varices by left gastric venous-caval shunt. *Arch Surg* 1970; **100**: 157–62.

121 Idezuki Y, Kakudo N, Sanjo K, Bandai Y. Sugiura procedure for management of variceal bleeding in Japan. *World J Surg* 1994; **18**: 216–21.

122 Dagenais M, Langer B, Taylor BR, Greig PD. Experience with radical esophagogastric devascularization procedures (Sugiura) for variceal bleeding outside Japan. *World J Surg* 1994; **18**: 222–8.

123 D'Amico G, Pagliaro L, Bosch J. The treatment of portal hypertension: a meta-analytic review. *Hepatology* 1995; **22**: 332.

124 Villanueva C, Balanzo J, Novella MT. Nadolol plus isosorbide mononitrate compared with sclerotherapy for the prevention of variceal bleeding. *N Engl J Med* 1996; **334**: 1624–9.

125 Teres J, Baroni R, Bordas JM, Visa J, Pera C, Rodes J. Randomized trial of portacaval shunt, stapling transection and endoscopic sclerotherapy in uncontrolled variceal bleeding. *J Hepatol* 1987; **4**: 159–67.

126 Henderson JM, Kutner MH, Millikan WJ, Jr, Galambos JT, Riepe SP, Brooks WS, *et al.* Endoscopic variceal sclerosis compared with distal splenorenal shunt to prevent recurrent variceal bleeding in cirrhosis: a prospective randomized trial. *Ann Intern Med* 1990; **112**: 262–9.

127 Maffei-Faccioli A, Gerunda G, Neri D, Merenda R, Zangrandi F, Medure F. Selective variceal decompression and its role relative to other therapies. *Am J Surg* 1990; **211**: 178–86.

128 Spina GP, Santambrogio R, Opocher E, Cosentino F, Zambelli A, Passoni GR, *et al.* Distal splenorenal shunt versus endoscopic sclerotherapy in prevention of variceal rebleeding. *Ann Surg* 1990; **211**: 178–86.

129 Spina GP, Henderson JM, Rikkers LF, Teres J, Burroughs AK, Conn HO, *et al.* Distal splenorenal shunt versus endoscopic sclerotherapy in the prevention of variceal bleeding: a meta-analysis of 4 randomized clinical trials. *J Hepatol* 1992; **16**: 338–45.

130 Rikkers LF, Jin G, Burnett DA, Buchi KN, Cormier RA. Shunt surgery versus endoscopic sclerotherapy for variceal hemorrhage: late results of a randomized trial. *Am J Surg* 1993: **165**: 27–33.

131 Conn HO. Transjugular intrahepatic portal-systemic shunts: the state of the art. *Hepatology* 1993; **17**: 148–58.

132 LaBerge JM, Ring EJ, Gordon RL, Lake JR, Doherty MM, Somber KA, *et al.* Creation of transjugular intrahepatic portosystemic shunts with the wallstent endoprosthesis: results in 100 patients. *Radiology* 1993; **187**: 413–20.

133 Simpson KJ, Chalmers N, Redhead DN, Finlayson NDC, Bouchier IAD, Hayes PC. Transjugular intrahepatic-portasystemic stent shunting for control of acute and recurrent upper gastrointestinal hemorrhage related to portal hypertension. *Gut* 1993; **34**: 968–73.

134 Haskal ZJ, Pentecost MJ, Soulen MC, Shlansky-Goldberg RD, Baum RA, Cope C. Transjugular intrahepatic portosystemic shunt stenosis and revision: early and midterm results. *Am J Roentgenol* 1994; **163**: 439–44.

135 Rossle M, Haag K, Ochs A, Sellinger M, Noldge G, Perarnau JM, *et al.* The transjugular intrahepatic portosystemic stent-shunt procedure for variceal bleeding. *N Engl J Med* 1994; **330**: 165–71.

136 Cabrera J, Maynar M, Granados R, Gorriz E, Reyes R, Pulido-Duque JM, *et al.* Transjugular intrahepatic portosystemic shunt versus sclerotherapy in the elective treatment of variceal hemorrhage. *Gastroenterology* 1996; **110**: 832–9.
137 Henderson JM, Kutner MH, Millikan WJ, Jr, Galambos JT, Riepe SP, Brooks WS, *et al.* Endoscopic variceal sclerosis compared with distal splenorenal shunt to prevent recurrent variceal bleeding in cirrhosis: a prospective, randomized trial. *Ann Intern Med* 1990; **112**: 262–9.
138 Jin G Jr, Rikkers LF. Selective variceal decompression: current status. *HPB Surg* 1991; **5**: 1.
139 Henderson JM, Gilmore GT, Hooks MA, Galloway JR, Dodson TF, Hood MM, *et al.* Selective shunt in the management of variceal bleeding in the era of liver transplantation. *Ann Surg* 1992; **216**: 248–55.
140 Iwatsuki S, Starzl TE, Too D, Marsh W, Lee RG, Fung JJ. Liver transplantation in the treatment of bleeding esophageal varices. *Surgery* 1998; **104**: 697–705.
141 Bismuth H, Adams R, Mathur S, Sherlock D. Options for elective treatment of portal hypertension in cirrhotic patients in the transplantation era. *Am J Surg* 1990; **160**: 105–10.
142 Wood RP, Shaw BW, Rikkers LF. Liver transplantation for variceal hemorrhage. *Surg Clin North Am* 1990; **70**: 449–61.
143 Henderson JM. Liver transplantation for portal hypertension. *Gastroenterology Clin North Am* 1992; **21**: 197.
144 Ringe B, Lang H, Tusch G, Pichlmayr R. Role of liver transplantation in management of esophageal variceal hemorrhage. *World J Surg* 1994; **18**: 233–9.
145 Ring EJ, Lake JR, Roberts JP, Gordon R, LaBerge J. Using transjugular intrahepatic portosystemic shunts to control variceal bleeding before liver transplantation. *Ann Intern Med* 1992; **116**: 304–9.
146 Woodle ES, Darcy M, White HM, Perdrizet GA, Vesely TM, Picus D, *et al.* Intrahepatic portosystemic vascular stents: a bridge to hepatic transplantation. *Surgery* 1993; **113**: 344–51.
147 Reichle FA, Fahmy WE, Golsorkhi M. Prospective comparative clinical trial with distal splenorenal and mesocaval shunts. *Am J Surg* 1979; **137**: 13–21.
148 Fischer JE, Bower RH, Ataman S, Welhng R. Comparison of distal and proximal splenorenal shunts: a randomized prospective trial. *Ann Surg* 1981; **194**: 531–44.
149 Langer B, Taylor BR, Mackenzie DR, Gilas T, Stone RM, Blendis I. Further report of a prospective randomized trial comparing distal splenorenal shunt with end-to-side portacaval shunt. *Gastroenterology* 1985; **88**: 424–9.
150 Millikan WJ, Warren WD, Henderson JM, Salam A, Smith RB, Galambos J, Rudman D. The Emory prospective randomized trial: selective versus nonselective shunt to control variceal bleeding: ten-year follow up. *Ann Surg* 1985; **160**: 67–74.
151 Harley HA, Morgan T, Redeker AG, Reynolds TB, Villamil F, Weiner JM, Yellin A. Results of a randomized trial of end-to-side portacaval shunt and distal splenorenal shunt in alcoholic liver disease and variceal bleeding. *Gastroenterology* 1986; **91**: 802–9.
152 Grace ND, Conn HO, Resnick RH, Groszmann RJ, Atterbury CE, Wright SC, *et al.* Distal splenorenal vs. portal-systemic shunts after hemorrhage from varices: a randomized controlled trial. *Hepatology* 1988; **8**: 1475–81.
153 Rikkers LF. Is the distal splenorenal shunt better? *Hepatology* 1988; **8**: 1705.
154 Mercado MA, Morales-Linares JC, Granados-Garcia J, Gomez-Mendez TJ, Chan C, Orozco H. Distal splenorenal shunt versus 10 mm low-diameter mesocaval shunt for variceal hemorrhage. *Am J Surg* 1996; **171**: 591–5.
155 Pagliaro L, Burroughs AK, Sorensen TIA, D'Amico R, Bosch J. Therapeutic controversies and randomized controlled trials: prevention of bleeding and rebleeding in cirrhosis. *Gastroenterol Int* 1989; **2**: 71.
156 Epstein M. Renal function abnormalities in cirrhosis. In: Zakim D, Boyer TD (eds). *Hepatology: a textbook of liver disease*. Philadelphia: W. B. Saunders, 1982; 449.

157 Lieberman FL, Reynolds TB. Plasma volume in cirrhosis of the liver: its relation to portal hypertension, ascites and renal failure. *J Clin Invest* 1967; **46**: 1297.
158 Epstein M, Pins DS, Schneider H, Levinson R. Determinants of deranged sodium and water homeostasis in decompensated cirrhosis. *J Lab Clin Med* 1976; **87**: 822–39.
159 Epstein M. Deranged sodium homeostasis in cirrhosis. *Gastroenterology* 1979; **76**: 622–35.
160 Lieberman FL, Dension EK, Reynolds TB. The relationship of plasma volume, portal hypertension, ascites, and renal sodium retention in cirrhosis: the overflow theory of ascites formation. *Ann NY Acad Sci* 1970; **170**: 202.
161 Runyon BA, Montano AA, Akriviadis EA, Antillon MR, Irving MA, McHutchison JG. The serum-ascites albumin gradient is superior to the exudate-transudate concept in the differential diagnosis of ascites. *Ann Intern Med* 1992; **117**: 215–20.
162 Stanley MM. Pathogenesis of ascites in cirrhosis unitary hypothesis. *ASAIO Trans* 1989; **35**: 161.
163 Llach J, Rimola A, Navasa M, Gines P, Salmeron JM, Gines A, *et al.* Incidence and predictive factors of first episode of spontaneous bacterial peritonitis in cirrhosis with ascites: relevance of ascitic fluid protein concentration. *Hepatology* 1992; **16**: 724–7.
164 Pelletier G, Lesur G, Ink O, Hagege H, Attali P, Buffet C, Etienne JP. Asymptomatic bacterascites: is it spontaneous bacterial peritonitis. *Hepatology* 1991; **14**: 112–15.
165 Arroya V, Gines P, Rodes J. Treatment of ascites in patients with cirrhosis of the liver. *J Hepatol* 1986; **2**: 504–12.
166 Salerno F, Badalamenti S, Incerti P, Tempini S, Restelli B, Bruno S, *et al.* Repeated paracentesis and IV albumin infusion to treat 'tense' ascites in cirrhotic patients. *J Hepatol* 1987; **5**:102–8.
167 Planas R, Gines P, Arroyo V, Llach J, Panes J, Vargas V, *et al.* Dextran-70 versus albumin as plasma expanders in cirrhotic patients with tense ascites treated with total paracentesis. *Gastroenterology* 1990; **99**: 1736–44.
168 Ferral H, Bjarnason H, Wegryn SA, Rengel GJ, Nazarian GK, Rank JM, *et al.* Refractory ascites: early experience in treatment with transjugular intrahepatic portosystemic shunt. *Radiology* 1993; **189**: 795–801.
169 Ochs A, Rossle M, Haag K, Hauenstein KH, Deibert P, Siegerstetter V, *et al.* The transjugular intrahepatic portosystemic stent-shunt procedure for refractory ascites. *N Engl J Med* 1995; **332**: 1192–7.
170 Shiffman ML, Jeffers L, Hoofnagle JH, Tralka TS. The role of transjugular intrahepatic portal systemic shunt for treatment of portal hypertension and its complications: a conference sponsored by the National Digestive Diseases Advisory Board. *Hepatology* 1995; **22**: 1591–7.
171 Crenshaw WB, Gordon FD, McEniff NJ, Perry LJ, Hartnell G, Anastopoulos H, *et al.* Severe ascites: efficacy of the transjugular intrahepatic portosystemic shunt in treatment. *Radiology* 1996; **200**: 185–92.
172 Gines P, Arroyo V, Vargas V, Planas R, Casafont F, Panes J, *et al.* Paracentesis with intravenous infusion of albumin as compared with peritoneovenous shunting in cirrhosis with refractory ascites. *N Engl J Med* 1991; **325**: 829–35.
173 Scholz DG, Nagorney DM, Lindor KD. Poor outcome from peritoneovenous shunts for refractory ascites. *Am J Gastroenterol* 1989; **84**: 540–3.
174 Moskovitz M. The peritoneovenous shunt: expectation and reality. *Am J Gastroenterol* 1990; **85**: 917–29.
175 Hillaire S, Labinaca M, Borgonovo G, *et al.* Peritoneovenous shunting of intractable ascites in patients with cirrhosis: improving results and predictive factors of failure. *Surgery* 1993; **113**: 373.
176 Millikan WJ, Jr, Henderson JM, Sewell CW, Guyton RA, Potts JR 3rd, Cranford CA, Jr, *et al.* Approach to the spectrum of Budd–Chiari syndrome: which patients require portal decompression? *Am J Surg* 1985; **149**: 167–76.
177 Valla D, Casadevall N, Lacombe C, Varet B, Goldwasser E, Franco D, *et al.* Primary myeloproliferative disorders and hepatic vein thrombosis: a prospective study of erythroid colony formation in vitro in 20 patients with Budd–Chiari syndrome. *Ann Intern Med* 1985; **103**: 329–34.

178 McDonald GB, Sharma P, Matthews DE, Shulman HM, Thomas ED. Veno-occlusive disease of the liver after bone marrow transplantation: diagnosis, incidence and predisposing factors. *Hepatology* 1984; **4**: 116–22.

179 Hadenque A, Poliquin M, Vilgrain V, Belghiti J, Degott C, Erlinger S, Benhamon J. The changing scene of hepatic vein thrombosis: recognition of asymptomatic cases. *Gastroenterology* 1994; **106**: 1042–7.

180 Murphy FB, Steinberg HV, Shires GT, Martin LG, Bernandino ME. The Budd–Chiari syndrome: a review. *Am J Roentgenol* 1986; **147**: 9–15.

181 Henderson JM, Warren WD, Millikan WJ, Jr, Galloway JR, Kawasaki S, Stahl RL, Hertzler G. Surgical options, hematological evaluation, and pathological changes in Budd–Chiari syndrome. *Am J Surg* 1990; **159**: 41–50.

182 Chang CH, Lee MC, Shieh MJ, Chang JP, Lin PJ. Transatrial membranotomy for Budd–Chiari syndrome. *Ann Thorac Surg* 1989; **48**: 409–12.

183 Orloff MJ, Girard B. Long-term results of treatment of Budd–Chiari syndrome by side-to-side portacaval shunt. *Surg Gynecol Obstet* 1989; **168**: 33–41.

184 Franco D, Vons C, Lecompte Y, Nuzzo G, Smadja C. Portoatrial shunt in Budd–Chiari syndrome. *Surgery* 1986; **99**: 378–80.

185 Klein AS, Cameron JL. Diagnosis and management of Budd–Chiari syndrome. *Am J Surg* 1990; **160**: 128–33.

186 Lopez RR Jr, Benner KG, Hall L, Rosch J, Pinson CW. Expandable venous stents for treatment of Budd–Chiari syndrome. *Gastroenterology* 1991; **100**: 1435–41.

187 Blum U, Rossle M, Haag K, Ochs A, Blum H, Havenstein K, Astinet F, Langen M. Budd–Chiari syndrome: technical hemodynamics and clinical results of treatment with transjugular intrahepatic portosystemic shunt. *Intervent Radiol* 1995; **197**: 805–11.

188 Bismuth H, Sherlock DJ. Portosystemic shunting versus liver transplantation for the Budd–Chiari syndrome. *Ann Surg* 1991; **214**: 581–9.

189 Shaked A, Goldstein RM, Klintmalm GB, Drazan K, Husberg B, Busuttil RW. Portosystemic shunt versus orthotopic liver transplantation for Budd–Chiari syndrome. *Surg Gynecol Obst* 1992; **174**: 453–9.

190 Webb LJ, Sherlocks S. The etiology, presentation and national history of extrahepatic portal venous obstruction. *QJM* 1979; **48**: 627–39.

191 Warren WD, Millikan WJ, Smith RB, Henderson JM, Salam AS, Galambos J. Noncirrhotic portal vein thrombosis physiology before and after surgery. *Ann Surg* 1980; **71**: 745–9.

192 Valla D, Casadevall N, Huisse MG, Tulliez M, Grange JD, Muller O, *et al*. Etiology of portal vein thrombosis in adults. A prospective evaluation of primary myeloproliferative disorders. *Gastroenterology* 1988; **94**: 1063–9.

193 Howard ER, Stringer MD, Mowat AP. Assessment of injection sclerotherapy in the management of 152 children with esophageal varices. *Br J Surg* 1988; **75**: 404–8.

194 Gauthier F, De Dreuzy O, Valayer J, Montupet PH. H-type shunt with an autologous venous graft for treatment of portal hypertension in children. *Pediatr Surg* 1989; **24**: 1041.

195 Warren WD, Henderson JM, Millikan WJ, Galambos JT, Bryan FC. Management of variceal bleeding in patients with non-cirrhotic portal vein thrombosis. *Ann Surg* 1988; **207**: 623–34.

196 Alvarez F, Bernard O, Brunelle F, Hadchouel P, Odievre M, Alagille D. Portal obstruction in children. II. Results of portasystemic shunts. *J Pediatr* 1983; **103**: 703–7.

197 Warren WD. Control of variceal bleeding: reassessment of rationale. *Am J Surg* 1983; **148**: 8.

198 Galloway JR, Henderson JM. Management of variceal bleeding in patients with extrahepatic portal vein thrombosis. *Am J Surg* 1990; **160**: 122–6.

15

Ischemic colitis

GERARD STANSBY, HUW J.W. THOMAS, AND ROBERT D. GOLDIN

INTRODUCTION

The intestinal mucosa is sensitive to ischemia, and ischemic colitis is the result of an event that leads to a reduction in colonic blood flow sufficient to cause ischemia or infarction of the colonic wall but not sufficient to produce full-thickness infarction and perforation. Ischemic colitis has a variable presentation and prognosis. It is often diagnosed only after a period of delay. It frequently occurs following surgery or in association with concurrent medical conditions, which may hamper diagnosis and management.

The term 'ischemic colitis' was first introduced by Marston *et al.* in 1966.[1] The condition may result from several different pathological processes. Firstly, it may occur because of an acute vascular event in the arterial supply to the colon (such as thrombosis of, or embolism into, the superior mesenteric artery [SMA]). Secondly, it may occur after periods of hypotension when the colon is one of the most susceptible organs. The term is often used for cases where full-thickness infarction is present acutely, but this usage is incorrect – not all ischemic colons have ischemic colitis, although the two may coexist if the involvement is patchy. Acute infarction of the colon and chronic mesenteric ischemia are different entities and will not be discussed here.

ETIOLOGY AND CLASSIFICATION

The colon differs from the small bowel in that it has no villi and therefore no countercurrent mechanism. It also appears that the capacity for autoregulation is lower in the colon than it is in the small bowel, as is the overall blood supplied per gram of tissue. The mucosa may also be relatively hypoperfused in conditions of low cardiac output because it is at the end of the microvascular arcades supplying the colonic wall. In addition, blood is shunted away from the splanchnic circulation as part of the physiological responses to shock and vasoconstrictors. It appears that a perfusion pressure of 50–60 mm Hg is a threshold level below which autoregulation is insufficient to maintain oxygen delivery in the colonic wall.[2]

Ischemic colitis is conventionally classified into two broad categories based on the etiology (Table 15.1): either found in association with occlusive vascular disease or occurring as a result of low cardiac output (or low perfusion) states. In practice, the two may often coexist. A third form has also been described due to mesenteric venous occlusive disease, where increased venous resistance leads to impaired microcirculatory arterial perfusion. Also, there are numerous case reports of drug-induced ischemic colitis, often occurring in young patients. Although ischemic colitis is often considered a condition of elderly people or patients with vascular disease, it also occurs in healthy adults under 60 years, and in one recent series of 24 patients, five (21%) patients were under 45 years.[3] Ischemic colitis in young patients seems to affect women more than men and it may have a generally self-limiting course.[4] It is predisposed to by rheumatoid arthritis, diabetes,

Table 15.1 *Etiology of ischemic colitis*

- *Occlusive*
  - Major vessel disease
    - Occlusion due to atheroma
    - Occlusion due to embolus
    - Occlusion due to trauma
    - Occlusion due to surgical ligation
    - Occlusion due to tumor invasion/compression/pancreatitis
    - Iatrogenic: post-mesenteric angiography
  - Small vessel disease
    - Vasculitis, e.g. polyarteritis nodosa,[10] systemic lupus erythematosus, Wegener's granulomatosis[11]
    - Radiation injury
    - Diabetes mellitus
  - Venous occlusion
    - Portal hypertension
      - Acute mesenteric venous thrombosis
    - Post-trauma
    - Post-hypotension/hypovolemia
    - Thrombophilia[12–14]
    - Sickle cell disease[15]
    - Intra-abdominal inflammation as in pancreatitis
    - Malignancy: venous compression/invasion
- *Non-occlusive*
  - Severe shock/low flow
    - Cardiogenic
    - Septic
    - Hypovolemic
  - In association with colonic obstruction or volvulus[16]
- *Drug-induced*
  - Cocaine[17–19]
  - Catecholamines
  - Oral contraceptives[20]
  - Phenobarbital
  - Nasal decongestants[21,22]
  - Dextroamphetamine[23]
  - Sumatriptan[24]

and the oral contraceptive pill, but many cases are idiopathic.[5, 6]. It has also been described in endurance athletes, such as marathon runners, where dehydration, hypercoagulability, and physiological shunting away from the splanchnic bed may be responsible,[7] and in people with sickle cell disease, where hypercoagulability and sludging in smaller vessels appears to be the etiology.[8] Occasionally, ischemic colitis also occurs in pancreatitis if there is occlusion of the mesenteric vessels.[9]

ANATOMICAL FACTORS IN ISCHEMIC COLITIS

The colon is normally supplied by the superior and inferior mesenteric arteries, with some contributions from the internal iliac arteries via the superior hemorrhoidal vessels. This anatomical arrangement usually allows for free collateralization between the celiac artery, SMA, and inferior mesenteric artery (IMA) territories such that occlusion of individual vessels does not necessarily result in ischemia. Important in the collateral pathways between the SMA and IMA territory is the marginal artery of Drumond and the arc of Riolan (meandering mesenteric artery). These collaterals may not always be adequate because of congenital variations, previous surgery such as bowel resection resulting in division of the marginal artery, or the suddenness of onset of the condition. This latter factor may be important in many cases of ischemic colitis where the potential capacity for collateralization requires time to develop. As a consequence, the vast majority of cases of ischemic colitis occur in a relatively acute fashion or follow acute events, such as surgery or myocardial infarction. The internal iliac arteries provide an important collateral supply to the IMA territory of the rectum and descending colon via the middle and inferior hemorrhoidal vessels. The watershed between the areas of colon supplied by the SMA and IMA is usually described to be between the proximal two-thirds and distal one-third of the transverse colon based on the embryological division between mid- and hind-gut. This anatomical arrangement explains the vulnerability of the splenic flexure to ischemia. However, in the arteriopath with an occluded IMA, the watershed point will be shifted distally towards the rectum. In addition, in up to 50% of people the marginal artery of the colon may be developed poorly, which may result in right-sided ischemic colitis, although right-sided ischemic colitis probably represents only about 5–10% of all cases.[25] Right colonic involvement appears to be associated with severe forms of ischemic colitis and occurs frequently in patients with chronic cardiac disease, such as aortic stenosis, and in patients with chronic renal failure requiring hemodialysis.[26]

PATHOPHYSIOLOGY

The ischemic colonic mucosa, even if not infarcted, loses its barrier function rapidly, allowing invasion by luminal bacteria and absorption of endotoxin. The mucosa may also slough in a patchy fashion, which can result in increased peristalsis and diarrhea mixed with blood. In the more severe forms of the condition, bacterial invasion leads to portal pyemia and death. In less severe cases, inflammatory cytokine release may contribute to multiple organ failure, with renal impairment and respiratory impairment possibly via the activation of neutrophil leukocytes.

The gut mucosa is rich in the enzyme xanthine dehydrogenase, which results in the production of reactive oxygen species and free radicals, such as superoxide. This causes oxidative tissue damage, which is actually most severe when reperfusion follows a period of ischemia. This reperfusion injury also results in the activation of polymorphonuclear leukocytes, which then result in systemic events that may potentially lead to tissue injury and inflammation remote to the gut, such as in the lung and kidneys. The more extreme manifestations of this are termed the systemic inflammatory response syndrome (SIRS). In addition, endothelin 1 (ET-1), a potent vasoconstrictor, is released following ischemia and reperfusion, and may lead to splanchnic vasoconstriction, thus exacerbating the situation further.

CLINICAL FEATURES

Clinical features are very variable. Mild degrees of ischemic colitis may in fact be subclinical and never diagnosed. Clinically evident ischemic colitis appears to affect 1–2% of patients after aortic surgery, but it is found more frequently if colonoscopy or biopsy evidence is sought. In a study by Welch and co-workers, where colonoscopy and biopsies were performed post-aortic surgery, a surprisingly high 30% of patients were found to have features of ischemic colitis on biopsy, although virtually all of these were asymptomatic.[27] If only the submucosa and mucosa are involved, then the presentation is more likely to be mild. If the muscularis is involved, then symptoms are likely to be worse and the colitis may fail to resolve or a stricture may subsequently form. If full-thickness infarction of the muscularis has occurred, then this will lead to perforation, peritonitis, and the need for surgery. This situation should not be classified as true ischemic colitis, but most series do include 10–20% of such cases, presumably reflecting the fact that full-thickness infarcts and true ischemic colitis can coexist in a patchy fashion.

DIFFERENTIAL DIAGNOSIS

Ischemic colitis typically presents with features of abdominal pain associated with bloody diarrhea (Table 15.2). These clinical features have a broad differential diagnosis, which includes infective colitis, pseudomembranous colitis dhue to *Clostridium difficile* toxin, inflammatory bowel disease, acute diverticulitis, and radiation colitis (Table 15.3). Perhaps the commonest diagnostic dilemma is between *C. difficile* colitis and ischemic colitis in the post-operative patient where *C. difficile* is probably more common. In one report of 120 patients having aortic surgery, five developed *C. difficile* colitis and only one developed ischemic colitis.[29]

Table 15.2 *Typical clinical features of ischemic colitis*

Elderly patient with vascular disease
Abdominal pain
Abdominal tenderness
Abdominal distension
Diarrhea, often bloodstained
Shock and toxemia

Table 15.3 *Main differential diagnosis of ischemic colitis*

Infective colitis
Inflammatory bowel disease (ulcerative colitis and Crohn's disease)
Pseudomembranous colitis (*Clostridium difficile*)
Acute diverticulitis
Escherichia coli 0157[28]
Radiation colitis

However, the diagnosis can usually be made by a combination of microbiology for *C. difficile* and its toxin and histological features on biopsy. In a report of biopsies from 25 patients with *C. difficile* colitis and 24 with ischemic colitis, hyalinization of the lamina propria was seen in 19 cases of ischemic colitis but none of *C. difficile* colitis ($P < 0.0001$), atrophic-appearing microcrypts were seen in 18 ischemic cases and six *C. difficile* cases ($P < 0.0006$), and lamina propria hemorrhage, full-thickness mucosal necrosis, and a diffuse microscopic distribution of pseudomembranes were also significantly more common in ischemic colitis than in *C. difficile* colitis.[30] *Escherichia coli* 0157 may cause a hemorrhagic colitis resembling ischemic colitis. In addition, *E. coli* may also be a cause of ischemic colitis in that it can result in fibrin thrombus in colonic vessels as a secondary event.[28]

PATHOLOGY

Macroscopic pathology

Ischemic bowel disease can take a number of macroscopic forms, including hemorrhagic infarction, ulceration (which can resemble that seen in inflammatory bowel disease), a lesion associated with a pseudomembrane (resembling pseudomembranous colitis), or a stricture (which can mimic a number of other pathological processes). These lesions are most common in the watershed areas, such as the splenic flexure, although any part of the bowel can be affected.

Microscopic pathology

Three phases of the evolution of an ischemic lesion can be identified: the acute, organizing, and healed phases. The *acute phase* is characterized by necrosis of the superficial epithelium and hemorrhage into the lamina propria, with the intestinal crypts being spared. These changes are reversible, but if the ischemia is extensive or prolonged, then they may progress to ulceration or even transmural infarction.

In the *organizing* phase, the ulceration and associated granulation tissue are not associated with marked inflammatory changes, which helps in making the distinction from those seen in inflammatory bowel disease. Another clue to their etiology is the presence of hemorrhage or iron deposition. The use of a Perl's stain for iron is very useful in identifying the latter.

In the *healed phase*, the large-bowel mucosa may show architectural distortion of the crypts. In the small bowel, there may be villous distortion and pyloric metaplasia. In severe cases, there is transmural fibrosis. In resection specimens, an attempt should be made to identify the underlying cause of the ischemia, although frequently it is not possible to find one. It is important to distinguish a primary vasculitis from the changes seen in areas of active inflammation, which are much more likely to be secondary. It should be noted that vasculitis may be confined to the intestinal tract, as is the case with enterocolic (lymphocytic) phlebitis.[31] The vessels should also be examined carefully for the presence of cholesterol emboli, which is another uncommon cause of intestinal ischemia.

In acute disease, the differential diagnosis from inflammatory bowel disease is based on the distribution and acute nature of the inflammatory infiltrate together with absence of crypt architectural distortion in ischemic enterocolitis. In the later stages, the distinction is more difficult because of the presence of chronic inflammatory changes and crypt distortion in both conditions. However, full-thickness fibrosis is uncommon in inflammatory bowel disease, and iron deposition is a clue to the ischemic etiology.

A number of conditions are associated with an ischemic component. These include necrotizing enterocolitis (seen in premature infants), neutropenic colitis (seen in immunosuppressed patients), Behçet's disease (gastrointestinal disease is seen in up to 15% of cases), pseudomembranous colitis (seen following antibiotic therapy), stercoral ulceration, radiation enterocolitis (seen weeks to years following radiotherapy), and drug-induced colitis (e.g. due to potassium chloride, hydrochlorthiazide, cocaine, or non-steroidal anti-inflammatory drugs) The distinction from ischemic bowel disease depends on a combination of clinical and pathological features. For example, in radiation enterocolitis, characteristic features include mucosal atrophy, telangectatic blood vessels, hyaline fibrosis, radiation fibroblasts, and a spectrum of ischemic lesions, including intimal fibrosis and fibrinoid necrosis. The distinction from pseudomembranous colitis can be difficult. The presence of a hyalinized lamina propria appears to be a specific and sensitive marker for ischemia in colon biopsies with pseudomembranes. The presence of atrophic crypts, hemorrhage into the lamina propria, full-thickness mucosal necrosis, and diffuse involvement of all the surface of all biopsies by pseudomembranes are further markers of ischemia.

INVESTIGATIONS IN ISCHEMIC COLITIS

The diagnosis of ischemic colitis requires a high index of clinical suspicion and confirmation by further investigations. These may include plain X-ray, stool culture, endoscopic examination, barium enema, computed tomography (CT) and/or magnetic resonance imaging (MRI), blood gas/pH measurement and blood lactate, and angiography/Duplex scanning.

Plain X-ray

Plain X-ray (Figure 15.1) of the abdomen may show fluid levels in the colon, toxic colonic dilation, intramural gas, or free gas if a perforation has occurred. In severe cases, intraportal air may be seen, a finding that suggests an outcome with a high mortality. In the postoperative patient, plain X-ray may help to exclude mechanical obstruction or indicate the need for laparotomy.

Stool culture

Stool samples should be sent for microscopy for ova cysts and parasites, culture, and sensitivity, and for analysis for *C. difficile* toxin.

Endoscopic examination

Sigmoidoscopy may reveal blood in the lumen, but the mucosa may appear normal at this point if the watershed zone is higher up. Additional information may be gleaned from mucosal biopsies for histopathology, even if the mucosa appears macroscopically normal. In general, such biopsies should be taken if ischemic colitis is suspected clinically. Colonoscopy

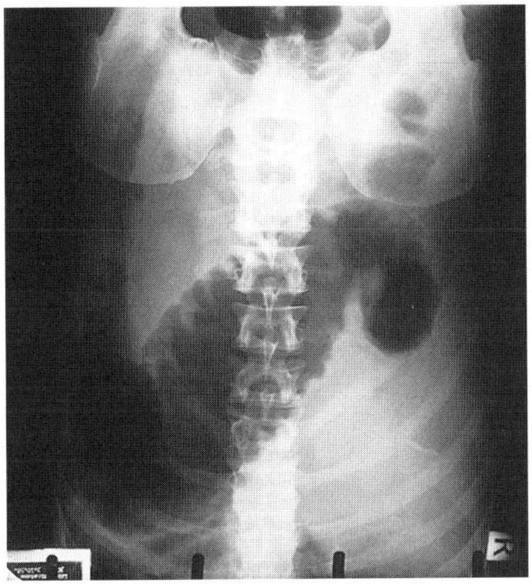

Figure 15.1 *Plain X-ray.*

(Plate 22) may be undertaken unprepared if the patient is unwell or with a gentle bowel preparation. Examination will determine the extent of the ischemia. In mild disease, the mucosa has a pale appearance with petechiae; in more severe disease, the mucosa may be blue or black with slough and ulceration. A recent meta-analysis suggested that colonoscopy can help to diagnose ischemic colitis but cannot separate transmural from mucosal ischemia, and that it had no impact on mortality in any of the prospective series.[32] Another report suggested that endoscopy could identify pseudomembranes significantly more often with *C. difficile* than with ischemia, while the endoscopic appearance of masses or polyps was seen exclusively in cases of ischemia.[30] Occasionally, however, colonoscopy has been implicated as the cause of ischemic colitis.[33, 34]

Barium enema

An instant enema (Figure 15.2) may reveal thumbprinting due to mucosal edema.[35] However, endoscopy with biopsy is preferable as a means of investigation if it is available.

Computed tomography scanning and magnetic resonance imaging

CT and MRI scans may be normal in the early stages of ischemic colitis or they may show nonspecific thickening or edema with a target appearance. CT can be used to confirm the clinical suspicion of ischemic colitis, to suggest ischemia when it is unsuspected, and to diagnose complications. In ischemic colitis, CT typically demonstrates circumferential, symmetric wall thickening with fold enlargement.[36] However, intrinsic colonic abnormalities cannot be used to diagnose or predict the development of infarction. A recent report of 54 cases evaluated by CT showed a wet appearance with heterogeneous areas of edema in 61% of patients, a dry appearance with mild homogeneous thickening in 33% of patients, and intramural air in 6% of patients.[37] Modern spiral CT may also allow assessment of patency of visceral vessels as may gadolinium-enhanced magnetic resonance angiography (MRA).

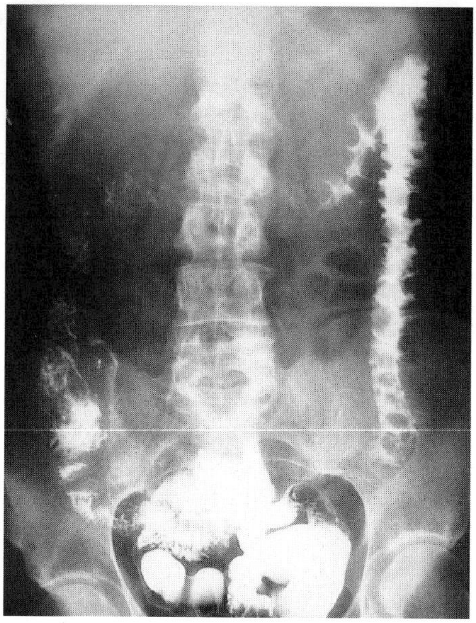

Figure 15.2 *Barium enema.*

Blood gas/pH measurement and blood lactate

In severe intestinal infarction, the patient will develop a metabolic acidosis with a large base excess and low pH. However, these metabolic changes are usually signs of advanced full-thickness infarction rather than ischemic colitis. L-lactate is produced by all cells as a product of glycolysis, but is produced in excess during conditions of hypoxia. Raised levels of L-lactate are therefore related to inadequate tissue perfusion. However, since L-lactate from the intestine is mostly removed by the liver, it is not usually helpful in the diagnosis of ischemic colitis, although it may be raised in colonic infarction. D-Lactate is produced by intestinal bacteria and may be more predictive of colonic ischemia after aortic surgery than L-lactate, although there has been no prospective study. [38]

Angiography/Duplex scanning

Angiography and Duplex scanning of mesenteric vessels are rarely helpful or diagnostic for ischemic colitis in the acute situation, and will not be discussed further.

MANAGEMENT

Conservative management is appropriate in most patients if they are clinically stable and there are no signs of full-thickness involvement or peritonitis. Patients should be given intravenous fluids and antibiotics and should be monitored carefully. If there are conditions predisposing to intestinal ischemia, such as hypercoagulability, they should be treated.

A number of pharmacological treatments have been shown experimentally to improve intestinal blood flow, such as glucagon and prostanoid infusions,[39] but these are not used widely clinically. Early colonoscopy should be carried out to confirm the diagnosis. The question of whether to use angiography and other investigations remains controversial. In the case of a patient who appears to be settling and who is systemically well, intervention is probably best avoided. In a patient who continues with significant bowel disturbance or systemic symptoms, then angiography may be considered in order to assess the visceral circulation. If there is a significant stenosis or occlusion of visceral arteries, then consideration of mesenteric angioplasty might be given, although the number of patients who may ultimately benefit will be small. Patients who have peritonitis or who deteriorate should have a laparotomy with resection of the ischemic segment; both ends of the bowel should be brought out as stomas. A primary anastomosis is probably always best avoided when the colon is involved. If attempted, then the anastomosis certainly should be covered by a defunctioning stoma. Re-look laparotomy the following day should also be considered. Although there is a high associated mortality, in patients who survive it will often be possible to reverse the stoma at a later date.[40]

Mortality rates in patients requiring emergency surgery for peritonitis and resection of ischemic bowel may be as high as 50%. This is due to the associated co-morbid medical conditions, including peripheral vascular disease, ischemic heart disease, and cerebrovascular disease. If the original surgery has involved aortic grafting, then there is also a high risk of subsequent graft infection.

Following an episode of ischemic colitis where the muscularis has been involved, a colonic stricture may develop as a late complication. This will typically present with features of subacute obstruction. On investigation, the stricture is smooth and may have a normal mucosal pattern. The recognition of the ischemic etiology of the stricture is important as there may be an increased instance of anastomotic dehiscence following bowel resection due to the poor vascular supply.[41] However, in cases where the etiology is non-occlusive, the blood supply will be normal and resection of the stricture and re-anastomosis can be carried out safely.

POSTOPERATIVE ISCHEMIC COLITIS

The single commonest cause of ischemic colitis is post-aortic surgery, where a combination of inferior mesenteric occlusion, postoperative hypoxia or hypotension, and a vulnerable, elderly patient are probably the main causes.[42]

Patients in shock who are operated on for ruptured aneurysms were found to be at greatest risk of developing postoperative intestinal ischemia.[43] Excluding patients in shock, operation for aneurysmal disease and for occlusive disease carried the same risk. Renal disease, emergency surgery, age, type of hospital, aortobifemoral graft, operating time, cross-clamping time, and ligation of one or both internal iliac arteries were also independent risk factors for developing ischemic colitis.[43] In another study, the importance of hypotension, duration of hypotension, temperature less than 35 °C, pH less than 7.3, fluid requirement of more than 5 l, and packed red cells of more than 6 units were found to be predictive of ischemic colonic complications following ruptured aortic aneurysm repair on multivariate analysis.[42]

Ischemic colitis occurs most frequently following aortic vascular surgery, although it can occur after any major surgical procedure. It has also been described following aortic stent grafting.[44] There are a number of reasons why ischemic colitis can occur after aortic surgery. There may be interruption to colonic blood supply due to division of the IMA as well as the possibility of hypotension and hypoxia in the perioperative period and stenosis of other visceral vessels due to the underlying atherosclerotic disease process. IMA ligation is probably the most important factor in the majority of cases. Other possible factors include embolization into the IMA territory during dissection of the aneurysm, injury to mesenteric vessels by retractors, and mesenteric compression by hematoma.

As with ischemic colitis in general, two patterns are usually described: full-thickness infarction and mucosal ischemia alone. Such colonic ischemia has long been recognized as an occasional clinical event in infrarenal aortic aneurysm surgery, usually associated with a high incidence of morbidity and mortality, especially if colonic resection is required. Its clinical incidence is of the order of 1–2% in most series.[29, 45] In ruptured aortic aneurysm repair, however, the incidence is much higher, up to 30%.[42, 46] Studies with colonoscopy or sigmoidoscopy have, however, shown an even higher incidence: 7–35% of elective cases[47] and up to 60% of survivors of ruptured aneurysm repair.[48] Fanti *et al.* reported a series of 105 patients who underwent rectosigmoidoscopy within 72 h of aortic surgery.[49] Colonic ischemia was found in 12 patients, but seven were asymptomatic and all were managed conservatively. Welch *et al.* studied a group of patients undergoing elective aortic surgery with pre- and postoperative colonoscopy with biopsy in order to assess the true incidence of the condition.[27] They studied 28 patients having aneurysm repair and 28 having reconstruction for occlusive aortoiliac disease. Postoperative colonoscopy and biopsy was carried out at 1 week. As might be expected, all patients had normal appearances and biopsies before surgery. Postoperatively, however, 30% had features of ischemic colitis upon biopsy. Interestingly, there was no difference in the incidence of this finding between those with occlusive and those with aneurysmal disease. Overall, there was an incidence of complications of 34%, mostly cardiac and respiratory, and four patients died. One patient died of a fatal colonic infarction. Two factors studied correlated significantly with the finding of histological evidence of ischemia at 1 week. These were prolonged cross-clamp time and postoperative diarrhea. Co-morbidity was also higher in the ischemia group. Ligation of a patent IMA did not correlate significantly, although the data showed a trend in this direction. This study demonstrates clearly that colonic ischemia occurs much more frequently after aortic surgery than suggested previously by studies relying on clinical endpoints or visual assessment on colonoscopy. This may be of clinical importance, since ischemia, even if it may ultimately resolve and not lead to infarction, may result in disruption of the gut mucosal barrier, allowing the passage of organisms and endotoxin into the portal circulation and thereby leading to septic complications and multi-organ failure. It is interesting to speculate that many cases of multi-organ failure, systemic inflammatory response syndrome, or postoperative problems with renal function or gas exchange may be due, at least partially, to decreased mucosal barrier function and endotoxin absorption due to (subclinical) colonic ischemia.

As postoperative ischemic colitis occurs in an unpredictable fashion, there are no clear guidelines as to how surgeons may avoid its development. Preoperative angiography may demonstrate the occasional patient with enlarged collaterals who is at risk, but the normal IMA is not well demonstrated at routine angiography and in any event does not give any guide as to which patients with a patent IMA will be at risk if it is ligated, since normal collaterals are not normally seen at routine angiography. Routine angiography clearly cannot be justified in all patients undergoing aortic surgery. Reimplantation of all patent IMAs has been suggested as one possible strategy. This approach was used by Seeger *et al.* in 151 aortic reconstructions, and no patients developed colonic infarction compared with 2.7% in the previous 186 patients where reimplantation was carried out based on stump pressure, Doppler, and inspection.[50] Methods have also been proposed whereby patients who may need IMA reimplantation can be identified intraoperatively and selective reimplantation practiced (Table 15.4). However, as most of these techniques are not available in the major-

Table 15.4 *Techniques for objectively assessing colonic circulation intraoperatively*

1. Doppler ultrasound[53]
2. Inferior mesenteric artery stump pressure
3. Photoplethysmography[54]
4. Sigmoid tonometry
5. Laser Doppler fluximetry[55–57]

ity of centers, most attention has been focussed on the use of IMA stump pressure. In a recent report, a prospective assessment of the IMA circulation was performed in a test group of consecutive patients operated on by one surgeon.[51] If mean IMA stump pressure was less than approximately 50 mm Hg, then it was reimplanted. The incidence of ischemic colitis in this group was not, however, less than in historical controls or in a concomitant control group under other surgeons. Arguments against routine reimplantation of IMAs include the possibility of increased bleeding from the anastomosis[52] and the fact that it is technically demanding and takes extra time. However, in some hands, at least, it appears to be safe, although its efficacy at preventing postoperative colonic infarction and ischemic colitis still remains controversial. Certainly, if preoperative angiography has shown occlusion or stenosis of the SMA, and if neither internal iliac is reconstructed or if the colon appears ischemic at the end of the procedure, then reimplantation of a patent IMA seems sensible.

Attention should be given to ensuring that at least one internal iliac artery is revascularized whenever possible. Johnston and Scobie showed that diarrhea and ischemic colitis were more common if this was not done.[52] Attempts should also be made to maintain blood pressure and oxygen levels, as any fall may be critical in patients with an already compromised colonic mucosal perfusion. It is also important that the surgeon remembers the simple measure of checking the colonic appearance visually at the end of the procedure – one suspects that this simple step is often neglected. If the colon appears ischemic, then revascularization or the need for a second look can be considered. Intraoperative assessment of the left colon by visual inspection is probably worth doing, but reliance on it may have a fundamental flaw. Firstly, the circulation and blood pressure are likely to be in an optimized state, and secondly, there will usually be a hyperemic period of increased flow once clamps have been released. What appears to be an adequate circulation may be less than adequate in a postoperative scenario of hypotension, hypovolemia, or hypoxia. This may be compounded by the use of inotropes or vasoconstrictors postoperatively.

REFERENCES

1 Marston A, Pheils MT, Thomas ML, Morson BC. Ischemic colitis. *Gut* 1966; **7**: 1–15.

2 Kvietys PR, Granger DN. Physiology, pharmacology and pathology of the colonic circulation. In: Shepherd AP, Granger DN (eds). *Physiology of the intestinal circulation*. New York: Raven Press, 1984.

3 Arnott ID, Ghosh S, Ferguson A. The spectrum of ischemic colitis. *Eur J Gastroenterol Hepatol* 1999; **11**: 295–303.

4 Newell AM, Deckert JJ. Transient ischemic colitis in young adults. *Am Fam Physician* 1997; **56**: 1103–8.

5 Judge JS, Hoffman NE, Levitt MD. Transient ischemic colitis in young adults. *Aust NZ J Surg* 1994; **64**: 721–2.

6 Matsumoto T, Iida M, Kimura Y, Nanbu T, Fujishima M. Clinical features in young adult patients with ischemic colitis. *J Gastroenterol Hepatol* 1994; **9**: 572–5.

7 Lucas W, Schroy PC 3rd. Reversible ischemic colitis in a high endurance athlete. *Am J Gastroenterol* 1998; **93**: 2231–4.

8 Moukarzel AA, Rajaram M, Sundeep A, Guarini L, Feldman F. Sickle cell anemia: severe ischemic colitis responding to conservative management. *Clin Pediatr (Phila)* 2000; **39**: 241–3.

9 Yamagiwa I, Obata K, Hatanaka Y, Saito H, Washio M. Ischemic colitis complicating severe acute pancreatitis in a child. *J Pediatr Gastroenterol Nutr* 1993; **16**: 208–11.

10 Okada M, Konishi F, Sakuma K, Kanazawa K, Koiwai H, Kaizaki Y. Perforation of the sigmoid colon with ischemic change due to polyarteritis nodosa. *J Gastroenterol* 1999; **34**: 400–4.

11 Storesund B, Gran JT, Koldingsnes W. Severe intestinal involvement in Wegener's granulomatosis: report of two cases and review of the literature. *Br J Rheumatol* 1998; **37**: 387–90.

12 Klestzick HN, McPhedran P, Cipolla D, Berry WA, DiCorato M, Denowitz J. The antiphospholipid syndrome and ischemic colitis. *Gastroenterologist* 1995; **3**: 249–56.

13 Yee NS, Guerry D 4th, Lichtenstein GR. Ischemic colitis associated with factor V Leiden mutation. *Ann Intern Med* 2000; **132**: 595–6.

14 Ludwig D, Stahl M, David-Walek T, Bruning A, Siemens A, Zwaan M, *et al*. Ischemic colitis, pulmonary embolism, and right atrial thrombosis in a patient with inherited resistance to activated protein C. *Dig Dis Sci* 1998; **43**: 1362–7.

15 Moukarzel AA, Rajaram M, Sundeep A, Guarini L, Feldman F. Sickle cell anemia: severe ischemic colitis responding to conservative management. *Clin Pediatr (Phila)* 2000; **39**: 241–3.

16 Scott B, Isbister WH. Ischemic colitis as a complication of left-sided large bowel obstruction: a case report. *Aust NZ J Surg* 1991; **61**: 73–5.

17 Niazi M, Kondru A, Levy J, Bloom AA. Spectrum of ischemic colitis in cocaine users. *Dig Dis Sci* 1997; **42**: 1537–41.

18 Papi C, Candia S, Masci P, Ciaco A, Montanti S, Capurso L. Acute ischemic colitis following intravenous cocaine use. *Ital J Gastroenterol Hepatol* 1999; **31**: 305–7.

19 Simmers TA, Vidakovic-Vukic M, Van Meyel JJ. Cocaine-induced ischemic colitis. *Endoscopy* 1998; **30**: S8–9.

20 Mann DE, Jr, Kessel ER, Mullins DL, Lottenberg R. Ischemic colitis and acquired resistance to activated protein C in a woman using oral contraceptives. *Am J Gastroenterol* 1998; **93**: 1960–2.

21 Lichtenstein GR, Yee NS. Ischemic colitis associated with decongestant use. *Ann Intern Med* 2000; **132**: 682.

22 Dowd J, Bailey D, Moussa K, Nair S, Doyle R, Culpepper-Morgan JA. Ischemic colitis associated with pseudoephedrine: four cases. *Am J Gastroenterol* 1999; **94**: 2430–4.

23 Dirkx CA, Gerscovich EO. Sonographic findings in methamphetamine-induced ischemic colitis. *J Clin Ultrasound* 1998; **26**: 479–82.

24 Knudsen JF, Friedman B, Chen M, Goldwasser JE. Ischemic colitis and sumatriptan use. *Arch Intern Med* 1998; **158**: 1946–8.

25 Schuler JG, Hudlin MM. Cecal necrosis: infrequent variant of ischemic colitis. Report of five cases. *Dis Colon Rectum* 2000; **43**: 708–12.

26 Flobert C, Cellier C, Berger A, Ngo A, Cuillerier E, Landi B, *et al*. Right colonic involvement is associated with severe forms of ischemic colitis and occurs frequently in patients with chronic renal failure requiring hemodialysis. *Am J Gastroenterol* 2000; **95**: 195–8.

27 Welch M, Baguneid MS, McMahon RF, Dodd PDF, Fulford PE, Griffiths GD, Walker MG. Histological study of colonic ischemia after aortic surgery. *Br J Surg* 1998; **85**: 1095–8.

28 Su C, Brandt LJ, Sigal SH, Alt E, Steinberg JJ, Patterson K, Tarr PI. The immunohistological diagnosis of *E. coli* O157:H7 colitis: possible association with colonic ischemia. *Am J Gastroenterol* 1998; **93**: 1055–9.

29 Valentine RJ, Hagino RT, Jackson MR, Kakish HB, Bengtson TD, Clagett GP. Gastrointestinal complications after aortic surgery. *J Vasc Surg* 1998; **28**: 404–11.

30 Dignan CR, Greenson JK. Can ischemic colitis be differentiated from *C. difficile* colitis in biopsy specimens? *Am J Surg Pathol* 1997; **21**: 706–10.

31 Saraga E, Bouzourenne H. Enterocolic (lymphocytic) phlebitis: a rare cause of intestinal ischemic necrosis: a series of six patients and review of the literature. *Am J Surg Pathol* 2000; **24**: 824–9.
32 Houe T, Thorboll JE, Sigild U, Liisberg-Larsen O, Schroeder TV. Can colonoscopy diagnose transmural ischemic colitis after abdominal aortic surgery? An evidence-based approach. *Eur J Vasc Endovasc Surg* 2000; **19**: 304–7.
33 Cremers MI, Oliveira AP, Freitas J. Ischemic colitis as a complication of colonoscopy. *Endoscopy* 1998; **30**: S54.
34 Wheeldon NM, Grundman MJ. Ischemic colitis as a complication of colonoscopy. *BMJ* 1990; **301**: 1080–1.
35 Reeders JW, Rosenbusch G, Tytgat GN. Radiological aspects of ischemic colitis. A review. *Diagn Imaging* 1981; **50**: 4–16.
36 Horton KM, Corl FM, Fishman EK. CT evaluation of the colon: inflammatory disease. *Radiographics* 2000; **20**: 399–418.
37 Balthazar EJ, Yen BC, Gordon RB. Ischemic colitis: CT evaluation of 54 cases. *Radiology* 1999; **211**: 381–8.
38 Welch M. D-Lactate as an early marker of intestinal ischemia after ruptured abdominal aortic aneurysm repair. *Br J Surg* 1999; **86**: 712.
39 Nakai M, Uchida H, Hanaoka T, Sugiyama S, Sano S, Shimizu N. Beneficial effects of prostaglandin E1 on ischemic colitis following surgery on the abdominal aorta. *Surg Today* 1998; **28**: 1146–53.
40 Longo WE, Ward D, Vernava AM 3rd, Kaminski DL. Outcome of patients with total colonic ischemia. *Dis Colon Rectum* 1997; **40**: 1448–54.
41 Simi M, Pietroletti R, Navarra L, Leardi S. Bowel stricture due to ischemic colitis: report of three cases requiring surgery. *Hepatogastroenterology* 1995; **42**: 279–81.
42 Levison JA, Halpern VJ, Kline RG, Faust GR, Cohen JR. Perioperative predictors of colonic ischemia after ruptured abdominal aortic aneurysm. *J Vasc Surg* 1999; **29**: 40–5.
43 Bjorck M, Bergqvist D, Troeng T. Incidence and clinical presentation of bowel ischemia after aortoiliac surgery – 2930 operations from a population-based registry in Sweden. *Eur J Vasc Endovasc Surg* 1996; **12**: 139–44.
44 Jaeger HJ, Mathias KD, Gissler HM, Neumann G, Walther LD. Rectum and sigmoid colon necrosis due to cholesterol embolization after implantation of an aortic stent-graft. *J Vasc Interv Radiol* 1999; **10**: 751–5.
45 Longo WE, Lee TC, Barnett MG, Vernava AM, Wade TP, Peterson GJ, *et al.* Ischemic colitis complicating abdominal aortic aneurysm surgery in the U.S. veteran. *J Surg Res* 1996; **60**: 351–4.
46 Bandyk DF, Florence MG, Johansen KH. Colon ischemia accompanying ruptured abdominal aortic aneurysm. *J Surg Res* 1981; **30**: 297–303.
47 Fry PD. Colonic ischemia after aortic reconstruction. *Can J Surg* 1988; **31**: 162–4.
48 Hagihara PF, Ernst CB, Griffen WO, Jr. Incidence of ischemic colitis following abdominal aortic reconstruction. *Surg Gynecol Obstet* 1979; **149**: 571–3.
49 Fanti L, Masci E, Mariani A, Chiesa R, Jannello A, Melissano G, *et al.* Is endoscopy useful for early diagnosis of ischemic colitis after aortic surgery? Results of a prospective trial. *Ital J Gastroenterol Hepatol* 1997; **29**: 357–60.
50 Seeger JM, Coe DA, Kaelin LD, Flynn TC. Routine reimplantation of patent inferior mesenteric arteries limits colon infarction after aortic reconstruction. *J Vasc Surg* 1992; **15**: 635–41.
51 Killen DA, Reed WA, Gorton ME, Muehlebach GF, Borkon AM, Piehler JM, Wathanacharoen S. Is routine postaneurysmectomy hemodynamic assessment of the inferior mesenteric artery circulation helpful? *Ann Vasc Surg* 1999; **13**: 533–8.
52 Johnston KW, Scobie TK. Multicenter prospective study of nonruptured abdominal aortic aneurysms. I. Population and operative management. *J Vasc Surg* 1988; **7**: 69–81.
53 Danse EM, Laterre PF, Van Beers BE, Goffette P, Dardenne AN, Pringot J. Early diagnosis of acute intestinal ischemia: contribution of colour Doppler sonography. *Acta Chir Belg* 1997; **97**: 173–6.

54 Garcia-Granero E, Alos R, Uribe N, Sala C, Lledo S. Intraoperative photoplethysmographic diagnosis of ischemic colitis. *Am Surg* 1997; **63**: 765–8.

55 Krohg-Sorensen K, Kvernebo K. Laser Doppler flowmetry in evaluation of colonic blood flow during aortic reconstruction. *Eur J Vasc Surg* 1989; **3**: 37–41.

56 Krohg-Sorensen K, Line PD, Haaland T, Horn RS, Kvernebo K. Intraoperative prediction of ischemic injury of the bowel: a comparison of laser Doppler flowmetry and tissue oximetry to histological analysis. *Eur J Vasc Surg* 1992; **6**: 518–24.

57 Sakakibara Y, Jikuya T, Saito EM, Mitsui T, Ijima H. Does laser Doppler flowmetry aid the prevention of ischemic colitis in abdominal aortic aneurysm surgery? *Thorac Cardiovasc Surg* 1997; **45**: 32–4.

16

Management of angiodysplasia of the gastrointestinal tract

PAUL A. THODIYIL AND AJAY K. KAKKAR

INTRODUCTION

In about 5% of gastrointestinal bleeding, the cause is not found by routine upper- and lower-gastrointestinal endoscopy. Gastrointestinal angiodysplasia is now recognized as one of the major causes of gastrointestinal bleeding, accounting for approximately 6.0% of cases of lower-intestinal bleeding and 1.2–8.0% of cases of hemorrhage from the upper gastrointestinal tract. Colonic angiodysplasia probably accounts for most of the obscure bleeding, followed by small-bowel angiodysplasia, which accounts for 30–40% in this subset of patients.[1] As a cause of severe rectal bleeding in the elderly, it is probably on a par with diverticular hemorrhage, accounting for about 15–20% of cases.[2, 3]

Galbadini coined the term 'angiodysplasia' in 1974 for submucosal ectatic microvasculature that principally involved the cecum and ascending colon in the elderly.[4] With no consensus on nosology for gastrointestinal vascular abnormalities, the term 'angiodysplasia' has been used to describe gastrointestinal vascular ectasia irrespective of etiology. However, it is preferable to restrict its use to localized gastrointestinal vascular ectasia not associated with one of the several systemic, metabolic, or hereditary diseases that cause diffuse intestinal and extraintestinal vascular ectasia.[5]

BACKGROUND

Intestinal angiodysplasia is primarily a disease of the elderly, with a mean age reported between 62 and 70 years,[6, 7] with 70% above the age of 65 years.[7] There is also a clear male preponderance (62.5%).[7] Documented coagulopathy and cardiac valvular disease have been found in 28% and 25%, respectively.[7]

Colonic angiodysplasia has a prevalence of 0.83% in consecutive screening colonoscopy of healthy, asymptomatic adults over the age of 50 years.[8] When specimen injection techniques are used, over 25% of patients aged over 60 years undergoing colonic resections for other reasons have angiodysplasia.[9] These lesions are small (mean size 4.0 mm) and multiple in 42%, with 62% of these located proximal to the hepatic flexure. They tend to have a benign course with a very low risk of bleeding (0% in 3 years).[8] The next common sites are the rectum (14%) and sigmoid colon (17%),[10] with the transverse and descending colons the least common sights. Among those presenting with bleeding, 78% of the colonic angiodysplasias are located in the right colon.[7] They are the most common cause of recurrent lower-intestinal bleeding in the elderly,[11] where despite endoscopic or operative treatment 12% have recurrent bleeding. The stomach and duodenum are the commonest sites of angiodysplasia in the upper gastrointestinal tract, with a 1–2% prevalence in patients undergoing upper-gastrointestinal endoscopy for all indications.[12] It is responsible for about 4% of acute gastrointestinal bleeding.[13] As with colonic angiodysplasias, it is principally a lesion of the elderly, with a mean age of 60 years. Approximately 15–20% of those with upper-gastrointestinal angiodysplasia also have similar lesions in the colon.[14]

Angiodysplasia has also been described in the pediatric population, where the mean age of clinical onset in a case series of nine patients was 2.3 years.[15] Colonic bleeding and anemia were present in all cases, most presenting before the first year of life. The left colon was the most frequently involved area. Colonic angiodysplasia in those aged under 21 years may have a congenital etiology, being associated frequently with Meckel's diverticulum.[16]

Angiodysplasia in the elderly is thought to be an acquired degenerative process resulting from chronic submucosal arteriovenous shunting secondary to mucosal ischemia.[17] There is increased expression of basic fibroblast growth factor and vascular endothelial growth factor, suggesting a role for angiogenic factors in pathogenesis.[18] There is also a deficiency of collagen type IV in the mucosal vessels in angiodysplasia that may be related to their susceptibility to ectasia and haemorrhage.[19]

Acquired angiodysplasias is also reported to occur with higher frequency in patients with renal failure, von Willebrand's disease, aortic stenosis, cirrhosis, and portal hypertension. However, not all of these associations have been subject to critical study.

Portal hypertension appears to predispose to angiodysplasia of the colon, with the relative risk of colonic angiodysplasia in alcoholic cirrhotic patients estimated to be 14.4 (95% confidence interval 2.8–75.3) compared with cirrhotic patients without esophageal varices.[20] Furthermore, unlike the angiodysplasia found in non-cirrhotic patients, colonic angiodysplasia in cirrhotic patients tends to be diffuse.[21]

Although there is thought to be an association between aortic stenosis and angiodysplasia,[22] a prospective case-control study of 40 patients with colonic angiodysplasia showed equal prevalence of aortic valve stenosis in both cases and controls. Furthermore, aortic valve disease was not an independent predictor of bleeding from these lesions.[23] A retrospective study by Oneglia and colleagues also showed a low prevalence (1.6%) of aortic valve stenosis in patients with gastrointestinal angiodysplasia.[24] However, what has not been studied is the prevalence of angiodysplasia in patients with aortic stenosis. It has been suggested that an acquired, reversible deficiency of plasma von Willebrand factor (equivalent to vWD-IIA) may explain the association with gastrointestinal hemorrhage in aortic stenosis.[25, 26]

Among patients with chronic renal failure, upper-gastrointestinal angiodysplasia accounted for 13% of acute bleeds. This contrasted with a 1.3% prevalence of angiodysplasia in those without renal failure ($P = 0.01$). The prevalence of angiodysplasia as a cause of bleeding was related both to the duration of renal failure ($P = 0.004$) and need for hemodialysis (P=0.0001). Furthermore, angiodysplasia was also the commonest cause of recurrent bleeding in this group.[27] The use of ambulatory peritoneal dialysis instead of hemodialysis appears to reduce the incidence of such recurrent bleeds.[28]

The pathological features of angiodysplasia consist of ectatic submucosal veins and capillaries associated with areas of overlying mucosal thinning and occasional ulcerations. There is a submucosal shunt, resulting in early venous drainage. These vessels may rupture or ulcerate, producing chronic intestinal bleeding and, less often, acute hemorrhage.

They appear macroscopically as telangiectactic areas on the mucosal surface, but they may be readily obscured by gas insufflation during endoscopy. The lesions are rarely more than a few millimeters in diameter, and their demonstration in resected specimens requires special vessel-injection techniques.[5] They are most often localized or segmental in distribution, but in some conditions they diffusely involve the gastrointestinal tract.

CLINICAL PRESENTATION

The vast majority of angiodysplasias appear to remain silent in life. Indeed, incidentally discovered lesions have an extremely low probability of bleeding.[8] When angiodysplasias do bleed, it is typically intermittent and over a prolonged period of time. Between 30 and 50% of those presenting with intestinal bleeding have no recurrent bleeds in the 2 years after the index episode. On the other hand, 50% of those patients with gastric, duodenal, or colonic lesions and manifesting with bleeding will have at least one frank, overt bleeding episode in the subsequent 1 or more years.[13, 29] Angiodysplasia may present with melena (42%), anemia (27%), rectal bleeding (26%), and hematemesis (5%).[6] About 20% will present as emergencies with life-threatening hemorrhage. In this case, the bleeding is typically dark red, suggesting proximal colonic origin. In the remainder, haemodynamic stability is maintained, typically leading to chronic anemia that may lead to iron or transfusion dependence. The median duration of symptoms is 50 weeks (range 1 day–36 years),[6] with a history of prior admissions for intestinal bleeding of undetermined etiology. These patients are also more likely than other gastrointestinal bleeders to be smokers.[30]

DIAGNOSIS

The problem faced by the clinician is most often that of a patient presenting with chronic transfusion-dependent anemia with or without overt rectal bleeding; less often, a patient over 50 years presents with sudden onset of rectal bleeding requiring blood transfusions. It is important in these patients to first perform a thorough upper-gastrointestinal endoscopy and colonoscopy.[31, 32] Even when the presentation is with rectal bleeding, an upper-gastrointestinal endoscopy will identify the 10–14% of patients where the source is proximal to the second part of the duodenum.[33, 34]

Angiodysplasia of the stomach and duodenum is almost invariably diagnosed at endoscopy, often with little contribution from angiography. Furthermore, about one-quarter of those with colonic angiodysplasia also have similar gastric and duodenal lesions. The endoscopic appearance of these lesions is variable. They are flat and rarely larger than 5 mm in diameter, with discrete borders and a red color (Plate 23). The prominent draining vein may also be visible. The lesions may be indistinguishable from bruising caused by the endoscope, or may not even be visible endoscopically.

The value of emergency colonoscopy has been much debated. Most data support its use, reporting diagnostic accuracy of up to 89% in identifying the bleeding lesion.[35, 36] In 44%, the examination was incomplete, failing to reach the cecum. While colonoscopy may identify diverticular bleeding and ischemic colitis, it is often very difficult to identify angiodysplastic lesions on colonoscopy as the inadequate views and gas insufflation invariably obscure these lesions. Where no other lesion has been identified, the absence of angiodysplasias in these circumstances does not exclude them as a cause and therefore calls for a visceral angiography.

The most useful and specific test confirming angiodysplasia is selective visceral angiography examining the celiac, superior mesenteric and inferior mesenteric artery territories.[4] This shows early-filling, tortuous, intramural veins lying on the antimesenteric part of the bowel, while on the capillary phase, ectatic capillaries appear as tiny lakes of contrast medium. These drain into submucosal veins that are already visible in the arterial phase (Figure 16.1)[37] and are the most common and sometimes the only angiographic sign. The proportion of patients identified to be bleeding from colonic angiodysplasia varies from 40% ($n = 131$)[6] to 66% ($n = 12$).[38] Of these, actual evidence of bleeding, such as contrast extravasation, is found in only 10–20%. In chronic gastrointestinal bleeding of obscure origin, angiography may be the only modality of investigation that can successfully locate a microvascular abnormality.[39] Selective visceral angiogram can provide a diagnosis in 87% of acute gastrointestinal bleeders and 74% of chronic bleeders where routine endoscopies were unhelpful.[40] In 40% of the latter group, angiodysplasia is the lesion identified.

In addition to its diagnostic value, angiography provides the potential for effective therapeutic intervention through embolization. Unfortunately, while embolization may be

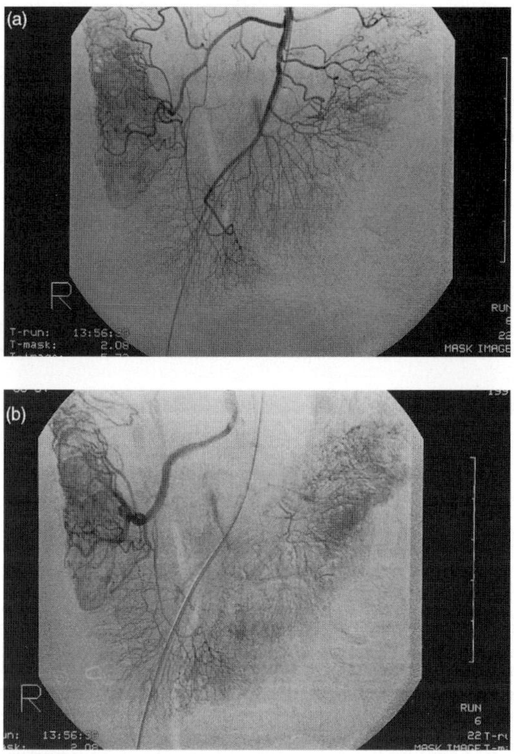

Figure 16.1 *This highly selective visceral angiogram demonstrates some of the classic features of angiodysplasia. (a) The tiny coalescing lakes of contrast medium are caused by ectatic capillaries. (b) This shows an early-filling and prominent submucosal vein that is already visible in the arterial phase. Courtesy of James Jackson, FRCR, Hammersmith Hospital, London.*

undertaken successfully in about 36% of acute bleeders and 10% of chronic bleeders, this technique cannot be extended to those with intestinal angiodysplasias.[40]

Angiography and colonoscopy are complementary. Among those with angiographically proven colonic angiodysplasias, performing an elective colonoscopy identified 68% of them. This figure increases to 80% when lesions are in the area examined at colonoscopy.[41, 42] About 9% were falsely negative, while 23% failed because of incomplete examination. Apart from identifying other disease, such as polyps, colonoscopy provides the therapeutic possibility of electrocoagulation.[41] A potential pitfall in angiographic diagnosis of angiodysplasia is malignant angiodysplasia, where the neovasculature of colonic tumors has an appearance indistinguishable from angiodysplasia.[43]

Visceral angiography can also be complemented by endoscopic examination of the small bowel. The diagnostic yield of push-enteroscopy is high in patients where initial esophagogastroduodenoscopy and colonoscopy have failed to identify a lesion. In a series of 83 such patients, Schmit *et al.*, using a 240-cm push-enteroscope, identified a potential bleeding lesion in 49 (59%) patients, of whom 33 (40%) had angiodysplasias.[44] Similarly, among 233 patients, push-enteroscopy had a positive diagnostic yield of 53%, of which 63% were vascular malformations.[45] Where the bleeding is from the small bowel, angiodysplasias are responsible for 40% of cases.[46]

The demonstration of angiodysplasias by any modality does not always imply that they are the source of the clinical presentation. In about 22–33% of patients where colonic angiodysplasias have been identified, a synchronous small-bowel lesion is also seen.[47] When affected patients have been evaluated by colonoscopy, concomitant lesions have been diagnosed in one-third of instances.[1] The angiodysplasia can only be diagnosed confidently as the source of blood loss if it is seen to be actively bleeding. As the angiodysplasias bleed only intermittently, the synchronous lesion should be removed at the same time as the colectomy.[47]

Barium studies are of little value. Nor are fecal occult blood tests, which are only intermittently positive. The value of computed tomographic angiography (CTA) and magnetic resonance angiography (MRA) in the diagnosis of intestinal angiodysplasia is unclear. It is possible that their resolution may not be high enough to image this microvascular abnormality. However, some studies have begun to show promise in these non-invasive techniques. Junquera and co-workers prospectively showed significant diagnostic agreement between CTA and a combination of colonoscopy plus visceral angiography ($\kappa = 0.68$, $P = 0.001$). The sensitivity, specificity, and positive predictive values for detection of colonic angiodysplasia were 70%, 100%, and 100%, respectively.[48] While magnetic resonance imaging (MRI) is useful in assessing vascular malformations in soft tissue,[49] it has no proven value in gastrointestinal angiodysplasias. A further new technology is wireless capsule endoscopy.[50] This is a small (11 $\times$ 30 mm) radiotelemetric device that, when swallowed, provides good views from mouth to colon and provides information on the location of surface abnormalities. While it cannot be used for biopsy or therapy, its value may lie in assessing small bowel inaccessible with the current enteroscopes. Clinical experience with this device is very limited but promising.[51]

Where a preoperative diagnosis has not been made, intraoperative enteroscopy may be used to examine the small bowel beyond the reach of routine enteroscopy.[52, 53]

MANAGEMENT

The general approach to managing obscure gastrointestinal bleeding is summarized in the algorithm shown in Figure 16.2. The clinical presentation is the main guide to the choice and

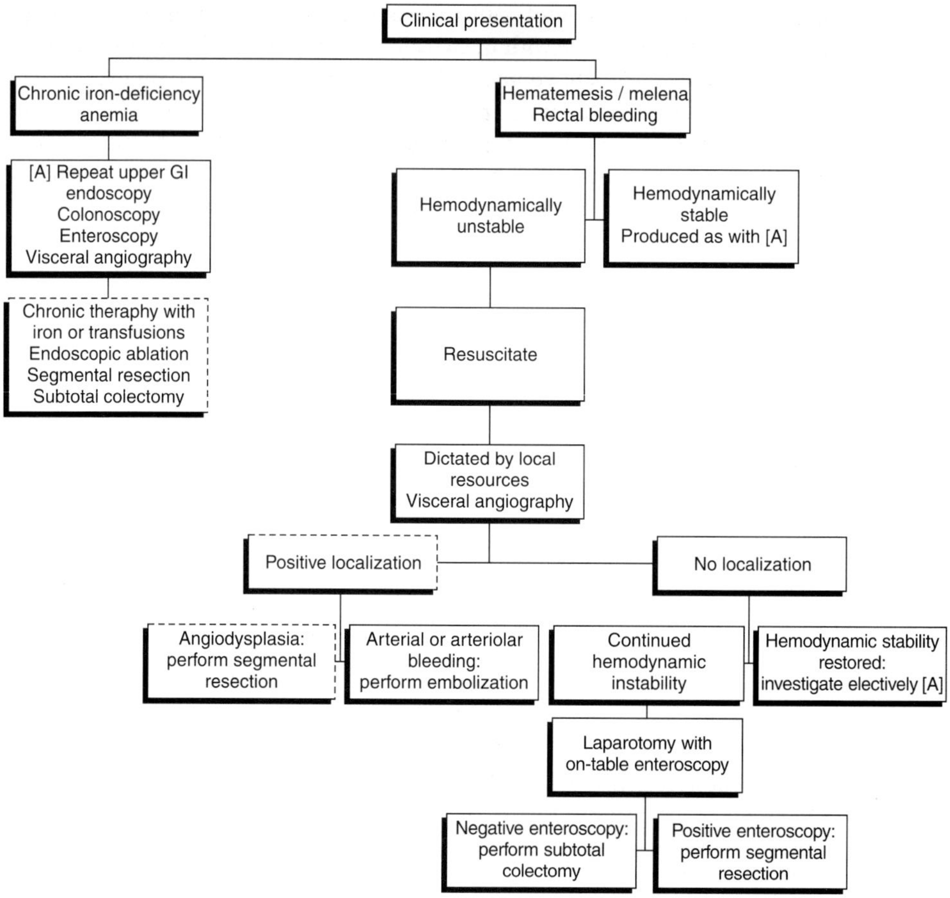

Figure 16.2 *Algorithm summarizing an approach to the management of obscure gastrointestinal bleeding where initial upper-gastrointestinal endoscopy has been normal. In the unstable patient, once adequately resuscitated, colonoscopy or visceral angiography may not only reduce the extent of operative resection, but could even make it unnecessary. Colonoscopy in the emergency setting has little proven therapeutic value in angiodysplasia, but it is valuable in diverticular bleeding.*

sequence of investigation and treatment. In patients presenting with rectal bleeding, it is imperative to first perform an upper-gastrointestinal endoscopy. In the emergency setting, where resources are available, visceral angiogram is potentially more valuable than colonoscopy. Angiograms not only allow embolization of bleeding diverticulae but also localize angiodysplasia for segmental resection. Where radiological localization has failed and there is continuing bleeding with hemodynamic instability, laparotomy with on-table enteroscopy may help guide a resection. Where this exonerates the small bowel, the risks of re-bleeding are minimized by undertaking a subtotal colectomy.

However, approximately 43% of patients with colonic angiodysplasias spontaneously stop bleeding with supportive conservative management.[47] Overall, endoscopic coagulation will effectively manage 50% of patients; 12.5% require surgical resection, and 9% need supportive observation in 9 per cent. Of these, 12.5% develop recurrent bleeding and 12.5% die of various causes. Of those treated non-operatively, 31% have further bleeding during a 42-month follow-up, of whom 5% consequently die,[7] with no report of death or recurrent bleeding among those operated on. The risk of recurrent bleeding is a problem with non-operative treatment, and surgical resection gives better results in this respect.

Intestinal angiodysplastic lesions are not generally amenable to vascular embolization. This disease occurs at the capillary and post-capillary venule level and, as such, the vessels are inaccessible for such treatment. However, in some circumstances where operative intervention may not be appropriate, embolization may be undertaken, but with considerable risk of failure or transmural necrosis.[54] The value of transcatheter infusion of vasopressin in angiodysplasia is doubtful, and it is probably no more beneficial than embolization. The case series of successful use of selective intra-arterial infusion of vasopressin in colonic hemorrhage consists mostly of patients with arterial hemorrhage from diverticulae, where it has been delivered via a selectively placed angiographic catheter at 0.2 U/min.[55, 56] However, the use of vasopressin in these situations has been superseded by transcatheter embolization, which has a lower re-bleeding rate.[57]

Medical management

Medical therapy can include observation, supportive care, or specific medications, including iron supplements and avoidance of non-steroidal anti-inflammatory drugs and anticoagulants. Blood transfusions may be necessary on a regular basis. The specific medical therapies that have been tried include estrogens, somatostatin, and aminocaproic acid. Such therapy is usually used for diffuse involvement of the bowel or where lesions are inaccessible. Patients with failed endoscopic or surgical management and who on account of co-morbidity are unsuitable for endoscopic or surgical intervention may also be offered these systemic therapies.[58]

The role of hormonal therapy is not clearly established. Uncontrolled case studies have reported reduction or cessation of bleeding in subjects managed with conjugated estrogens.[59–62] However, prospective, randomized, controlled trials assessing the efficacy of hormonal therapy are limited.[1]

A double-blind cross-over trial ($n = 10$) using 0.05 mg ethinylestradiol and 1 mg norethisterone daily significantly reduced transfusion requirements.[63] However, a cohort study of 64 patients failed to demonstrate any difference in transfusion requirements between a control group and those receiving daily hormonal therapy (norethynodrel with mestranol or conjugated oestrogens).[64] What is not clear is whether such hormonal treatment may be of benefit to specific groups of patients, such as those with chronic renal failure[65] or in some of the rare constitutional thrombocytopathies.[66]

The somatostatin receptor analogue octreotide has, in a number of uncontrolled studies,[67, 68] been shown to reduce transfusion requirements in patients with portal hypertension and von Willebrand's disease.[69] There is no evidence that it results in regression of angiodysplasias.[68] In patients with portal hypertension and gastrointestinal angiodysplasia, surgical or medical reduction of portal pressures may ameliorate the intestinal blood loss.[70]

Endoscopic therapy

Endoscopic treatment has generally replaced surgery as the first line of definitive treatment for angiodysplasias in most patients. Not only does it usually establish the diagnosis, but elective endoscopic coagulation is also an effective and safe method of treatment allowing most patients to be managed without an operation. Monopolar or bipolar diathermy, argon plasma coagulator, neodymium : yttrium aluminum garnate (Nd : YAG) laser,[71] sclerotherapy,[72] and band ligation[73, 74] have all been tried successfully. While the safety profiles are good, recurrent bleeding and perforation of the right colon are recognized problems, especially where monopolar electrocoagulation and lasers are used.

In the upper gastrointestinal tract, endoscopic Nd : YAG laser obliteration appears to be the most effective in reducing recurrent bleeding rates and transfusion requirements.[75, 76] Monopolar electrocautery and injection sclerotherapy are both associated with recurrent bleeding in up to 50% of patients.[13, 77]

Push-enteroscopy with cauterization of accessible small-intestinal angiodysplasias significantly reduces chronic transfusion requirements.[78] In a series of 55 patients, Askin and Lewis showed that the cauterized group (follow-up 30 $\pm$ 18 months) significantly improved, requiring 2.40 $\pm$ 2.97 units/month before treatment and 0.32 $\pm$ 0.91 units/month after cauterization ($P = 0.0001$, paired *t*-test).[78] Similar results were obtained by Morris *et al.*, where patients required a median of one (range one to five) examinations to treat small-bowel lesions (median of two, range one to seven) identified at enteroscopy.[46]

With enteroscopy and cautery, Schmit and colleagues were able to achieve good outcome (neither recurrence of anemia nor blood transfusion requirements) in 75% (25/33) of patients with small-bowel angiodysplasia.[44]

The value of emergency colonoscopy in the management of colonic angiodysplasia is unclear. Two prospective studies showed a potentially useful role of urgent, prepared colonoscopy for diverticular hemorrhage.[2] However, only 20% of these patients had active diverticular bleeding identified and dealt with endoscopically. Ohyama and co-workers showed that in emergency colonoscopy, hemostasis can be achieved in 66%, but the proportion that were angiodysplasias is uncertain.[35] Incidental colonic angiodysplasia requires no endoscopic treatment as their risk of bleeding is very low.[8] In the elective obliteration of colonic angiodysplasia, short-term results appear to be good, but the 2-year risk of recurrent bleeding is as high as 40–50% irrespective of the technique used. Furthermore, there is little difference in transfusion requirements compared with those receiving no specific therapy.[75, 77] The use of Nd : YAG lasers in the right colon is associated with colonic perforation in up to 15% of patients.[75, 79]

Operative management

While endoscopic therapy remains the first line of treatment, operative intervention will become necessary in 56% of those with colonic angiodysplasia on account of failed endoscopic treatment or life-threatening hemorrhage.[47] Operative intervention may also become necessary where angiographically demonstrated lesions are not seen on endoscopy. However, operative treatment is not always curative.

It is important for the clinician to bear in mind that the absence of angiodysplastic lesions on endoscopy does not exclude their presence. At laparotomy, the lesions can neither be felt nor seen by the surgeon. As such, the extent of resection should be guided by the extent of angiodysplasia on angiography and endoscopy. A right hemicolectomy is the most common resection (83%) and carries a 9% perioperative mortality in this group of patients.[47] Re-bleeding rates after surgical resection have been reported as high as 19–25% within the first 5 years.[47, 80] They usually result from other intestinal angiodysplasia or an unsuspected second lesion. In about 40% of cases, no cause is identified for the recurrent bleeding.[47]

The failure rate of bowel resection is much higher (80%) in patients with aortic stenosis. In these patients, replacement with a bioprosthetic valve may bring about dramatic cessation of bleeding.[81] Where the angiodysplasia involves the small bowel, localizing the segment for

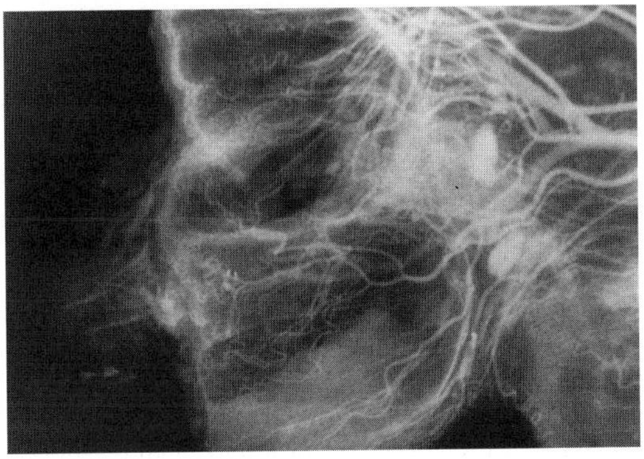

Figure 16.3 *This specimen angiogram was performed by injecting barium into the ileocolic artery of this right hemicolectomy specimen. Note the lakes of contrast medium and the draining vein. Courtesy of Paul Tait, FRCR, Hammersmith Hospital, London.*

resection can be challenging. Preoperative selective placement of arterial catheters and intraoperative injection of methylene blue through them are of value in intraoperative localization.[82, 83]

It is essential that the specimen is treated in a manner that allows demonstration of the angiodysplastic lesions. The colon has to be removed with its vessels undamaged. In the operating suite, the specimen should have barium gelatin (or a suitable alternative) injected into the main vascular pedicle that readily fills the minute vascular abnormalities, rendering them visible to the eye or under the microscope.[4, 38, 84] A specimen angiography is then carried out to show the morphology of the vascular lesion (Figure 16.3).[85] This would confirm whether the offending lesion has been removed. This preparation also allows the pathologist to take tissue blocks directly from affected areas of the bowel.[38]

SUMMARY

Gastrointestinal angiodysplasia is a common and challenging condition to treat. Most patients can be managed successfully with iron replacement, supported if necessary with endoscopic obliteration. Operative intervention is most appropriately reserved for the few with severe hemorrhage and hemodynamic instability or those with large chronic transfusion dependence.

REFERENCES

1 Foutch PG. Angiodysplasia of the gastrointestinal tract. *Am J Gastroenterol* 1993; **88**: 807–18.

2 Jensen DM, Machicado GA, Jutabha R, Kovacs TO. Urgent colonoscopy for the diagnosis and treatment of severe diverticular hemorrhage. *N Engl J Med* 2000; **342**: 78–82.

3 Jensen DM, Machicado GA. Diagnosis and treatment of severe hematochezia. The role of urgent colonoscopy after purge. *Gastroenterology* 1988; **95**: 1569–74.

4 Galbadini JJ. Case records of the Massachusetts General Hospital. Weekly clinicopathological exercises. Case 36–1974. *N Engl J Med* 1974; **291**: 569–75.

5 Athanasoulis CA, Galdabini JJ, Waltman AC, Novelline RA, Greenfield AJ, Ezpeleta ML. Angiodysplasia of the colon: a cause of rectal bleeding. *Cardiovasc Radiol* 1977; **1**: 3–13.

6 Thompson JN, Salem RR, Hemingway AP, Rees HC, Hodgson HJ, Wood CB, *et al*. Specialist investigation of obscure gastrointestinal bleeding. *Gut* 1987; **28**: 47–51.
7 Gupta N, Longo WE, Vernava AM 3rd. Angiodysplasia of the lower gastrointestinal tract: an entity readily diagnosed by colonoscopy and primarily managed nonoperatively. *Dis Colon Rectum* 1995; **38**: 979–82.
8 Foutch PG, Rex DK, Lieberman DA. Prevalence and natural history of colonic angiodysplasia among healthy asymptomatic people. *Am J Gastroenterol* 1995; **90**: 564–7.
9 Boley SJ, Sammartano R, Adams A. On the nature and etiology of vascular ectasia of the colon: degenerative lesions of aging. *Gastroenterology* 1977; **72**: 650.
10 Hochter W, Weingart J, Kuhner W, Frimberger E, Ottenjann R. Angiodysplasia in the colon and rectum. Endoscopic morphology, localisation and frequency. *Endoscopy* 1985; **17**: 182–5.
11 Boley SJ, DiBiase A, Brandt LJ, Sammartano RJ. Lower intestinal bleeding in the elderly. *Am J Surg* 1979; **137**: 57–64.
12 Gunnlaugsson O. Angiodysplasia of the stomach and duodenum. *Gastrointest Endosc* 1985; **31**: 251–4.
13 Clouse RE, Costigan DJ, Mills BA, Zuckerman GR. Angiodysplasia as a cause of upper gastrointestinal bleeding. *Arch Intern Med* 1985; **145**: 458–61.
14 Cappell MS. Spatial clustering of simultaneous nonhereditary gastrointestinal angiodysplasia. Small but significant correlation between nonhereditary colonic and upper gastrointestinal angiodysplasia. *Dig Dis Sci* 1992; **37**: 1072–7.
15 De la Torre Mondragon L, Vargas Gomez MA, Mora Tiscarreno MA, Ramirez Mayans J. Angiodysplasia of the colon in children. *J Pediatr Surg* 1995; **30**: 72–5.
16 Hemingway AP, Allison DJ. Angiodysplasia and Meckel's diverticulum: a congenital association? *Br J Surg* 1982; **69**: 493–6.
17 Baum S, Athanasoulis CA, Waltman AC, Galdabini J, Schapiro RH, Warshaw AL, *et al*. Angiodysplasia of the right colon: a cause of gastrointestinal bleeding. *Am J Roentgenol* 1977; **129**: 789–94.
18 Junquera F, Saperas E, de Torres I, Vidal MT, Malagelada JR. Increased expression of angiogenic factors in human colonic angiodysplasia. *Am J Gastroenterol* 1999; **94**: 1070–6.
19 Roskell DE, Biddolph SC, Warren BF. Apparent deficiency of mucosal vascular collagen type IV associated with angiodysplasia of the colon. *J Clin Pathol* 1998; **51**: 18–20.
20 Naveau S, Bedossa P, Poynard T, Mory B, Chaput JC. Portal hypertensive colopathy. A new entity. *Dig Dis Sci* 1991; **36**: 1774–81.
21 Naveau S, Leger-Ravet MB, Houdayer C, Bedossa P, Lemaigre G, Chaput JC. Nonhereditary colonic angiodysplasias: histomorphometric approach to their pathogenesis. *Dig Dis Sci* 1995; **40**: 839–42.
22 Galloway SJ, Casarella WJ, Shimkin PM. Vascular malformations of the right colon as a cause of bleeding in patients with aortic stenosis. *Radiology* 1974; **113**: 11–15.
23 Bhutani MS, Gupta SC, Markert RJ, Barde CJ, Donese R, Gopalswamy N. A prospective controlled evaluation of endoscopic detection of angiodysplasia and its association with aortic valve disease. *Gastrointest Endosc* 1995; **42**: 398–402.
24 Oneglia C, Sabatini T, Rusconi C, Gardini A, Paterlini A, Buffoli F, *et al*. Prevalence of aortic valve stenosis in patients affected by gastrointestinal angiodysplasia. *Eur J Med* 1993; **2**: 75–8.
25 Warkentin TE, Moore JC, Morgan DG. Aortic stenosis and bleeding gastrointestinal angiodysplasia: is acquired von Willebrand's disease the link? *Lancet* 1992; **340**: 35–7.
26 Hertzberg MS, Facey DA, Favaloro EJ, Koutts J. Type 2B von Willebrand's disease and angiodysplasia. *Thromb Haemost* 1999; **82**: 1557–8.
27 Chalasani N, Cotsonis G, Wilcox CM. Upper gastrointestinal bleeding in patients with chronic renal failure: role of vascular ectasia. *Am J Gastroenterol* 1996; **91**: 2329–32.
28 Chang IT, Chen TW, Ng YY, Yang WC. Recurrent intestinal angiodysplastic bleeding in a patient on hemodialysis ceasing spontaneously with CAPD. *Perit Dial Int* 1998; **18**: 342–3.

29 Trudel JL, Fazio VW, Sivak MV. Colonoscopic diagnosis and treatment of arteriovenous malformations in chronic lower gastrointestinal bleeding: clinical accuracy and efficacy. *Dis Colon Rectum* 1988; **31**: 107.

30 Cappell MS, Gupta A. Changing epidemiology of gastrointestinal angiodysplasia with increasing recognition of clinically milder cases: angiodysplasia tend to produce mild chronic gastrointestinal bleeding in a study of 47 consecutive patients admitted from 1980–1989. *Am J Gastroenterol* 1992; **87**: 201–6.

31 Heald RJ. Defining the clinical problem of massive rectal bleeding. *Dis Colon Rectum* 1974; **17**: 432–8.

32 Moody FG. Rectal bleeding. *N Engl J Med* 1974; **290**: 839–41.

33 Meyer-Wyss B, Rederer M. The role of emergency endoscopy in acute lower intestinal hemorrhage. *Helv Chir Acta* 1991; **58**: 33–7.

34 Wilcox CM, Alexander LN, Cotsonis G. A prospective characterization of upper gastrointestinal hemorrhage presenting with hematochezia. *Am J Gastroenterol* 1997; **92**: 231–5.

35 Ohyama T, Sakurai Y, Ito M, Daito K, Sezai S, Sato Y. Analysis of urgent colonoscopy for lower gastrointestinal tract bleeding. *Digestion* 2000; **61**: 189–92.

36 Tada M, Shimizu S, Kawai K. Emergency colonoscopy for the diagnosis of lower intestinal bleeding. *Gastroenterol Jpn* 1991; **26** (suppl 3): 121–4.

37 Boley SJ, Sprayregen S, Sammartano RJ, Adams A, Kleinhaus S. The pathophysiologic basis for the angiographic signs of vascular ectasias of the colon. *Radiology* 1977; **125**: 615–21.

38 Tarin D, Allison DJ, Modlin IM, Neale G. Diagnosis and management of obscure gastrointestinal bleeding. *BMJ* 1978; **2**: 751–4.

39 Allison DJ. Gastrointestinal bleeding. Radiological diagnosis. *Br J Hosp Med* 1980; **23**:358–65.

40 Allison DJ, Hemingway AP, Cunningham DA. Angiography in gastrointestinal bleeding. *Lancet* 1982; **2**: 30–3.

41 Salem RR, Wood CB, Rees HC, Kheshavarzian A, Hemingway AP, Allison DJ. A comparison of colonoscopy and selective visceral angiography in the diagnosis of colonic angiodysplasia. *Ann R Coll Surg Engl* 1985; **67**: 225–6.

42 Richter JM, Hedberg SE, Athanasoulis CA, Schapiro RH. Angiodysplasia: clinical presentation and colonoscopic diagnosis. *Dig Dis Sci* 1984; **29**: 481.

43 Lu C, Fukuya T, Landas S, Ellison K. Angiodysplasia of the colon mimicking adenocarcinoma. *Am J Roentgenol* 1993; **160**: 898.

44 Schmit A, Gay F, Adler M, Cremer M, Van Gossum A. Diagnostic efficacy of push-enteroscopy and long-term follow-up of patients with small bowel angiodysplasias. *Dig Dis Sci* 1996; **41**: 2348–52.

45 Descamps C, Schmit A, Van Gossum A. 'Missed' upper gastrointestinal tract lesions may explain 'occult' bleeding. *Endoscopy* 1999; **31**: 452–5.

46 Morris AJ, Mokhashi M, Straiton M, Murray L, Mackenzie JF. Push enteroscopy and heater probe therapy for small bowel bleeding. *Gastrointest Endosc* 1996; **44**: 394–7.

47 Steger AC, Galland RB, Hemingway A, Wood CB, Spencer J. Gastrointestinal haemorrhage from a second source in patients with colonic angiodysplasia. *Br J Surg* 1987; **74**: 726–7.

48 Junquera F, Quiroga S, Saperas E, Perez-Lafuente M, Videla S, Alvarez-Castells A, *et al.* Accuracy of helical computed tomographic angiography for the diagnosis of colonic angiodysplasia. *Gastroenterology* 2000; **119**: 293–9.

49 Witte CL, Witte MH. An imaging evaluation of angiodysplasia syndromes. *Lymphology* 2000; **33**: 158–66.

50 Iddan G, Meron G, Glukovsky A, Swain P. Wireless capsule endoscopy. *Nature* 2000; **405**: 417.

51 Appleyard M, Glukhovsky A, Swain P. Wireless-capsule diagnostic endoscopy for recurrent small-bowel bleeding. *N Engl J Med* 2001; **344**: 232–3.

52 Lau WY, Wong SY, Yuen WK, Wong KK. Intraoperative enteroscopy for bleeding angiodysplasias of small intestine. *Surg Gynecol Obstet* 1989; **168**: 341.

53 Lopez MJ, Cooley JS, Petors JG. Complete intraoperative small-bowel endoscopy in the evaluation of occult gastrointestinal bleeding using the sonde endoscope. *Arch Surg* 1996; **131**: 272.

54 Uflacker R. Transcatheter embolisation for treatment of acute lower gastrointestinal bleeding. *Acta Radiol* 1987; **28**: 425.

55 Shaff MI, Becker H. Diagnosis and control of diverticular bleeding by arteriography and vasopressin infusion. A report of 2 cases. *S Afr Med J* 1979; **56**: 72–4.

56 Baum S, Athanasoulis CA, Waltman AC. Angiography in the diagnosis and therapy of hemorrhage from the large bowel. *Radiologe* 1975; **15**: 427–33.

57 Browder W, Cerise EJ, Litwin MS. Impact of emergency angiography in massive lower gastrointestinal bleeding. *Ann Surg* 1986; **204**: 530–6.

58 Lewis BS. Medical and hormonal therapy in occult gastrointestinal bleeding. *Semin Gastrointest Dis* 1999; **10**: 71–7.

59 Manning RJ. Estrogen/progesterone treatment of diffuse antral vascular ectasia. *Am J Gastroenterol* 1995; **90**: 154–6.

60 Moshkowitz M, Arber N, Amir N, Gilat T. Success of estrogen-progesterone therapy in long-standing bleeding gastrointestinal angiodysplasia. Report of a case. *Dis Colon Rectum* 1993; **36**: 194–6.

61 Bellucci S, Zini JM, Bitoun P, Dupuy E, Drouet L, Tobelem G, *et al.* Diffuse severe digestive angiodysplasia in Bernard-Soulier syndrome. Improvement of bleeding by oestroprogestative therapy. *Thromb Haemost* 1995; **74**: 1610–12.

62 Hermans C, Goffin E, Horsmans Y, Laterre E, Van Ypersele de Strihou C. Watermelon stomach. An unusual cause of recurrent upper GI tract bleeding in the uraemic patient: efficient treatment with oestrogen–progesterone therapy. *Nephrol Dial Transplant* 1996; **11**: 871–4.

63 Van Cutsem E, Rutgeerts P, Vantrappen G. Treatment of bleeding gastrointestinal vascular malformations with oestrogen–progesterone. *Lancet* 1990; **335**: 953.

64 Lewis BS, Salomon P, Rivera-MacMurray S, Kornbluth AA, Wenger J, Waye JD. Does hormonal therapy have any benefit for bleeding angiodysplasia? *J Clin Gastroenterol* 1992; **15**: 99–103.

65 Richardson JD, Lordon RE. Gastrointestinal bleeding caused by angiodysplasia: a difficult problem in patients with chronic renal failure receiving hemodialysis therapy. *Am Surg* 1993; **59**: 636–8.

66 Bellucci S, Caen JP. Gastrointestinal angiodysplasia in constitutional thrombocytopathies. *Int J Hematol* 1997; **65**: 419–20.

67 Nardone G, Rocco A, Balzano T, Budillon G. The efficacy of octreotide therapy in chronic bleeding due to vascular abnormalities of the gastrointestinal tract. *Aliment Pharmacol Ther* 1999; **13**: 1429–36.

68 Rossini FP, Arrigoni A, Pennazio M. Octreotide in the treatment of bleeding due to angiodysplasia of the small intestine. *Am J Gastroenterol* 1993; **88**: 1424–7.

69 Bowers M, McNulty O, Mayne E. Octreotide in the treatment of gastrointestinal bleeding caused by angiodysplasia in two patients with von Willebrand's disease. *Br J Haematol* 2000; **108**: 524–7.

70 Bellaiche G, Nouts A, Brassier D, Ley G, Slama JL. Severe iron-deficiency anemia due to diffuse antral vascular ectasia in a cirrhosis patient. Cure after surgical portacaval shunt. *Gastroenterol Clin Biol* 1998; **22**: 832–4.

71 Feydy P, Aazib L, Driguez E, Ghannoun Z, Brasseur J, Coevoet B. Treatment with laser Nd YAG of digestive angiodysplasias in dialysed patients with chronic kidney failure. *Gastroenterol Clin Biol* 1993; **17**: 772–3.

72 Bemvenuti GA, Julich MM. Ethanolamine injection for sclerotherapy of angiodysplasia of the colon. *Endoscopy* 1998; **30**: 564–9.

73 Baniukiewicz A, Laszewicz W. An endoscopic treatment for gastric angiodysplasia, using an endoscopic ligating device designed to treat esophageal varices. *Endoscopy* 1996; **28**: 787.

74 Jo Y, Matsumoto T, Aoyagi K, Yano Y, Kawasaki A, Fujishima M. Endoscopic band ligation device for bleeding gastric angiodysplasia. *Gastrointest Endosc* 1999; **50**: 599.

75 Rutgeerts P, Van Gompel F, Geboes K, Vantrappen G, Broeckaert L, Coremans G. Long term results of treatment of vascular malformations of the gastrointestinal tract by neodymium YAG laser photocoagulation. *Gut* 1985; **26**: 586.

76 Bown SG, Swain CP, Storey DW, *et al*. Endoscopic laser treatment of vascular anomalies of the upper gastrointestinal tract. *Gut* 1985; **26**: 1338–48.

77 Hutcheon DF, Kabelin J, Bulkley GB, Smith GW. Effect of therapy on bleeding rates in gastrointestinal angiodysplasia. *Am Surg* 1987; **53**: 6.

78 Askin MP, Lewis BS. Push enteroscopic cauterization: long-term follow-up of 83 patients with bleeding small intestinal angiodysplasia. *Gastrointest Endosc* 1996; **43**: 580–3.

79 Buchi KN. Vascular malformations of the gastrointestinal tract. *Surg Clin North Am* 1992; **72**: 559.

80 Salem RR, Thompson JN, Rees HC, Hemingway AP, Spencer J, *et al*. Outcome of surgery in colonic angiodysplasia. *Gut* 1985; **26**: A1155.

81 King RM, Pluth JR, Giuliani ER. The association of unexplained gastrointestinal bleeding with calcific aortic stenosis. *Ann Thorac Surg* 1987; **44**: 514.

82 Thompson JN, Hemingway AP, McPherson GA, Rees HC, Allison DJ, Spencer J. Obscure gastrointestinal haemorrhage of small-bowel origin. *BMJ* (*Clin Res Ed*) 1984; **288** (6431): 1663–5.

83 Athanasoulis CA, Moncure AC, Greenfield AJ, Ryan JA, Dobson TF. Intraoperative localisation of small bowel bleeding sites with combined use of angiographic methods and methylene blue injection. *Surgery* 1980; **87**: 77–84.

84 Koga H, Iida M, Nagai E, Aoyagi K, Matsumoto T, Takesue M, *et al*. Jejunal angiodysplasia confirmed by intravascular injection technique in vitro. Report of a case and review of the literature. *J Clin Gastroenterol* 1996; **23**: 139–44.

85 Allison DJ. Who's aware of angiodysplasia? *World Med* 1981; **16**: 59–60.

Index

Page numbers in *italics* refer to figures and tables.